LOTUS JAPONICUS HANDBOOK

Lotus japonicus Handbook

Edited by

Editorial Director

A.J. MÁRQUEZ
University of Seville, Spain

Section Editors

J. STOUGAARD
M. UDVARDI
M. PARNISKE
H. SPAINK
G. SAALBACH
J. WEBB
M. CHIURAZZI
A. J. MÁRQUEZ

Springer

A C.I.P. Catalogue record for this book is available from the Library of Congress.

ISBN-13 978-90-481-6945-0 (PB)
ISBN-13 978-1-4020-3735-1 (e-book)

Published by Springer,
P.O. Box 17, 3300 AA Dordrecht, The Netherlands.

www.springeronline.com

Printed on acid-free paper

Printed in the Netherlands.

Contents

Editors

EDITORIAL DIRECTOR

Antonio J. Márquez
Profesor Titular
Plant Biochemistry and Mol. Biol. Dept.
University of Seville, SPAIN
cabeza@us.es

TECHNICAL EDITOR

Megan McKenzie
Science Writer
Corvallis Oregon, USA
megan@mckenzie.us.com

SECTION EDITORS

Lotus japonicus, a general introduction

Jens Stougaard, University of Aarhus, DENMARK
Antonio J Márquez, University of Seville, SPAIN

Symbiotic processes

Michael Udvardi, MPI of Molecular Plant Physiology, Golm, GERMANY
Martin Parniske, University of Munich, GERMANY

Cell and molecular biology protocols

Herman Spaink, Leiden University, THE NETHERLANDS
Martin Parniske, University of Munich, GERMANY

Functional genomics

Michael Udvardi, MPI of molecular Plant Physiology, Golm, GERMANY
Gerhardt Saalbach, Risoe National Laboratory, DENMARK

Mutants, gene-tagging, and genetic analysis

Jens Stougaard, University of Aarhus, DENMARK
Judith Webb, IGER, Aberystwyth, UNITED KINGDOM

Transformation and reverse genetic analysis

Maurizio Chiurazzi, Institute of Genetics and Biophysics, CNR, Naples, ITALY

Primary and secondary metabolism

Antonio J Márquez, University of Seville, SPAIN
Judith Webb, IGER, Aberystwyth, UNITED KINGDOM

Lotus japonicus literature

Antonio J Márquez and Marco Betti, University of Seville, SPAIN

Preface

Legumes are very important plants (VIPs). They account for about one-third of the world's primary crop production, human dietary protein, and processed vegetable oil, and are a major source of feed for livestock and raw materials for industry. Legumes are a key component of sustainable agricultural systems because of symbiotic nitrogen fixation, which provides these plants and associated or subsequent crops with a free and renewable source of nitrogen. Legumes are also able to establish beneficial symbioses with soil fungi that enable them to mine phosphorous and other essential nutrients from the soil more effectively. However, despite their importance in agriculture, increases in legume yield through breeding over the past few decades have lagged behind those of cereals. While classical plant breeding can and will lead to further improvements in legume genotypes in the future, the genomics revolution offers alternative and complementary approaches that can aid and accelerate plant breeding. Genomics and functional genomics, together with the more classical scientific disciplines of genetics, biochemistry, physiology, and molecular and cell biology, have already accelerated discoveries in legume molecular and systems biology.

Legumes have played a key role in biological research, a good example of which is Mendel's work on the common garden pea that provided the groundwork for modern genetics. Mendel's work also highlighted the importance of choosing a tractable model species for scientific research. While diploid pea was a fine model for classical genetics, the large size of the pea genome together with other less than optimal features have hindered the isolation and characterisation of genes with important roles in legume biology and agriculture. For this reason *Lotus japonicus* was chosen as a model species for legume research some ten years ago. Since then, many groups within Europe and around the world have adopted *L japonicus* as a model and have developed numerous resources and protocols to facilitate basic and applied research on this species. Over 200 research papers focusing on *L japonicus* have now been published, many of them groundbreaking. Amongst the most important discoveries that have come from Lotus research have been the isolation of genes with crucial roles in symbiosis development, which have provided amazing new insight into the nature and evolution of signalling in plant-microbe symbioses.

This handbook represents the first effort to compile basic descriptions and methods for research in *L japonicus.* We wish to thank all the authors for their efforts in making the book as complete and up-to-date as possible. We are also grateful to the EU, which has provided funding to promote Lotus research, including the research and training networks HPRN-CT2000-00086 (coordinated by Michael Udvardi) and MRTN-CT-2003-505227 (coordinated by Martin Parniske). In fact, this book was conceived within the framework of the first of these networks. Finally, we wish to thank Megan McKenzie, our Technical Editor for her extraordinary professional work, as well Junta de Andalucía (Spain) for support in preparing the camera-ready version of this handbook for publication.

THE EDITORS

COLOUR PLATES

Colour plate 1. Different plant organs of Lotus japonicus *plants. a-d) Developmental stages of* Lotus japonicus *flowers relevant to hand pollination (reproduced with permission from Jiang and Gresshoff, 1997). Flowers in a and b are too young and inappropriate for fertilisation as female flower. The correct age of the flower is shown in c: optimal flowers are 6-7 mm in length and the standard petal is still fused, they show no curvature of the style as well as complete extension of the anther filament and ripening of the pollen. The flower in d is too old, pollen has been released from the anther and self-pollination has already taken place. e) Flowers from Gifu (yellow) and Filicaulis (red) ecotypes. (a courtesy of K. Szczyglowski) f) Seedpod containing up to 20 seeds. g) Trifoliate leaf. h) 2 months old* Lotus japonicus *plant. See chapter 1.1.*

Colour plate 1.

Colour plate 2. Formation of nodules in Lotus japonicus *plants infected with* Mesorhizobium loti. *A-C) Cleared tissues showing successive stages of nodule development. A: Cortical cell division, B: Bump, C: Developing nodule. D)* Lotus japonicus *nodule 2 weeks after infection with* Mesorhizobium loti. *DAPI staining. E) Brightfield micrograph of* L japonicus *nodule 10 days after inoculation with* M loti *strain NZP2253 carrying a hemA:LacZ reporter gene fusion. Roots were stain for β-galactosidase activity, cleared, and examined using brightfield microscopy. F) Brightfield micrograph of* L japonicus *roots showing the progression of the IT towards the developing nodule primordium. L japonicus roots were inoculated with* M loti *strain NZP2253 carrying a hemA:LacZ reporter gene fusion, stained for β-galactosidase activity, and examined using brightfield microscopy. G) Fully mature nitrogen-fixing nodule formed on* L japonicus *roots.* *See chapter 2.1.*

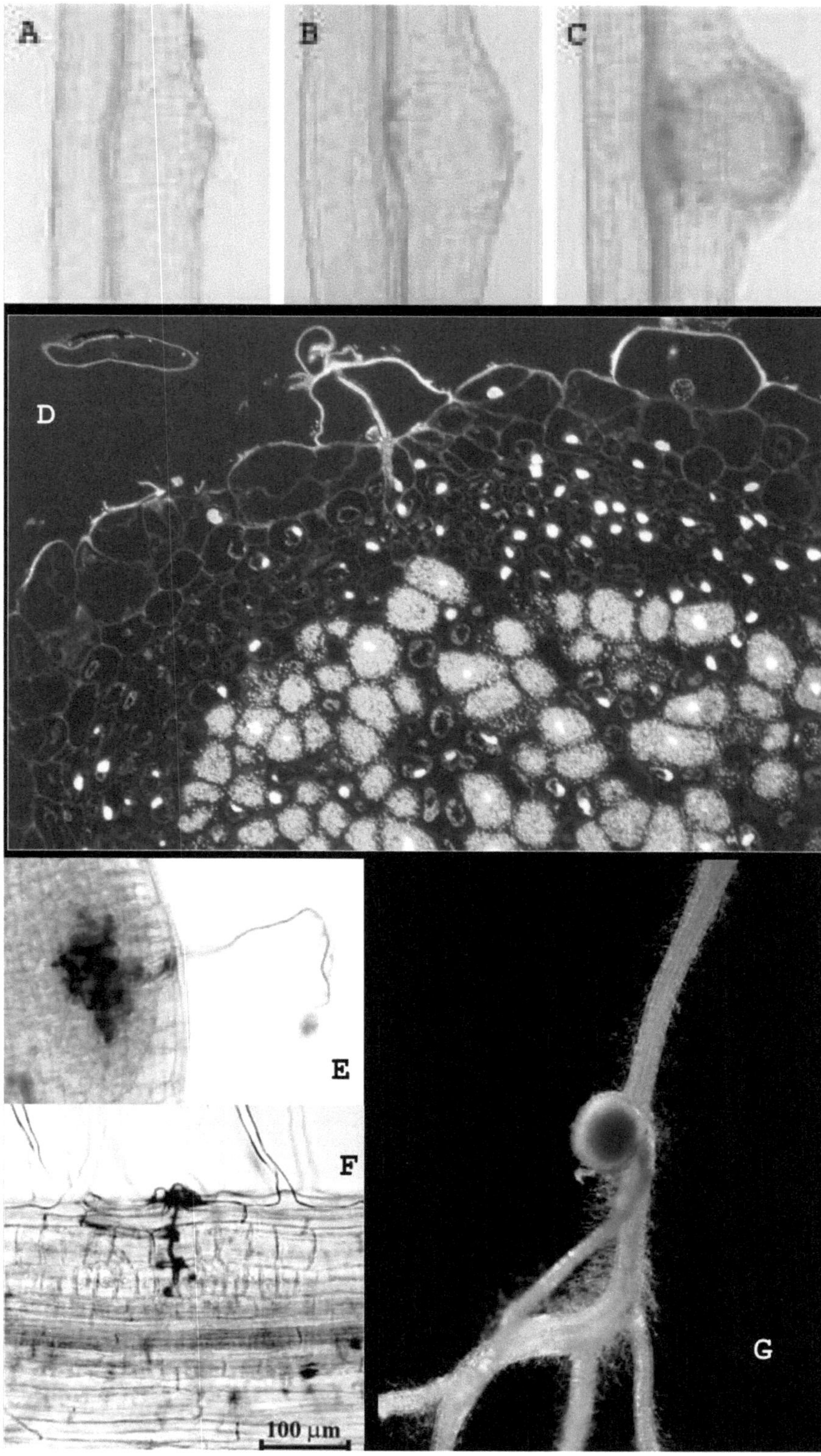

Colour plate 2.

Colour plate 3. A-F) Double staining of L. japonicus *hairy roots expressing MsENOD12B-gusA (X-gluc as substrate, blue precipitate) during nodulation induced by* M. loti *R7A constitutively expressing β-gal (M-gal as substrate, red precipitate). Bar in panels A, B, D, and E=125 μm. Bar in panel C=95 μm. Bar in panel F=30 μm (See chapter 3.1). G) Immunolocalisation of MnSOD in* Lotus japonicus *nodules by fluorescent detection of MnSOD primary antibodies. Images represent longitudinal sections of 13 days post inoculation (dpi) nodules. Alexa488-conjugated anti-rabbit immunoglobulin Gs were used as secondary antibodies. B1, B2, and B3, are the confocal microscope images showing the signal in the infected cells of the central tissue (background subtracted). B4, B5, and B6, are overlays of B1, B2, and B3, and the corresponding transmitted light images. B3 and B6 are a high magnification of the central tissue. No signal was observed in the controls (B1, B4). Bars = 30 μm (B1, B2, B4, and B5) and 3 μm (B3 and B6) (See chapter 3.2).*

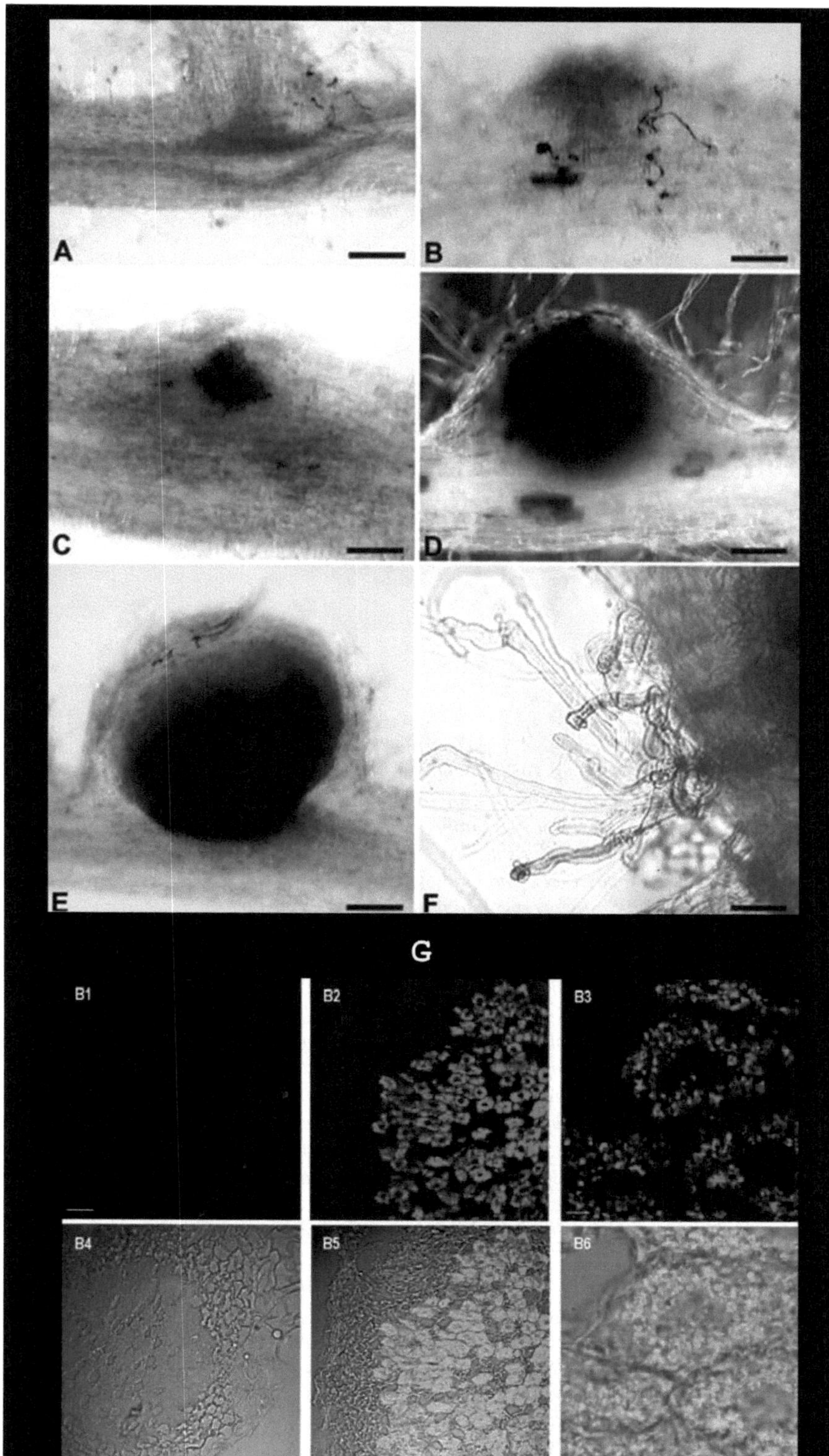

Colour plate 3.

Colour plate 4. A-G) Hairy roots induced by A. rhizogenes *LBA 1334 in* L japonicus *seedlings. Panel A: there is no swelling observed at the base of the hypocotyl of a Lotus seedling that has not been infected with LBA 1334 4 days after sectioning of the main root. In contrast, a slight swelling and root hairs appears in the area of wounding in a seedling 4 days after sectioning of the main root at the base of the hypocotyl and subsequent infection with LBA 1334 (panel B). Panels C, E, and G show GUS expression as blue precipitate using 5-bromo-4chloro-3-indolyl-β-glucoronic acid as substrate in roots transformed with CaMV35S-gus:intron cloned in various binary vectors. GUS histological staining was performed as described in Scarpella et al (2000). Panel C shows that all emerging hairy roots of a Lotus seedling were tranformed with LBA1334 carrying GUS as reporter gene cloned in pCAMBIA1301. Note that GUS is also expressed in loci that have not yet differentiated into hairy roots. Depending on the binary vector used, 5-10% of the transformed plants will show GUS expression in all emerging roots, the rest of the transformed plants wil present a combination of GUS-stained and non-stained roots. Staining was performed 7 days after transfer of the seedling to HRE medium. Panel E shows blue roots of seedlings transformed with LBA1334 carrying gus in a pPZP derivative and panel D, roots emerging in control seedlings, infected with LBA 1334, that do not show blue staining. GUS histological staining was performed 10 days after transfer to HRE. As many as 20 roots can be initiated, however only a few (2-10 roots) grow rapidly and produce lateral roots at this stage of development. Panel F shows the results of GUS staining of a control plant tranformed with LBA 1334 carrying pBin19 and panel G, the staining of a plant tranformed with LBA 1334 carrying pBin19 CaMV35S-gus:intron and showing GUS activity in nodulated hairy roots. Note that in both cases nodules are concentrated on the most developed root, which in 60-70 % of the cases, is transformed. Predominance of 1 or 2 roots is apparent after 2-3 weeks growth in HRE, clay particles or Jensen-agar plates. Staining of plants in panels F and G was performed 5 weeks after inoculation with* M. loti *strain R7A and growth of roots in clay particles. Bars in panels A, B, C= 1mm; in panels D, E= 4.5 mm; in panels F, G= 10 mm (see chapter 6.2). H) GUS staining obtained in hairy roots co-transformed with the T-DNA vector carrying the 35S-gusAint cassette (see chapter 6.1).*

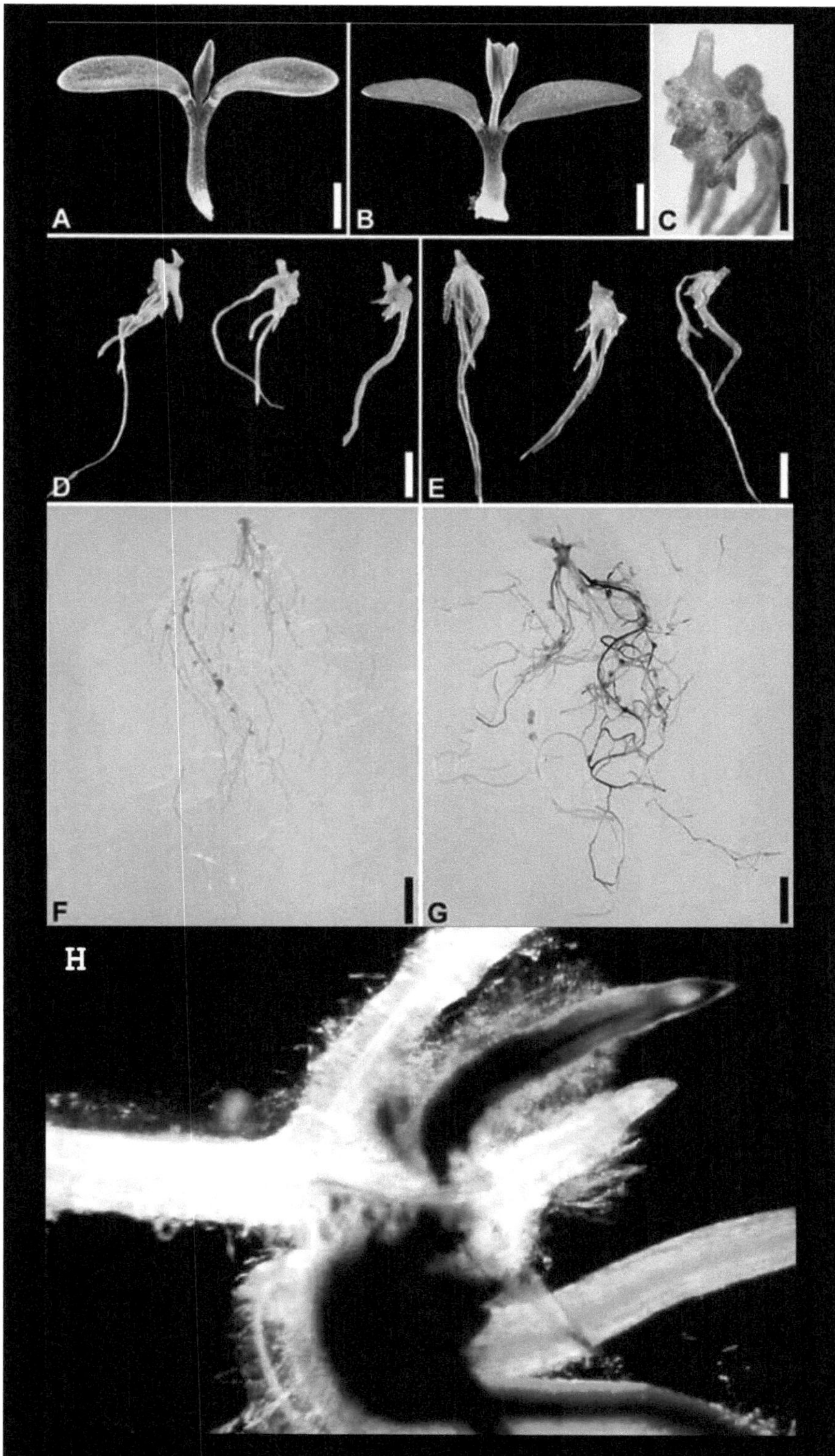

Colour plate 4.

SECTION 1

LOTUS JAPONICUS, A GENERAL INTRODUCTION

Chapter 1.1

LOTUS JAPONICUS AS A MODEL SYSTEM

Eloísa Pajuelo[1] and Jens Stougaard[2*]

[1]*Department of Microbiology and Parasitology, Faculty of Pharmacy. University of Sevilla, Prof. García González s/n, 41012-Sevilla, SPAIN;* [2]*Laboratory of Gene Expression, Department of Molecular Biology, University of Aarhus, Gustav Wieds Vej 10, 8000-Aarhus C, DENMARK; *Corresponding author.*

Email: stougaard@mb.au.dk

Phone: +45 8942 5011 Fax: +45 8620 1222

Keywords: Model legume, Lotus, symbiosis, ecotype, microsynteny

Lotus japonicus *was proposed as a model legume twelve years ago, because of a number of characteristics making this system very amenable for legume research. These characteristics include small plant, large and abundant flowers, easy hand pollination, high seed production, short generation time, easy cultivation, amenable to plant transformation and regeneration from tissue culture. At present, a set of genetic resources and tools has rapidly become available, including ecotypes, mutant lines, genetic maps, RIL lines, transformation procedures, EST sequences, and a whole genome-sequencing project. In these twelve years, research on* L. japonicus *has greatly contributed to the understanding of both symbiotic processes, i.e. with Rhizobium and mycorrhiza, making possible the cloning of several key genes involved in both symbioses. Now Lotus is regarded as one of the most useful plants for legume study and researchers who have interests in nodulation and other aspects of legume biology use it worldwide. In this introductory chapter we deal with the most general aspects and possibilities of the biology of* L. japonicus, *plant growth conditions, culture media, nitrogen supply and symbiotic partners; plant tissue culture; genetic transformation and regeneration of transgenic plants, contribution to the understanding of symbiotic processes and the role of this model plant for other research topics and exploiting microsynteny.*

INTRODUCTION

The family Leguminosae (Fabaceae), which is the third largest Angiosperm family, contains a number of important crop plants and woody trees, producing protein, carbohydrates, and oil for human consumption and animal fodder. Legumes are 'pioneer' plants that are able to grow in nutrient-poor soils by virtue of their ability to establish symbioses with nitrogen-fixing rhizobia bacteria, and with nutrient scavenging soil fungi, which provide the plants with phosphorous and other essential

A.J. Márquez (ed.), Lotus japonicus Handbook, 3-24.

nutrients. These natural features favoured the adoption of legumes in ancient agriculture, and make them an important part of sustainable agricultural systems to this day.

Among the plant families, symbiotic interaction with rhizobia is unique to legumes. The only known exception is *Parasponia andersonii* belonging to the elm family. Another symbiosis providing nitrogen to host plants is between gram-positive bacteria *Frankia* and several woody plant species forming actinorhizal symbiosis. Phylogenetic analysis based on DNA sequencing of the *rbcL* genes places all plants involved in rhizobial or actinorhizal symbiosis in the same clade (Rosid I), thus indicating that predisposition for nodulation evolved only once (Soltis et al, 1995, Doyle, 1998). However, different morphology and biochemistry of several types of nodules, indicate multiple origins of nodulation, perhaps even within the legume family (Doyle, 1998; Doyle and Luckow, 2003). The symbiotic interaction of legume plants with mycorrhizal fungi, though it is not exclusive to this family, is also an important interaction, which provides phosphorus and other nutrients to the plant. Obtaining sufficient nutrients from the soil is a common problem facing land plants, and 80% to 90% of all land plants are assisted in nutrient mining through association with symbiotic arbuscular mycorrhizal fungi. Both rhizobial and mycorrhizal symbiosis have been investigated previously, but modern molecular tools have only recently become available to study the genetics and genomics of root symbiosis. Increasing our knowledge of the biology and genetics of the legumes may improve this important agricultural resource as well as complement Arabidopsis, which does not develop rhizobial symbiosis or fungal symbiosis leading to vesicular arbuscular mycorrhiza.

BIOLOGY

With close to 18,000 different species belonging to the subfamilies Mimosoideae, Caesalpinioideae and Papilionoideae the legume family is a large and diverse plant family that includes several species of agricultural interest. The importance of the family from applied, botanical, and basic scientific perspectives merits a focused effort on a model legume. The poor ability of *Arabidopsis* to enter into symbiotic interaction with mycorrhiza and the inability to form nitrogen fixing root nodules in symbiosis with *Rhizobium* or it allies has been a further incentive to develop a model legume. A clear illustration of the need for "family specific" model plants is the relatively low levels of microsynteny between *Arabidopsis* and *L. japonicus* suggesting that the structure of legume genomes might be quite different from *Arabidopsis*. Recent bioinformatic based studies of genome evolution supports this view and indicates that legumes may not have undergone one of the genome duplication events inferred for *Arabidopsis*. Genomic research on models such as *Lotus japonicus*, should allow a fundamental understanding of the nodulation process and identification of genes involved in nodule formation and mycorrhizal symbiosis. Together with soybean, bean, cowpea, and *Sesbania*, *L. japonicus* belongs to legumes forming determinate nodules. Cortical cell divisions leading to the formation of the determinate nodule primordium initiate in the outer cortex. A meristem starts dividing to form the nodule primordium, and afterwards the meristematic activity ceases. Nodule size then increases via cell expansion.

Molecular characterisation of legume symbiotic genes and comparative genome analysis toward plants forming only mycorrhiza and plants like Arabidopsis with only limited capacity for symbiosis would contribute to the description of the molecular evolution of endosymbiosis.

Many cultivated legumes are either tetraploid, self-incompatible, or have a pod morphology making access to the seeds difficult, such as spiral pods in *Medicago*. *L. japonicus* is diploid, self-fertile, develops straight seedpods with two valves and seeds are arranged along a simple linear axis. Mature seedpods break along the rim, making seed harvest very easy. Seeds have no dormancy period and in the Lotus model, pod morphology is similar to the morphology of the most important grain legumes like soybean, pea, and cowpea making it a prime model for comparative studies of seed development.

L. japonicus (Regel) Larsen was first recognized at the ancient capital of Japan, Kyoto, centuries ago. The natural habitat for *L. japonicus* is in East- and Central-Asia including the region around Japan, Korea, China, and stretching into Afghanistan. In the 1950s, Professor Isao Hirayoshi (Kyoto University) collected *L. japonicus* plants growing on a riverbank in Gifu. Professor William F. Grant (McGill University, Montreal) collected its progeny as the accession B-129. In 1992, Kurt Handberg and Jens Stougaard of University of Aarhus (Denmark) obtained B-129 and established Lotus as a valuable tool for modern legume research. Many ecotypes of Lotus can be found growing in its natural habitat. Researchers in Japan have taken advantage of this and dozens of accessions have recently been collected (www.beansbase.agr.miyazaki-u.ac.jp/). Resources for investigating genotypic variability contributing to actual and potential traits of agronomic importance, such as seed yield, plant height, cold tolerance, and disease resistance is thus available for identification by mapping of quantitative trait loci (QTLs). In this context the possibility of obtaining fertile F1 plants and segregating F2 populations after crossing *L. japonicus* to diploid relatives like *L. filicaulis*, *L. frondosus*, *L. schoelleri*, *L. burttii* and others (Grant et al, 1962, Sz-Borsas et al, 1972, Sandal et al. 2002) is an additional source of biodiversity.

Ecotypes

A number of *L. japonicus* ecotypes have been collected. This material is a rich source of biological (phenotypical) and genetic variability. At present, 75 *L. japonicus* ecotype lines from different areas of Japan can be requested on the *L. japonicus* seed center at Miyazaki University Japan web site, http://www.beansbase.agr.miyazaki-u.ac.jp. The ecotype Miyakojima (MG-20) found on the island of the same name, has been used as a crossing partner with Gifu in order to establish populations for map-based cloning, since it has the highest degree of polymorphisms, about 6%, among other 15 ecotypes tested (Kawaguchi, 2000). Miyakojima displays several traits that distinguish it from other Japanese accessions: low concentrations of anthocyanin in the stem and petals, few trichomes, a more upright habit, broad leaflets and petals, and large black seeds. The first two traits, which are controlled by single recessive genes, serve as useful markers for following mutant crosses. The Kazusa DNA Research Institute is currently sequencing the MG-20 genome.

Morphological description of plant developmental stages

L. japonicus is a small prostrate perennial that flowers profusely. Flowers are yellow and relatively big, generally two per peduncle. It is self-fertile with a straight, cylindrical, and dehiscent pod containing up to 20 seeds.

L. japonicus grows relatively slowly during the first weeks after germination. The primary plant is small, allowing high-density germination, mutant plant screening on agar plates, nodulation tests, selection of transformants on selective conditions, etc. However, the mature plant has a bushy growth with many branches up to 30 cm long and many leaves, which makes it very useful for biochemical, chemical and physiological studies, since abundant material can be harvested. The mature plant exhibits abundant indeterminate flowering, with large yellow flowers, which are usually arranged in pairs and less frequently one or three flowers per peduncle. Mature flowers on the ecotype Gifu are about 8-12 mm in length. Under greenhouse conditions, *L. japonicus* has a generation time of approximately 3 months. Fertilisation and seed set occur without manipulation of the flowers. Pod formation in fertilised flowers occurs within 3-4 days. Mature seedpods are about 3 cm in length and contain up to 20 seeds. Mature seed weight is around 1.0-1.2 mg. Because *L. japonicus* is a perennial plant, new branches and flowers continue to develop after removing the old branches.

Colour plate 1 illustrates the main morphological characteristics of *L. japonicus* plants.

Cross-pollination

In order to use *L. japonicus* for genetic studies and map based cloning a cross-pollination method is required. The method described by Jiang and Gresshoff (1997) is generally used. Forceps are used to cut the tip of the petals and emasculate the flowers, gently removing the anthers. Pollen from the donor flower is deposited onto the stigma of the emasculated young flower, whose stigma should be slightly bent.

In Figure 1, from Jiang and Gresshoff (1997), the correct age of the flower is shown in C whereas the flowers in A and B are too young and inappropriate for fertilisation as female flower. The flower in D is too old (pollen has been released from the anther and self-pollination has already taken place). Optimal flowers for cross-pollination are 6-7 mm in length and the standard petal is still fused, they show no curvature of the style as well as complete extension of the anther filament and ripening of the pollen. The flowers used for crossing can be protected against drying out with small bags. It is important that the humidity is high during the development of the pod from the cross. *L. japonicus* ecotypes are generally easy to cross. *L. japonicus* can also be crossed to related diploid Lotus species like *L. filicaulis*.

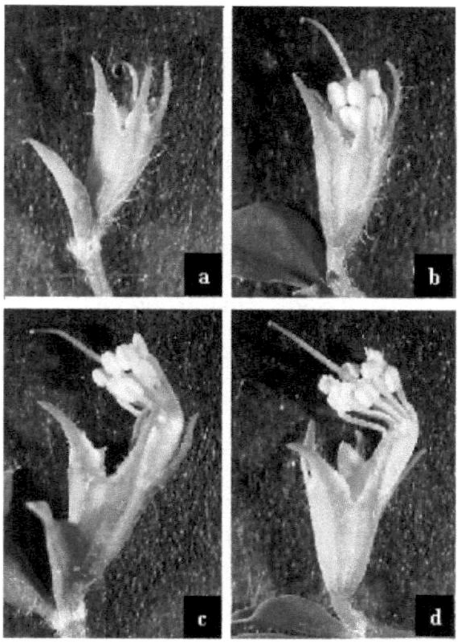

Figure 1. Developmental stages of Lotus japonicus *flowers relevant to hand pollination (reproduced with permission from Jiang and Gresshoff, 1997). See colour plate 1 for this figure in colour.*

As a legume model plant, *L. japonicus* offers a suit of advantages; high transformability, self pollinating proliferation, small genome size (ca. 450 MB), small plant size, large seed set and short life cycle (ca. three months) (Table 1). In addition to its model status, *L. japonicus* belongs to the Loteae and represents the genus *Lotus* consisting of approximately 180 distributed worldwide (Kirkbride Jr. J.H. 1999). Among the genus *Lotus,* one of the species of greater interest from an agricultural point of view is the tetraploid *Lotus corniculatus* L. (birdsfoot trefoil) 2n=24. *L. uliginosus/L. pedunculatus* Schkuhr (Big trefoil) (2n=12), *L. tenius/L. glaber* (Waldts&Kit) (Narrow-leaf birdsfoot trefoil) (2n=12 or 24) and *L. subbiflorus* (2n=12 or 24) are also used in agriculture (Swanson et al., 1990). All of these cultivated species are either self-incompatible or have marginal to low self-fertility. *L. corniculatus* (birdsfoot trefoil) is an herbaceous perennial dehiscent species native to Europe, where it is widely distributed, and has extensively colonized in North America and South America. *L. corniculatus* has become a successful forage crop during the past century both in Europe and America. While *L. corniculatus* has been reported as having diploid ecotypes (Grant 1995), it is generally known as a tetraploid species (2n=24). The anti-bloating properties due to tannin content, ability to grow in unfertile, acid or alkaline soils, and drought/flooding tolerance are the main agricultural advantages. In addition, the species has become widely used to prevent roadside erosion.

Growth	Growth in a variety of conditions	
	Short generation time	3-4 months
	Primary small plant, but bushy plant after secondary shoot formation, branching	
	Perennial	
	Straight seed pods, indeterminate flowering	
	Small seeds (1.2 g per 1,000 seeds)	
Propagation	Self-fertile	
	Large and abundant flowers	
	High seed production	Up to 20 seeds per pod, 6000 seeds per plant
	Vegetative propagation from nodal sections	
	Easy hand pollination	
Genome characteristics	Small genome	450 Mbp
	Diploid	2n=12
Plant tissue culture	Regeneration from callus	
	Amenable to plant transformation with *A. tumefaciens* and *A. rhizogenes* and regeneration of transgenic plants.	
	Hygromycine, kanamycine and phosphinothricin selection possible	
	Transgene stable in subsequent generations	
Symbiosis	Rhizobial and mycorrhizal symbiosis	*Glomus intraradices, Gigaspora margarita*?
	Determinate nodules	
	Fast growing *Mesorhizobium loti*	R7A, NZP2235, MAFF303099 TONO, NGR234
Tannin content	Low levels of tannin accumulation	No tannins in leaves
Galactomannan content in seed endosperm	Endospermic seeds	Galactomannan with a high degree of Gal substitution (Man/Gal=1.23)

Insect interaction	Cyanogenic and nitrile glucosides content: Lotaustralin and linamarin Rhodiocyanosides A and D	Lotaustralin:linamarin ratio approx. = 10
Spider mite interaction	Emission of volatile compounds (terpenoids) to attract mites	(*E*)-*β* -ocimene
Nematode interaction	Root-knot and cyst nematodes	Sensitive to root-knot nematodes: *M. incognita, Meloidogynes sp.* Resistant to cyst-nematode: *H. glycines* (hypersensitive response)

Table 1. L. japonicus *model plant features (taken from Handberg and Stougaard, 1992).*

PLANT GROWTH CONDITIONS, CULTURE MEDIA, NITROGEN SUPPLY, AND SYMBIOTIC PARTNERS

L. japonicus is a species able to grow under a variety of conditions. Many growth systems have been developed depending on the final use of the plant, for instance, microscopical observation, biochemical analysis, genetic screening, etc. Usually, the best conditions for plant cultivation are temperatures between 18-22°C and to avoid seedpod dehiscence, humidity around 70%. Photoperiods of 16 h light and 8 h dark are common plant growth chamber conditions. Among the inert materials used as support for plant growth are clay granules (leca), a 50% mix of perlite:vermiculite, or even vermiculite alone. *L. japonicus* is somewhat sensitive to over-watering. Watering from below is generally better than flooding from above and a "semidry" air-filled root support is better than a dense water soaked support. The pillow system is another system for plant growth and nodulation experiments and consists of bags filled with sand: vermiculite and submerged in plant growth solution (Szczyglowski et al., 1998). Seeds are planted in between two bags, and roots develop in between the bags. This system is also good for root and nodulation studies, since it is easy to pull out the plant without distorting roots and nodules. When grown on peat, plants can develop a bigger size, with stems up to 30 cm long and many branches and side shoots, giving a bushy appearance. This is the best condition for seed production. After a round of flowering and seed set, old branches can be removed and plants can regenerate new branches and start flowering again.

Plants are sufficiently small during the first 4-6 weeks to be grown on plates under sterile conditions on plant growth medium-agar. In this case, covering the root of the plants with dark paper or keeping the lower part of the plant under the dark is recommended. These conditions are especially good for nodulation studies, since it is easy to see nodules without taken the plant out of the plate. Magenta boxes filled with vermiculite or perlite:vermiculite are also suitable. In all these closed systems, the addition of ethylene inhibitors as L-α-(2-aminoetoxivinil)glycine (AVG) has been recommended in nodulation experiments (Pacios-Bras et al., 2000).

An alternative growth system is hydroponic culture. In this case, seeds are placed on a mesh and submerged in plant growth solution, under suitable temperature and photoperiod conditions. A similar system is aeroponic cultures, where plants are also germinated on a mesh, and plant growth solution is misted around the roots. Plants develop a big root system under these conditions, which is useful for experiments requiring a large amount of root tissue.

Plant growth solutions

Broughton & Dilworth solution (1971) often diluted one fourth and also Jensen solution have been used extensively for *L. japonicus* nodulation experiments, since they are N-free. Plants inoculated with compatible *Rhizobium* can grow well without additional nitrogen. A detailed study on nitrogen metabolism in this plant is included elsewhere in this book. For non-inoculated plants, Hornum solution has also been used (Handberg and Stougaard, 1992). Plants grow better in Hornum solution containing both ammonium and nitrate as nitrogen sources. Nitrate alone can also be used as nitrogen source, when the plant is not inoculated. Ammonium alone is not an optimal nitrogen source for *Lotus*, and growth in the presence of ammonium is slower. Nevertheless, it has also been used as nitrogen source in the form of ammonium succinate, which supports plant growth reasonably well. Seedlings are sensitive and may turn yellow on high nutrient concentrations. The best growth is often obtained by germination and seedling growth on water or low nutrients followed by application of nutrient solution after plants start to develop trifoliate leaves.

Frequently used culture solutions

- Hornum solution (100x stock): 40 g/l NH_4NO_3, 30 g/l KNO_3, 30 g/l $MgSO_4.7H_2O$, 10 g/l $NaH_2PO_4.H_2O$, 2 g/l Fe-EDTA (9% Fe), 120 mg/l $MnSO_4$.H2O, 120 mg/l H_3BO_3, 40 mg/l $CuSO_4.5H_2O$, 40 mg/l $ZnSO_4$.7H2O and 8 mg/l $Na_2MoO_4.2H_2O$. Diluted in tap water and pH adjusted to 6.8.
- Broughton & Dilworth solution: $CaCl_2$ 1 mM; KH_2PO_4 0.5 mM; Fe-citrate 10 μM; $MgSO_4$ 0.25 mM; K_2SO_4 0.25 mM; $MnSO_4$ 1 μM; H_3BO_3 2 μM; $ZnSO_4$ 0.5 μM; $CuSO_4$ 0.4 μM; $CoSO_4$ 0.2 μM; Na_2MoO_4 0.2 μM.
- Jensen medium: 0.1 g/l $CaHPO_4$; 20 mg/l K_2PO_4; 20 mg/l $MgSO_4.7H_2O$; 20 mg/l NaCl; 10 mg/l Fe-citrate; 100 μg/l Na_2MoO4.$4H_2O$; 200 μg/l $MnSO_4.4H_2O$; 10 μg/l $CuSO_4.5H_2O$; 25 μg/l $ZnSO_4.7H_2O$; 300 μg/l H_3BO_3.

Nodal propagation

The perennial character of this plant enables easy vegetative propagation: individual plants can be multiplied from nodal cuttings with a rooting frequency of 90%, even in the absence of hormones. For clonal propagation of individual plants, nodal sections 2 cm long with one trifoliate leaf are firmly pressed into vermiculite watered with Hornum solution and covered with a transparent plastic bag for a few days. Growth stops for a few days until a small root develops and then resumes, so it can be possible to get flowers in 1.5 months. Alternatively, we have also obtained plants from nodal sections in hydroponic cultures. A 5-6 cm long branched cutting is placed on a piece of polyurethane provided with a hole, which is floating on the Hornum nutrient solution, so the section can be submerged in the solution inside a plastic tray filled with solution and covered with a transparent tap. The cuttings quickly develop roots under these conditions, and can be transferred to peat or vermiculite afterwards.

Symbiotic partners

Both "fast-growing" *Mesorhizobium loti* and "slow-growing" *Bradyrhizobium* sp. (*Lotus*) strains can nodulate *L. japonicus*. Fast-growing strains induce nodules effective in nitrogen fixation while Bradyrhizobium sp. induce only ineffective nodules. Among the fast growing *M. loti* strains, the following are commonly used: R7A, NZP2235, JRL501 (Niwa et al., 2001; Kawaguchi et al., 2002), MAFF 303099 (complete genome sequence NC002678) (Kaneko et al., 2000; Stracke et al, 2002), and TONO (Kawaguchi et al., 2000). The fast-growing broad host range NZP2037 induces effective nodules but the frequency of nodulation is relatively low. Many of these strains are available with reporter genes such as *lacz*, *GFP* or *GUS*, in order to follow the different steps of the nodulation process, as visualisation of infection and nodule colonisation. Nodules are visible 7-10 days post-inoculation, clustered along the taproot. Three weeks after inoculation, nodules are about 1 mm in diameter. If inoculation is done with a GUS-carrying strain, extensive blue coloration in the nodule interior caused by GUS expression and infection thread structures can be visualised after staining. New rounds of nodulation can occur in the younger part of the root, so nodules at different developmental stages can be found in the same plant. The broad host range *Rhizobium* sp. NGR 234 has also been reported to nodulate *L. japonicus*, (Hussain et al., 1999). *Rhizobium etli* (Banba et al., 2001) nodulates but nitrogen fixation efficiency is poor and nodules senesce rapidly.

The major Nod-factor secreted by *M. loti* consists of a pentameric N-acetylglucosamine, which carries a cis-vaccenic acid and a carbamoyl group at the non-reducing end of the molecule and a 4-O-acetyl fucose at the reducing terminal residue (López-Lara et al., 1995, Niwa et al., 2001).

PLANT TISSUE CULTURE

Protocols for plant tissue culture and regeneration of plants were given in Handberg and Stougaard (1992). Calli can be obtained from different plant tissues, including hypocotyls, leaves, roots, and petals cultivated on B5 medium containing 2,4-D and kinetin. However, regeneration of the plant is better when done from hypocotyl explants. Shoot induction is achieved in B5 supplemented with 0.2 mg/ l BAP and 10 mM ammonium sulphate, and then 3-4 cm long shoots are transferred to soil mixture in the greenhouse for rooting using the conditions for cuttings (see chapter from Tirichine et al Ibid).

GENETIC TRANSFORMATION AND REGENERATION

Plant transformation and regeneration is a prerequisite for the molecular analysis of genes involved in nodulation and symbiotic nitrogen fixation. The molecular analysis of plant genes involved in nodulation has been hindered by the inability to produce high numbers of transgenic legume lines, since some legume plants are very recalcitrant to plant transformation and regeneration of stable transformants. Furthermore, efficient regeneration of transgenic plants is required for studies such as promoter/enhancer trapping and insertional mutagenesis (Thykjaer et al., 1995; Schauser et al., 1998; Martirani et al., 1999; Webb et al., 2000).

Species, Genotype	DNA Delivery	Explant	Selection		Citation
			Marker	Agent	
Bird's foot trefoil (*L. corniculatus* Leo)	At (LBA4404)	Leaves	*aphIV*	Hyg	Webb et al. (1996)
L. japonicus Gifu	At (LBA4404, C58C1, GV2260)	Hypocotyls	*nptII*, *hpt*	Kan, Hyg	Handberg and Stougaard (1992)
	Ar (9402, AR10)	Seedlings	*nptII*	Kan	Stiller et al. (1997)
	At (AGL1)	Hypocotyls	*bar*	Phosphinothricin (PPT)	Lohar et al. (2001)
	At	Root	*hpt*	Hyg	Lombari et al (2003)

Table 2. Summary of legume transformation systems yielding transformed plants. At = A. tumefaciens; *Ar* = A. rhizogenes. *(Modified from Somers et al., 2003).*

Table 2 summarises *Lotus* transformation systems. The first *L. japonicus* transformation-regeneration protocol was reported in 1992 by Handberg and Stougaard, along with the proposal to use this species as a model legume. Transformation of *L. japonicus* has been achieved by *Agrobacterium*-mediated hypocotyl transformation (Handberg and Stougaard, 1992; Thykjaer et al., 1995; Stiller et al., 1997). In these methods, selection for transgenics is accomplished using antibiotic-resistant selectable markers such as *nptII* and *hptII* genes. This method was further optimised and the time to produce whole plants reduced to about 4 months by Stiller et al. (1997). A collection of wild-type *A. rhizogenes* strains was tested for infectiveness and the most virulent strains were selected for further use. To open the possibility of searching for mutant phenotypes, a regeneration system from hairy roots cultures was developed enabling the regeneration of large numbers of transgenic plants in about 5-6 months. However, the whole regeneration process although effective is relatively expensive and time-consuming. Additionally, about 10-20% of the transgenic lines regenerated through these schemes are sterile, probably due to somaclonally-induced abnormalities. Somaclonal variation and sterility were significantly reduced by use of the bar gene as selectable marker and selection with phosphinothricin (PPT) (Lohar et al., 2001). The bar gene encodes Phosphinothricin Acetyl Transferase that detoxifies phosphinothricin (PPT). Transgenic *L. japonicus* plants resistant to PPT were positive upon PCR by bar gene-specific primers and in most cases PPT resistance segregated as a single dominant allele indicating a single T-DNA insertion into the plant genome. All regenerated plants were fertile and void of visible somaclonal abnormalities. More recently a new transformation-regeneration procedure has been reported (Lombari et al., 2003) in which dedifferentiated root explants are a source of large numbers of cells susceptible to *Agrobacterium*-mediated transformation resulting in a tenfold increase in the number of transformants within 4 months. *L. japonicus* is easily transformed by *A. rhizogenes* and hairy roots can form nodules 3-4 weeks after emergence. Composite plants with a transgenic root system can thus be used for complementation of plant mutants and other transgene studies.

L. JAPONICUS: UNDERSTANDING SYMBIOTIC PROCESSES

Since *L. japonicus* was proposed as a model legume a set of genetic resources and tools have rapidly become available, including ecotypes, mutant lines, genetic maps, RIL lines, transformation procedures, EST sequences and sequence of the gene rich regions of the genome (VandenBosch and Stacey, 2003). Techniques such as plant tissue culture, nodulation assay, crossing and histology have already been established for *L. japonicus* research (Jiang and Gresshoff, 1993; Szczyglowski et al., 1998; Hussain et al., 1999). Now, Lotus is regarded as one of the most useful plants for legume study and researchers who have interests in nodulation and other aspects of legume biology use it worldwide. Other resources developed for use in *L. japonicus* like forward and reverse genetics, gene tagging, a high-density genetic linkage map, establishment of inbred populations for map-based cloning of genes, functional genomics projects for the completion of the genome sequence of *L. japonicus*, microarrays, proteomics, metabolomics, and exploiting microsynteny are described in individual chapters of this book.

Forward genetics in *Lotus* has been approached in several laboratories (Schauser *et al.,* 1998; Martirani et al., 1999; Webb *et al.* 2000, T. Aoki, unpublished data). T-DNA tagged lines have been obtained and the maize *Activator* transposon was reported to be mobile in *L. japonicus* and could be used as a mutagen (Thykjaer et al., 1995; Schauser *et al.* 1999). This has allowed the cloning of the *Nin* (nodule inception) nodulation gene. The molecular characterization of the *nin* mutation is considered a scientific milestone in the knowledge of nodulation, since it was the first plant gene for nodule formation to be cloned using molecular genetic methods. The *Nin* gene encodes a regulator protein that triggers the initiation of infection threads and the onset of cortical cell division (Schauser *et al.* 1999). The putative protein has a DNA binding/dimerization domain showing strong homology to Mid (Minus dominance) proteins from *Chlamydomonas reinhardtii*, which is involved in regulation of sexual reproduction in response to nitrogen starvation. *Ljcbp1* has been isolated by promoter tagging as a transgenic line showing GUS reporter activity in roots only after infection with *M. loti* or after colonisation by *G.*

intraradices and *G. mossae*. Sequencing of the tagged locus revealed an open reading frame encoding a protein with similarity to calmodulins, and this protein undergoes a mobility shift in the presence of calcium (Webb et al., 2000).

A collection of mutants has been developed from the standard accession Gifu by chemical mutagenesis. Symbiotic as well as other types of mutants were isolated from these populations (Imaizumi-Anraku et al., 1997), (Schauser et al., 1998), (Wegel el al., 1998), (Szczyglowski et al., 1998), (Bonfante et al., 2000), (Senoo et al., 2000), (Kawaguchi et al., 2002), (Orea et al., 2002). In *Lotus*, mapping work is in progress to assign more than 100 monogenic recessive symbiotic nodulation mutations into their respective *loci*. For molecular characterisation of EMS mutants from these collections, a map-based cloning procedure was established and genetic linkage maps mainly based on micro satellites and amplified fragment length polymorphism (AFLP) are available. Two Lotus populations are being used in mapping studies: (a) progeny of an interspecific cross between *L. japonicus* ecotype Gifu and *L. filicaulis*, and (b) progeny of a cross of the two ecotypes Gifu and Miyakojima. (Hayashi *et al.* 2001; Sandal *et al.* 2002, Nakamura *et al.* 2002; Sato et al., 2001). Several BAC and YAC libraries have been constructed (Men et al., 2001; Kawaguchi et al., 2001; Asamizu et al., 2003). Reverse genetics has been accomplished using a TILLING strategy (Perry et al., 2003).

Developing the *Lotus* genetics has already allowed the positional cloning of four genes that condition nodulation phenotypes: The *Lotus* SYMRK gene encodes a receptor like kinase with an LRR-domain, that plays an essential role in both AM and *Rhizobium* symbioses (Stracke *et al.* 2002). The *har1* gene is involved in long-distance signalling of nodulation auto-regulation (shoot controlled auto-regulation of nodule number) and also lateral root development. Map-based cloning of the mutation indicated that the mutated gene encodes a receptor kinase with a LRR repeat, a transmembrane domain and an intracellular serine/threonine kinase domain with homology to CLAVATA1 (Krusell et al., 2002; Nishimura et al., 2002a).

The earliest visible cellular change in response to Nod factors is root hair deformation 24 hours after inoculation. This is not observed in the mutants *Ljnfr1* and *Ljnfr5*. Short-term analysis of the response indicates that there is no depolarization nor extracellular alkalinization within minutes after application of purified Nod factors. There was also no expression of genes downstream in the signalling pathway (*Nin, Enod2, LjCBP1*). *Nfr1* and *Nfr5* encode receptor kinases involved in the recognition of Nod factors in *L. japonicus* (Radutoiu et al., 2003; Madsen et al., 2003). Both proteins have LysM domains in the extracellular moiety, which is probably involved in the recognition of the backbone of Nod-factor, a transmembrane domain and a Ser/Thr kinase (PK) domain in the intracellular part. However, the PK domain of *Lj*NFR5 is atypical since it lacks the activation domain. The simplest model is that NFR1 and NFR5 act together in the recognition of the Nod-factor. Besides these four genes whose cloning has been achieved by map-based cloning, molecular characterisation of three other symbiotic loci involved in Nod-factor signal perception or immediate downstream signal transduction are in a very advanced stage.

In a few cases, it is possible to find the mutated gene based on phenotypic similarities. One gene, *LjBzf*, was identified by making use of the phenotypic similarity of a mutant (*Ljsym77*) in this gene to the Arabidopsis *hy5* mutant (Nishimura et al., 2002b). *LjBzf* encodes a basic leucine zipper (bZIP)-type zinc finger protein with different roles in shoot (photomorphogenic development) and roots (lateral root formation and number of nodules).

Functional genomics. Comprehensive genome and gene information are being generated as a result of the genome sequencing initiative conducted at the Kazusa DNA Research Institute, Japan, and the expressed sequence tags (ESTs) sequencing programs started in several laboratories (Asamizu *et al.* 2000, 2004; Sato *et al.* 2001; Cyranoski 2001; Colebatch et al., 2002). Efforts to identify global gene expression patterns associated with nodule transport, metabolism, and maintenance were addressed via cDNA array technology and transcriptome analysis (Colebatch et al., 2002 and 2004; see also VandenBosch and Stacey, 2003).

Proteomics is also a powerful tool for identifying genes involved in a plant process analysing the spectrum of proteins expressed in the cell using two-dimensional gel electrophoresis and liquid chromotography followed by mass spectrometry. Proteins present in the peribacteroid space and peribacteroid membrane of *Lotus* nodules have been identified (Weinkoop and Saalbach, 2003). This approach led to the identification of a large number of proteins: 94 known proteins matching database and 24 ESTs of unknown identity were detected at the PBM. A receptor protein kinase different to other recently described receptor kinases involved in nodule formation, and a potential plant defence protein (HIR protein) were found.

Metabolome analysis. Metabolite profiling, using gas chromatography-mass spectrometry has also commenced and has revealed quantitative changes in flavonoids after mycorrhizal infection. Comparisons were made of wild-type leaf, stem, and root tissues of *L. corniculatus* and *L. japonicus*. This comparison showed that although shoot profiles are similar, the roots of these two species are different. GC-MS analysis has also been used to profile changes in metabolites in nodules compared to roots (Colebatch et al., 2004) and also to study nodulation of mutant, non-N-fixing plants and wild-type nodules containing mutant rhizobia (Udvardi, this book).

L. JAPONICUS AS A MODEL FOR OTHER CURRENT RESEARCH TOPICS

More recently *L. japonicus* was also used in model studies for plant-nematode interactions (Lohar and Bird, 2003). In order to complete their life cycles both root-knot and cyst nematodes must interact with their host plants and induce specialised feeding structures by redirecting plant developmental pathways. Although much attention has been paid to root penetration, migration and feeding sites, little is known about host plant factors involved in the plant-nematode parasitic interaction. Parallels have been observed with rhizobial nodule development on legumes. Surprisingly, the *L. japonicus* hypernodulating mutant *har1*, which is defective in a CLAVATA1-like receptor kinase gene, was found to be hyperinfected by *M. incognita*. However, another hypernodulated Lotus mutant exhibited the same level of *M. incognita* infection as wild-type plants. In contrast to the response to root-knot nematode, the Gifu ecotype is resistant to *H. glycines*, and elicits a hypersensitive response. A model legume would greatly facilitate a better understanding of the differences between parasitic (nematode) and mutualistic (rhizobia and mycorrhizae) symbioses (Parniske, 2000). Conditions for efficient parasitism by root-knot nematode (*Meloidogyne spp.*) and by soybean cyst nematode (*Heterodera glycines*) have been established. Thus *L. japonicus* is emerging as a powerful model legume for studying compatible and incompatible plant-nematode interactions.

Other groups (Forslund et al., 2004) involved in the study of plant-insect interaction have introduced *L. japonicus* as a genetic model system to study cyanogenic glucosides, which are precursors of toxic hydrogen cyanide providing plants with an immediate chemical defence response to herbivores and pathogens that cause tissue damage (Møller and Seigler, 1999; Zagrobelny et al., 2003). *L. japonicus* contains the two cyanogenic glucosides, linamarin and lotaustralin, derived from Val and Ile, respectively. The glucosidase activity is high, causing rapid cyanide release upon tissue damage. The *L. japonicus* genome has been found to contain two CYP79 orthologs assigned as CYP79D3 and CYP79D4, which show different expression patterns at the tissue level. Metabolic engineering of cyanogenic glucoside synthesis in *L. japonicus* proceeds following two parallel approaches. Transgenic plants over-expressing pathway enzymes from cassava or sorghum are being analyzed for altered patterns of glucoside accumulation.

L. japonicus also serves as a useful system to study indirect plant-herbivore defense mechanisms in a legume. When plants are infested by herbivores, they start emitting a specific blend of volatiles (herbivore-induced volatiles) that attract carnivorous natural enemies of herbivores thus potentially protecting the plant (Paré and Tumlinson, 1999; Kessler and Baldwin, 2001; Pichersky and Gershenzon, 2002). Low-molecular-weight terpenoids such as monoterpenes (C-10), sesquiterpenes (C-15), and homoterpenes (C-11 or C-16) are among the most common plant volatile compounds induced and emitted upon herbivory and they are formed by the activity of families of terpenoid synthases (TPS). The acyclic monoterpene hydrocarbon compound (E)-β-ocimene is one of the most common volatile chemicals released from plants in response to herbivory. Shoots of *L. japonicus* (ecotype Gifu B-129) infested with spider mites released a blend of volatiles, which contributed significantly to the attraction of predator mites (Ozawa et al., 2000✦). Arimura et al., (2004) describe the identification, cloning, and functional characterization of a cDNA for (*E*)-β-ocimene synthase from *L. japonicus* ecotype Miyakojima MG-20. Increased emissions of (*E*)-β-ocimene as well as other volatiles from spider mite-infested *L. japonicus* plants seems to be associated with increased transcript levels of (*E*)-β-ocimene synthase, suggesting that herbivorous spider mites induce monoterpene volatile emissions by a mechanism that involves induced TPS gene expression.

Proanthocyanidins (or condensed tannins) are polymerized units of flavan-3-ols arising from a poorly understood branch of the flavonoid pathway. Interest in these plant metabolites has risen dramatically in recent years because of their anti-oxidant and protein-precipitating properties and their role in improving forage quality, human health, and plant resistance. Condensed tannins are found in a range of higher plant species. Trefoil can produce condensed tannins (CT) that can be beneficial or detrimental to grazing animals, depending upon the amount produced. These secondary compounds are of particular value in forage crops because their presence in the foliage of herbage legumes at levels of up to 3 to 4% dry weight is regarded as being a beneficial agronomic trait. The accumulation of tannins in shoot tissues is beneficial for grazing livestock from the perspective of the prevention of pasture bloat. High tannin concentrations have disadvantages that include inhibiting enzyme activities, inhibiting digestion of soluble carbohydrates and hemicellulose, decreasing palatability by producing an astringent reaction in the mouth and possibly degrading the gut wall of the rumen. This phenomenon mitigates against the effective use of a number of potential forage crop species that have good levels of primary productivity, often in non-optimal environments, but are of limited nutritive value because of accumulation of high levels of antinutritive secondary compounds. The nonbloating characteristic of birdsfoot trefoil, *Lotus corniculatus*, is attributed to the production of condensed tannins. The condensed tannin concentrations of birdsfoot trefoil were low enough to still be beneficial to protein availability without affecting palatability to grazing animals (Gebrehiwot et al., 2002). Although *L. japonicus* is not used in agriculture, it has also been studied for tannin accumulation. Condensed tannins were distributed throughout all the tissues of the plant although at a very low level, with the notable exception that unlike in *L. corniculatus,* leaf tissue was found to be free of tannins (Morris et al., 1993). Recently, the first characterization of a regulatory gene associated with leaf proanthocyanidin biosynthesis has been reported (Gruber et al., 2003.). A novel set of highly homologous basic helix-loop-helix (*myc*-like) genes from the forage trefoils (*Lotus* species) has been isolated. Expression of *TAN*1 correlates well with proanthocyanidin content in leaves of several forage legumes. A homologue of this gene is interrupted in a *L. japonicus* mutant (*tan*1) that accumulates leaf proanthocyanidin, opening up the possibility of genetically manipulating levels of tannin accumulation in crop and forage legumes. For plants with very high tannin content, the opposite strategy has been taken, using antisense technology, to decrease the levels of condensed tannins in crop species. Transgenic birdsfoot trefoil (*Lotus corniculatus* L.) plants harbouring antisense dihydroflavonol reductase sequences have been obtained with decreased levels of tannins (Colliver et al., 1994; Robbins et al., 1998).

Seedpod shattering in cultivable *Lotus* species remains an agricultural problem. Shattering of the pod results in yield reduction and causes the high price of *Lotus corniculatus* seed for the farmer. Some *Lotus* species have been crossed to *Lotus corniculatus* in order to transfer desirable traits, such as pod indehiscence. Complete pod-shattering resistance has not been reported for any species in the *L. corniculatus* group. Such species exhibiting partial to complete resistance to pod shattering include *L. conimbricensis* Brot., *L. edulis* L., *L. ornithopodioides* L. and *L. tetragonolobus* (*Tetraglonobus purpureus*) Moench. Attempts to transfer the indehiscent character by hybridization between diploids and the tetraploid *L. corniculatus* have met with little success because of the chromosome number differences between the species (Grant, 1996). Three genes have been identified which control shatter in the model plant *Arabidopsis thaliana*: FUL, SHP1, and SHP2. Two SHP homologous genes have been identified in the model legume *L. japonicus* that are expressed in pod walls. A FUL-like gene was also identified in *L. japonicus*, but it shared less similarity to its *Arabidopsis* counterpart (Bilyeu and Beuselink, 2001). Success in producing indehiscent *L. japonicus* can be rapidly applied to the closely related crop species *L. corniculatus*, and has broad implications for other dehiscent species.

EXPLOITING MICROSYNTENY: THE ROLE OF A MODEL PLANT

The effort to identify and characterize genes and their functions in model species is proving to be particularly valuable. By focusing on relatively few species as models, the conservation of gene function in plants can be examined and those similarities and differences exploited for agronomic benefit in crops. Knowledge of the syntony (that is, the existence of chromosomal regions from different species containing orthologous genes more or less in the same order) between model legumes and economically important legumes would help us to clone genes more easily from the latter, for which little genetic information is available. In addition, syntenic relationships will help us to understand how legume crops evolved. The existence of a wide region of macrosyntony between pea and *L. japonicus* (including SMRK, SMHT, ENOD40, CHS and PR5 in *L. japonicus* and PsSym19, SMHT, ENOD40 and CHS together with PsSym10 in pea) has helped the cloning of Pssym19, the orthologue of SMRK (Stracke et al., 2004) and *PsSym10*, the orthologue of *LjNfr5*. Clustering of all these genes and the existence of macrosyntony between these two legumes is very significant. A specific chapter on this subject is included in this book. The similarity between mutant phenotypes in crop and model plants can also aid the cloning of important crop genes, as was the case for the mutant *sym35* from *Pissum sativum* (Borisov et al., 2003), whose mutation affected the orthologue gene of *Nin* from *L. japonicus*.

Discovery of the whole signal cascade between macro and microsymbionts and comparison of genes with the orthologues in non-legumes, if they exist, will shed light on how older genes could have been recruited from functions that are more ancient and redirected to participate in the more recent legume-*Rhizobium* interaction (Szczyglowski and Amyot, 2003; Rodríguez-Llorente et al., 2004; Tangsenco et al., 2004). This would open new possibilities for engineering of non-legume plants and their interactions with endosymbionts.

Since *L. japonicus* was proposed as a model legume a decade ago, it has contributed greatly to the description of new genes involved in the symbiotic process (Harris, 2002; Downie and Parniske, 2002; Szczyglowski and Amyot, 2003; Morris and Walker, 2003). Newly available tools and approaches will make it possible to obtain a clearer picture of the molecular evolution of symbiosis. Clearly, studies on *L. japonicus* are likely to prove invaluable both to understanding the basic biology of legumes and to improve leguminous crops that have large and/or polyploid genomes.

ACKNOWLEDGEMENTS

Authors want to thank M Udvardi for critical reading of the manuscript.

REFERENCES

Arimura GI, Ozawa R, Kugimiya S, Takabayashi J, and Bohlmann J. (2004) **Herbivore-Induced Defense Response in a Model Legume: Two-Spotted Spider Mites Induce Emission of (E)-{beta}-Ocimene and Transcript Accumulation of (E)-{beta}-Ocimene Synthase in *Lotus japonicus*.** *Plant Physiology* 135, 1976-1983.

Asamizu E, Nakamura Y, Sato S, Tabata S. (2004) **Characteristic of the *Lotus japonicus* gen repertoire deduced from large-scale expressed sequence tag (EST) analysis.** *Plant Molecular Biology* 54, 504-514.

Asamizu E, Kato S, Sato S, Nakamura Y, Kaneko T, Tabata S. (2003) **Structural analysis of a *Lotus japonicus* genome. IV. Sequence features and mapping of seventy-three TAC clones which cover the 7.5 Mb-regions of the genome.** *DNA Research* 10, 115-122.

Asamizu E, Nakamura Y, Sato S, Tabata S. (2000) **Generation of 7137 non redundant expressed sequencing tags from a legume, *Lotus japonicus*.** *DNA Research* 7, 127-130.

Banba M, Siddique AB, Kouchi H, Izui K, Hata S. (2001) ***Lotus japonicus* forms early senescent root nodules with *Rhizobium etli*.** *Molecular Plant-Microbe Interactions* 14 , 173-80.

Bilyeu K, Beuselink P. (2001) **Genes Controlling Pod Dehiscence in *Lotus japonicus*.** Meeting Abstract.

Bonfante P, Genre A, Faccio A, Martini I, Schauser L, Stougaard J, Webb J, Parniske M. (2000) **The *Lotus japonicus* Ljsym4 gene is required for the successful symbiotic interaction of root epidermal cells.** *Molecular Plant-Microbe Interactions*, 13: 1109-1120.

Borisov AY, Madsen LH, Tsyganov VE, Umerhara Y, Voroshilova VA, Batagov AO, Sandal N, Frederiksen A, Schauser L, Ellis N, et al. (2003) **The *sym35* gene required for root nodule development is an ortholog of *Nin* from *Lotus japonicus*.** *Plant Physiology* 131, 1009-1017.

Broughton and Dilworth M. (1971) **Control of leghemoglobin synthesis in snake beans.** *Biochemical Journal* 125, 1075-1080.

Colebatch G, Kloska S, Trevaski B, Freund S, Altmann J, Udvardi MK. (2002) **Novel aspects of symbiotic nitrogen fixation uncovered by transcript profiling with cDNA arrays.** *Molecular Plant-Microbe Interactions* 15, 411-420.

Colebatch G, Debrosses G, Ott J, Krussell L, Montanari O, Kloska S, Kopka J, Udvardi MK. (2004). Global changes in transcription orchestrate metabolic differentiation during symbiotic nitrogen fixation. *The Plant Journal,* 39, 487-512.

Colliver SP, Robbins MP, Morris P. (1994) **An 'antisense' strategy for the genetic manipulation of condensed tannin and isoflavonoid phytoalexin accumulation in transgenic *Lotus corniculatus* L.** *Acta Horticulturae* 381, 148-151.

Cyranoski D. (2001) **Japanese legume project may help to fix the nitrogen problem.** *Nature* 409, 272.

Downie JA, Parniske M. (2002) **Fixation with regulation.** *Nature* 420, 369-370.

Doyle JJ. (1998) **Phylogenetic perspectives on nodulation: evolving views of plants and symbiotic bacteria.** *Trends in Plant Science* 3, 473-478.

Doyle JJ, Luckow MA. (2003) **The Rest of the Iceberg. Legume Diversity and Evolution in a Phylogenetic Context.** *Plant Physiology* 131, 900-910.

Forslund K, Morant M, Jorgensen B, Olsen CE, Asamizu E, Sato S, Tabata S, Bak S. (2004) **Biosynthesis of the nitrile glucosidesrhodiocyanoside A and D and the cyanogenic glucosides lotaustrali and linamarin in Lotus japonicus.** *Plant Physiology* 135, 71-84.

Gebrehiwot L, Beuselinck PR, Roberts CA. (2002) **Seasonal variations in condensed tannins concentration of three *Lotus* species**. *Agronomy Journal* 94, 1059-1065.

Grant WF, Bullen MR, Nettancourt D. (1962) **The cytogenetics of Lotus. I. Embryo-cultured interspecific diploid hybrids closely related to L. Corniculatus**. *Canadian Journal of Genetics and Cytology* 4, 105-128.

Grant WF. (1995) **A chromosome atlas and interspecific - intergeneric index for Lotus and Tetragonolobus (Fabaceae)**. *Canadian Journal of Botany* 73, 1787-1809.

Grant WF. (1996) **Seed pod shattering in the genus *Lotus* (Fabaceae): a synthesis of diverse evidence.** *Canadian Journal of Plant Science* 76, 447-456.

Gruber M, Ray H, Skadhauge B. (2003) **Characterization of proanthocyanin regulatory genes from legumes.** *Plant & Animal Genomes XI Conference.* Poster: P792.

Handberg K, Stougaard J. (1992). ***Lotus japonicus*, an autogamous diploid legume species for classical and molecular genetics**. *The Plant Journal* 2, 487-496.

Harris J. (2002**) Shedding light on an underground problem**. *Proceedings of the National Academy of Sciences USA* 99, 14616-14618.

Hayashi M, Miyahara A, Sato S, Kato T, Yoshikawa M, Pedrosa A, Ouda R, Imaizumi-Anraku H, Bachmair A, Sandal N, Stougaard J. (2001). **Construction of a genetic linkage map of the model legume *Lotus japonicus* using an intraspecific F2 population**. *DNA Research* 8, 301-310.

Hussain A, Jiang Q, Broughton W, Gresshoff P. (1999). ***Lotus japonicus* nodulates and fixes nitrogen with the broad host range *Rhizobium* sp. NGR234**. *Plant Cell Physiology* 40, 894-899.

Imaizumi-Anraku H, Kawaguchi M, Koiwa H, Akao S, Syono K. (1997). **Two ineffective nodulating mutants of *Lotus japonicus*: different phenotypes caused by the blockage of the endocytotic bacterial release and nodule maturation**. *Plant Cell Physiology* 38, 871-881.

Jiang Q, Gresshoff P. (1993). ***Lotus japonicus*: a model plant for structure function analysis in nodulation and nitrogen fixation**. In: *Current Topics of Plant Molecular Biology*, VoL. II,. (Gresshoff, PM, ed), CRC Press, Boca Raton, Florida. pp: 97-110.

Jiang Q, Gresshoff P. (1997) **Classical and molecular genetics of the model legume *Lotus japonicus***. *Molecular Plant-Microbe Interactions,* 10, 59-68.

Kaneko T, Nakamura Y, Sato S, Asamizu E, Kato T, sasamoto S, Watanabe A, Idesawa K, Ishikawa A, Kawashima K. (2000). **Complete genome estructure of the nitrogen fixing symbiotic bacterium *Mesorhizobium loti.*** *DNA Research* 7, 331-338, 381- 406.

Kawaguchi M. (2000) ***Lotus japonicus* "Miyakojima" MG-20: an early flowering accession suitable for indoor handling**. *Journal of Plant Research* 113, 507-509.

Kawaguchi M, Motomura T, Imaizumi-Anraku H, Akao S, Kawasaki S. (2001**) Providing the basis for genomics in *Lotus japonicus*: The accessions Miyakojima and Gifu are appropriate crossing partners for genetic analyses**. *Molecular Genetics and Genomics* 266, 157-166.

Kawaguchi M, Imaizumi-Anraku H, Koiwa H, Niwa S, Ikuta A, Syono K, Akao S. (2002). **Root, root hair, and symbiotic mutants of the model legume *Lotus japonicus***. *Molecular Plant-Microbe Interactions* 15, 17-26.

Kessler A, Baldwin IT. (2001) **Defensive function of herbivore induced plant volatile emissions in nature**. *Science* 291, 2141-2144.

Kirkbride, JH, Jr. (1999). ***Lotus* systematics and distribution**. *Special Publication in Crop Science Societies of America* 28, 10.

Krusell L, Madsen LH, Sato S, Aubert G, Genua A, Szczyglowski K, Duc G, Kaneko T, Tabata S, de Bruijn F, Pajuelo E, Sandal N, Stougaard J. (2002) **Shoot control of root development and nodulation is mediated by a receptor-like kinase**. *Nature* 420, 422-426.

Lohar DP, Schuller K, Buzas DM, Gresshoff PM, Stiller J. (2001). **Transformation of *Lotus japonicus* using the herbicide resistance bar gene as a selectable marker**. *Journal of Experimental Botany* 52, 1697-1702.

Lohar DP, Bird DM. (2003) ***Lotus japonicus*: a new model to study root-parasitic nematodes**. *Plant & Cell Physiology* 44, 1176-1184.

López-Lara I, Van der Berg JDJ, Thomas-Oates JE, Glushka JJLBJ, Spaink HP. (1995) **Structural identification fo the lipo-chitin oligosaccharide nodulation signals of *Rhizobium loti***. *Molecular Microbiology* 15, 627-638.

Lombari P, Ercolano E, El Alaoui H, Chiurazzi M. (2003) **A new transformation-regeneration procedure in the model legume *Lotus japonicus*: root explants as a source of large numbers of cells susceptible to *Agrobacterium*-mediated transformation**. *Plant Cell Reports* 21, 771-777.

Madsen EB, Madsen LH, Radutoiu S, Olbryt M, Rakwalska M, Szczyglowski K, Sato S, Kaneko T, Tabata S, Sandal N, Stougaard J. (2003) **A receptor kinase gene of the LysM type is involved in legume perception of rhizobial signals**. *Nature* 425, 637-640.

Martirani L, Stiller J, Mirabella R, Alfano F, Lamberti A, Radutoiu SE, Iaccarino M, Gresshoff PM, Chiurazzi M. (1999). **T-DNA Tagging of Nodulation-and Root-Related Genes in *Lotus japonicus*: Expression Patterns and Potential for Promoter Trapping and Insertional Mutagenesis**. *Molecular Plant-Microbe Interactions* 12 , 275-284.

Men AE, Meksem K, Kassem MA, Lohar D, Stiller J, Lightfoot D, Gresshoff PM. (2001). **Bacterial Artificial Chromosome Library of *Lotus japonicus* Constructed in an *Agrobacterium tumefaciens*-Transformable Vector**. *Molecular Plant-Microbe Interactions* 14, 422-425.

Moeller BL, SeiglerD. (1999). **Biosynthesis of cyanogenic glucosides, cyanolipids and related compounds.** In: *Plant amino acids. Biochemistry and Biotechnology* (Singh BK, Ed.) Marcel Dekker, Inc, New York. pp. 563-609.

Morris P, Carron TR, Robbins MP, and Webb J. (1993). **Distribution of condensed tannins in flowering plants of *Lotus corniculatus* var *japonicus* and tannin accumulation by transformed root cultures**. *Lotus Newsletter* 24 (Beuselinck PR, ed.).

Morris ER and Walker JC. (2003). **Receptor-like kinases: the key to response**. *Current Opinion in Plant Biology* 6, 339-342.

Nakamura Y, Kaneko T, Asamizu E, Kato T, Sato S, and Tabata S. (2002) **Structural Analysis of a Lotus japonicus Genome. II. Sequence Features and Mapping of Sixty-five TAC Clones Which Cover the 6.5-Mb Regions of the Genome**. *DNA Research* 9, 63-70.

Nishimura R, Hayashi M, Wu GJ, Kouchi H, Imaizumi-Anraku H, Murakami Y, Kawasaki S, Akao S, Ohmori M, and Nagasawa M. (2002a) **HAR1 mediates systemic regulation of symbiotic organ development**. *Nature* 420, 426-429.

Nishimura R, Ohmori M, Fujita H, and Kawaguchi M. (2002b). **A *Lotus* basic leucine zipper protein with a ring-finger motif negatively regulates the developmental program of nodulation.** *Proceedings of the National Academy of Sciences USA* 99, 15206-15210.

Niwa S, Kawaguchi M, Imaizumi-Anraku H, Chechetka SA, Ishizaka M, Ikuba A, and Kouchi H. (2001) **Responses of a model legume *Lotus japonicus* to liopchitin**

oligosaccharide nodulation factors purified from *Mesorhizobium loti* JRL501. *Molecular Plant-Microbe Interactions* 14, 848-856.

Orea A, Pajuelo P, Pajuelo E, Quidiello C, Romero JM, and Márquez AJ. (2002) **Isolation and characterisation of photorespiratory mutants from *Lotus japonicus* affected in chloroplastic glutamine synthetase.** *Physiologia Plantarum*, 115, 352-361.

Ozawa R, Shimoda T, Kawaguchi M, Arimura G, Nishioka T, and Takabayashi, J. (2000) ***Lotus japonicus* infested with herbivorous mites emits volatile compounds that attract predatory mites**. *Journal of Plant Research* 113, 427-433.

Pacios-Bras C, Jorda MA, Wijfjes AHM, Harteveld M, Stuurman N, Thomas-Oates JE, and Spaink, HP. (2000) **A *Lotus japonicus* nodulation system based on heterologous expression of the fucosyl transferase NodZ and the acetyl transferase NodL in *Rhizobium leguminosarum*.** *Molecular Plant-Microbe Interactions* 13, 475-479.

Paré PW and Tumlinson JH. (1999) **Plant volatiles as a defense against herbivores**. *Plant Physiology* 121, 325-332.

Parniske M. (2000) **Intracellular accomodation of microbes by plants: a common developmental program for symbiosis and disease?** *Curr. Opin. Plant Biology* 3, 320-328.

Perry JA, Wang TL, Welham TJ, Gardner S, Pike JM, Yoshida S, and Parniske M. (2003). **A TILLING Reverse Genetics Tool and a Web-Accessible Collection of Mutants of the Legume *Lotus japonicus***. *Plant Physiology* 131, 866-871.

Pichersky E and Gershenzon J. (2002) **The formation and function of plant volatiles: perfumes for pollinator attraction and defense**. *Current Opinion in Plant Biology* 5, 237-243.

Radutoiu S, Madsen LH, Madsen EB, Felle HH, Umerhara Y, Groenlund M, Sato S, Nakamura Y, Tabata S, Sandal N, and Stougaard J. (2003) **Plant recognition of symbiotic bacteria requires two LysM receptor like kinases.** *Nature* 425, 585-592.

Robbins MP, Bavage AD, Strudwicke C, and Morris P. (1998) **Genetic Manipulation of Condensed Tannins in Higher Plants. II. Analysis of Birdsfoot Trefoil Plants Harboring Antisense Dihydroflavonol Reductase Constructs**. *Plant Physiology* 116, 1133-1144.

Rodríguez-Llorente ID, Pérez-Hormaeche J, El Mounadi K, Dary M, Caviedes MA, Cosson V, Kondorosi A, Ratet P, and Palomares AJ. (2004) **From pollen tubes to infection threads: recruitment of Medicago floral pectic genes for symbiosis**. *The Plant Journal* 39, 587-598.

Sandal N, Krussell L, Radotoiu S, Olbryt M, Pedrosa A, Stracke S, Sato S, Kato T, Tabata S, Parniske M, Bachmair A, Ketelsen T, and Stougaard J. (2002) **A genetic linkage map of the model legume *Lotus japonicus* and strategies for fast cloning of new loci.** *Genetics* 161, 1673-1683.

Sato S, Kaneko T, and Nakamura Y. (2001) **Structural analysis of *Lotus japonicus* genome. I. Sequence features and mapping of fifty-six TAC clones which cover the 5.4 Mb regions of the genome**, *DNA Research* 8, 311-318.

Schauser L, Handberg K, Sandal N, Stiller J, Thykaer T, Pajuelo E, Nielsen A, and Stougaard J. (1998) **Symbiotic mutants deficient in nodule establishment after T-DNA transformation of *Lotus japonicus*.** *Molecular General Genetics* 259, 414-423.

Schauser L, Roussis A, Stiller J, and Stougaard J. (1999) **A plant regulator controlling development of symbiotic root nodules**. *Nature* 402, 191-195.

Schultze and Kondorosi (1998) **Regulation of symbiotic root nodule development**. *Annual Review of Genetics* 32, 33-57.

Senoo K, Soleiman MZ, kawaguchi M, Imaizumi-Anraku H, Akao S, Tanaka A, and Obata H. (2000) **Isolation of two different phenotypes of mycorrhizal mutants in the model legume plant *Lotus japonicus* after EMS-treatment**. *Plant & Cell Physiology* 41, 726-732.

Somers DA, Samac DA, and Olhoft PM. (2003) **Recent Advances in Legume Transformation.** *Plant Physiology* 131, 892-899.

Soltis DE, Soltis PS, Morgan DR, Swensen SM, Mullin BC, Dowd JM, and Martin PG. (1995). **Chloroplast gene sequence data suggest a single origin of the predisposition for symbiotic nitrogen fixation in angiosperms**. *Proceedings of the National Academy of Sciences USA* 92, 2647-2651.

Stiller J, Martirani L, Tuppale S, Chian R-J, Chiurazzi M, and Gresshoff PM. (1997) **High frequency transformation and regeneration of transgenic plants in the model legume *Lotus japonicus***. *Journal of Experimental Botany* 48, 1357-1365.

Stougaard J. (2001) **Genetics and Genomics of root symbiosis**. *Current Opinion in Plant Biology* 4, 328-335.

Stracke S, Kistner C, Yoshida S, Mulder L, Sato S, Kaneko T, Tabata S, Sandal N, Stougaard J, Szczyglowski K, and Parniske M. (2002) **A plant receptor-like kinase required for both bacterial and fungal symbiosis**. *Nature* 417, 959-962.

Stracke S, Sato S, Sandal N, Koyama M, Kaneko T, Tabata S, and Parniske M. (2004) **Exploitation of colinear relationships between the genomes of *Lotus japonicus, Pisum sativum* and *Arabidopsis thaliana*, for positional cloning of a legume symbiosis gene**. *Theoretical and Applied Genetics* 108, 442-449.

Swanson EB, Somers DA, and Tomes DT. (1990). **Birds foot trefoil (*Lotus corniculatus* L.)**. In: *Biotechnology in Agriculture and Forestry*, Vol 10, Legumes and oilseed crops I. (Bajaj, YPS, Ed.), Springer-Verlag, Berlin.

Sz-Borsos, Somaroo BH and Grant WF. (1972) **A new diploid species of Lotus (Leguminosae) in Pakistan**. *Canadian Journal of Botany* 50, 1865-1870.

Szczyglowski K, Shaw RS, Wopereis J, Copeland S, Hamburger D, Kasiborski B, Dazzo FB, and de Brujin FJ. (1998) **Nodule organogenesis and symbiotic mutants of the model legume *Lotus japonicus***. *Molecular Plant-Microbe Interactions* 11, 684-697.

Szczyglowski K and Amyot L. (2003) **Symbiosis, inventiveness by recruitment?** *Plant Physiology* 131, 935-940.

Tansengco ML, Imaizumi-Anraku H, Yoshikawa M, Takagi S, Kawaguchi M, Hayashi M, and Marooka Y. (2004).**Pollen development and tube growth are affected in the symbiotic mutant of *Lotus japonicus crinkle***. *Plant & Cell Physiology* 45, 511-520

Thykjaer T, Stiller J, Handberg K, Jones J, and Stougaard J. 1995. **The maize transposable element *Ac* is mobile in the legume *Lotus japonicus***. *Plant Molecular Biology* 27, 981-93.

VandenBosh KA and Stacey G. (2003) **Summaries of legume Genomics Projects from around the Globe. Community resources for Crops and Models**. *Plant Physiology* 131, 2-35.

Webb KJ, Skøt L, Nicholson MN, Jørgensen B, and Mizen S. (2000) ***Mesorhizobium loti* increases root-specific expression of a calcium-binding protein homologue identified by promoter tagging in *Lotus japonicus***. *Molecular Plant-Microbe Interactions* 13, 606-616.

Wegel E, Schauser L, Sandal N, Stougaard J, and Parniske M. (1998) **Mycorrhiza mutants of *Lotus japonicus* define symbiotically independent steps during symbiotic infection**. *Molecular Plant-Microbe Interactions* 11, 933-936.

Wienkoop S and Saalbach G. (2003) **Proteome analysis: novel proteins identified at the peribacteroid membrane from *Lotus japonicus* root nodules**. *Plant Physiology* 131, 1080-1090.

Zagrobelny M, Bak S, Rasmussen AV, Jorgensen B, Maumann CM, and Lingberg MB. (2003) **Cyanogenic glucosides and plant-insect interactions**. *Phytochemistry* 56, 293-306.

Chapter 1.2

LOTUS-RELATED SPECIES AND THEIR AGRONOMIC IMPORTANCE

Pedro Díaz*, Omar Borsani, and Jorge Monza

*Laboratorio de Bioquímica; Departamento de Biología Vegetal; Facultad de Agronomía; CP12900 Montevideo; URUGUAY; *Corresponding author.*

Email: pediaz@fagro.edu.uy

Phone: +598 23 54 0229 Fax: +598 23 59 0436

Keywords: *L. corniculatus, L. uliginosus, L. glaber, L. subbiflorus,* botanical features, pastures, environmental adaptation, plant breeding.

More than 180 species within the genus Lotus occur worldwide. Four have been domesticated and improved through selection and plant breeding: Lotus corniculatus, L. uliginosus, L. glaber *and* L. subbiflorus. *Since the model legume* L. japonicus *is related taxonomically to these species, knowledge can be transferred to the agronomical arena. The slow progress observed in Lotus cultivar improvements to date could be explained by the polyploid nature of some of these species, a feature not present in* L. japonicus. *This chapter reviews briefly the taxonomical relationships among these species. Secondly, it illustrates how Lotus species are currently used to improve pastures for which other forage legume species are not suitable. Finally, it touches on beneficial microorganism-plant interactions and the benefits of using Lotus species as animal fodder.*

INTRODUCTION

One of the principal protein sources of the human diet comes from animal origin. Beef and sheep meat production is based on natural, cultivated pastures and feedlot system with nutrient supplement. Cultivated pastures can be composed of a single cultivated species or a mixture of forage species. These pastures include grass species as rye grass, cocks-foot, oat, and *Festuca* sp. Other important botany components in these pastures are legume species such as trefoils, lucerne, and Lotus sp.

An alternative generalised technology to improve the production and the nutritive value of natural grasslands is legume surface over sowing with phosphate fertilisation. In the natural improved grassland from South America it has been observed that the introduction of legume pastures promotes sustainable development

A.J. Márquez (ed.), Lotus japonicus Handbook, 25-37.

of native vegetation and a higher productive level (Berreta *et al.*, 2000). Forage legumes play an important role in this beneficial effect through their contribution to the nitrogen content of soil derived from nitrogen fixation, and the phosphorous availability generated by the fertilisation of the natural grasslands.

There is a potential use of Lotus species related to their ability to grow in acid soils or soils with low fertility and their tolerance to aluminium, manganese and sodium chloride (Blumenthal and McGraw, 1999). *Lotus corniculatus* is also used as bioremediator in boron and selenium contaminated soils (Banuelos *et al.*, 1992) and its use for other contamination problems must be considered.

Besides its agronomical and environmental attributes, there is a great deal of interest in Lotus because some species are extremely amenable to tissue culture, in particular *Lotus corniculatus* L., and *L. japonicus* (Regel) K. Larsen (Webb *et al.*, 1990; Handberg and Stougaard, 1992; Jiang and Gresshoff, 1997; Aoki *et al.*, 2002). In spite of the fact that *L. japonicus* is not used as forage legume, this species is a good model legume in which a wealth of genetic, biochemical and molecular biology studies have been directed to elucidate bacteria-plant symbiosis, mycorrhizal interactions and nitrogen metabolism (Parniske, 2000; Udvardi, 2000; Orea *et al.*, 2002). More studies should be carried out in *L. japonicus* plants in order to elucidate existing problems in the Lotus agriculturally used species, and in this way improve their field performance.

DISTRIBUTION AND SOWING

The genus *Lotus* includes more than 180 species that are found world wide except in very cold regions and the low land tropical areas of Southeast Asia and South and Central America. Different species of Lotus are currently used to improve pastures and hay quality where other forage legume species are not suitable (Papadopulous and Kelman, 1999).

There are four species of Lotus that have been domesticated and improved by selection and plant breeding: *L. corniculatus* (birds foot trefoil), *L. uliginosus* Schkuhr. (greater lotus), *L. glaber* Mill. (narrow-leaf trefoil) and *L. subbiflorus* Lagasca (hairy birdsfoot trefoil). This last species is the only annual Lotus with agricultural importance and is principally sown in South America (Asuaga, 1994a). *L. glaber* and *L. uliginosus* were denoted for many years as *L. tenuis* and *L. pedunculatus,* respectively.

The natural distribution of agronomical important Lotus species is mainly in the Mediterranean basin (Europe and North Africa). *L. corniculatus* is undoubtedly the species considered to have the greatest agricultural importance and the widest distribution. The other species like *L. glaber, L. uliginosus, L. subbiflorus* and *L. japonicus* have earned recognition for their intrinsic value. There are more than 175 other Lotus species about which we know little yet. The high number of Lotus species provides the genus with wide genetic diversity resulting in adaptations to different environmental conditions. This genetic diversity is the key in genetic breeding programs for various important agronomic characters.

The principal world regions where Lotus species are sown are South America, North America, and Europe, with 1.85, 1.39 and 1.38 million hectares respectively. Ten countries sow the 95 % of Lotus species in the world and more than 90% of this area is planted with *L. corniculatus* (Figure 1). Within other agronomic Lotus species, in Argentina *L. glaber* is sown in the Salado region (Vignolio *et al*, 1994) and in USA mainly in western and northeastern states (Blumenthal and McGraw, 1999). *L. uliginosus* is sown in New Zealand and coastal southeast Australia. In Uruguay, 50,000 ha of *L. subbiflorus* are sown annually which makes this country the main user of this species for pastures improvement (Asuaga, 1994a).

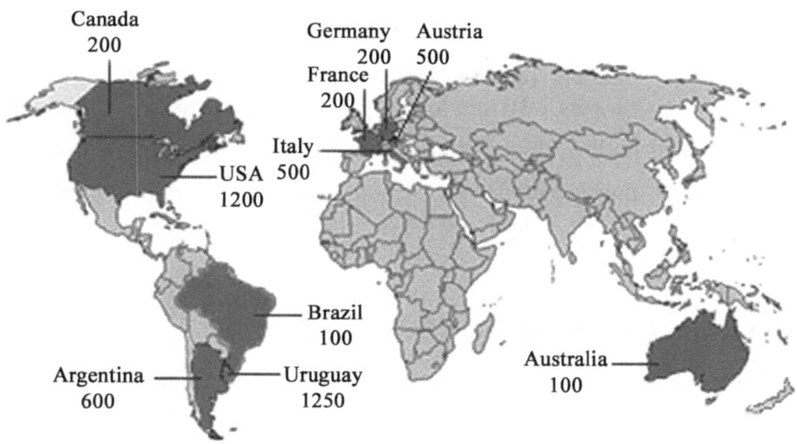

Figure 1. Countries with more than 100 thousand hectares sown with Lotus species. Under the name of the country, the sowing area in thousands of hectares is indicated.

BOTANICAL CHARACTERISTICS AND TAXONOMY

The genus Lotus comprises both annual and perennial species adapted to a wide range of ecological habitats. The agronomical species and also *L. japonicus,* are included in the subgenus *Edentolotus,* characterised for its non dentate style, section *Xantolotus,* composed with legume terete and laterally compressed yellow flowers (Arambarri, 1999; Izaguirre and Beyhaut, 1998).

Plants have erect or decumbent stems and pentafoliate leaves with two of the leaflets at the petiole base resembling stipules (Figure 2). Leaves are green to grey-green. Inflorescences with eight flowers (four for *L. subbiflorus*) are umbel-like cymes at the end of long axillary branches. All the agricultural important Lotus species mentioned above are self-sterile plants and must cross-pollinate mainly by insects. *L. japonicus* has the same morphological characteristics but the auto-compatible plants are self-pollinated.

The seedpods are about 2 cm long, contain 20 seeds, change from green to brown as they ripen and become coiled at dehiscence. Lotus species have seeds with round to oval shape, *L. corniculatus* and *L. glaber* present greenish to dark brown seeds as does *L. japonicus,* and *L. uliginosus* and *L. subbiflorus* present yellowish-brown to dark brown seeds.

Key to identifying *L. japonicus* and commercial Lotus species

1. Perennial plants: glabrous and subglabrous. Banner 8-14 mm long, wings 7-12 mm long, keel 7-12 mm long, style 4-7 mm long, straight pods 15-40 mm long.
    a. Stems solid, sometimes hollow just at the base. Rhizomatous and arhizomatous plants.
        i. Leaflets of long upper leaves 2-4 times longer than wide, flowers 10-17 mm long.
            1. Stems green..............................*L. corniculatus*
            2. Stems with dark pigmentation.............*L. japonicus*
        ii. Leaflets of long upper leaves 3-5 times longer than wide, flowers 7-10 mm long.....................................*L. glaber*
    b. Stems hollow, rhizomatous plants............................*L. uliginosus*
2. Annual plants: pubescent. Banner 6 mm long, wings 5 mm long, keel 6 mm long, style 4 mm long, straight pods 9-14 mm long..*L. subbiflorus*

ADAPTATIONS TO ENVIRONMENTAL CONDITIONS

The agronomical Lotus species are best adapted to temperate and humid environments. *L. corniculatus* winter hardiness depends on cultivar origin. North American and European cultivars are most tolerant to low temperatures but are adversely affected by severe winters with fluctuating temperatures (Beuselinck and Grant, 1995). *L. glaber* is less winter hardy than *L. corniculatus.* However, *L. glaber* proved to be tolerant to low temperatures and ice-sheeting of water in central-east Argentina (Vignolio *et al.*, 1999). *L. corniculatus* is more tolerant to low winter temperatures than *L. uliginosus*

Lotus species are adapted to infertile and acidic soils; they can be regarded as pioneer forage legumes and are moderately tolerant to soil salinity. Among the agronomical Lotus species, *L. uliginosus* presents more tolerance to acidic soil followed by *L. corniculatus*, independent of pH values. In soils with high aluminium concentration *L. uliginosus* and especially cv 'Maku' a tetraploid cultivar, shows more tolerance than *L. corniculatus* (Wheeler *et al.*, 1992). *L. corniculatus* and *L. uliginosus* agronomic performance respond to improve soil pH, drainage, and fertility. Application of phosphorous and potassium on infertile soils enhances forage yield, seed yield and winter hardiness (Russelle *et al.*, 1991). *L. uliginosus* also has a high tolerance to manganese (Wheeler and Dodd, 1995). Different Lotus species and cultivars exhibit different degree of tolerance to edaphic conditions such as low pH and aluminium soil content, but every species are more tolerant to these factors than lucerne or trefoils (Wheeler *et al.*, 1992).

In flooded soils, *L corniculatus* showed the best performance followed by *L glaber* and *L. uliginosus*. *L. corniculatus* tolerates poor soil drainage and is more tolerant to flooding than lucerne, though not highly tolerant. *L. glaber* also tolerates poor soil drainage and swampy conditions. This species has colonised the flooding pampa grasslands in central-east Argentina (Vignolio *et al.*, 1999). Besides, *L. uliginosus* tolerates periods of water logging better than most forage legumes on account of root suberisation, production of adventitious roots, thickening of submerged stems and transpiration of high quantities of water (Shiferaw *et al.*, 1992).

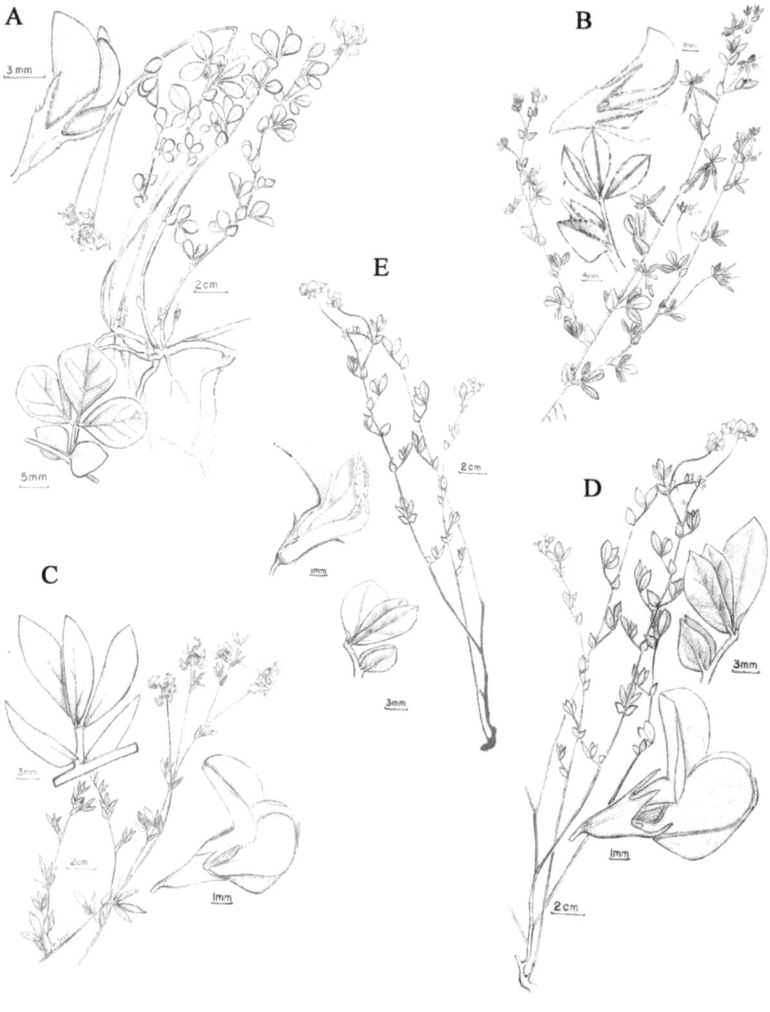

Figure 2. Botanical drawing of agriculturally important Lotus species and L. japonicus. *(A)* L. uliginosus; *(B)* L. subbiflorus; *(C)* L. glaber; *(D)* L. corniculatus *(reproduced with permission from Izaguirre and Beyhaut, 1998. Las leguminosas del Uruguay y regiones vecinas Ed. Hemisferio Sur); and (E)* L. japonicus.

Other edaphic factor that affects legume productivity-survival in the pasture is water soil content. *L. corniculatus* is more tolerant to dry soils than both *L. glaber* and *L. uliginosus*, although *L. corniculatus* is not as well adapted to high temperature or drought as *Medicago sativa* (Blumenthal and McGraw, 1999, Peterson *et al.*, 1992). On the other hand, *L. glaber* is moderately drought tolerant because of its shallow rooting and therefore can withstand high summer temperatures. *L. uliginosus* is low tolerant to drought, much less than *L. corniculatus*. *L. subbiflorus* is an annual species with a very short cycle finished at beginning of summer, so the plants tolerance to the dry season has not yet been evaluated.

L. uliginosus has certain shade tolerance and is sown as ground cover in pine (*Pinus radiata*) plantations in New Zealand (Gadgil *et al*., 1986). Moreover, *L. glaber* and *L. corniculatus* are intolerant to shading from other plants, particularly during establishment.

SEEDLING ESTABLISHMENT AND REGROWTH

Lotus species show poor seedling vigour and therefore a clean seedbed is essential for a good establishment. Because of their slow establishment ability, Lotus species does not compete well with broad-leaved weeds or aggressive grass weeds. The land for Lotus establishment must be well laboured. A uniform and firm seedbed is required since the small seeds need shallow sowing.

Lotus seeds are normally drilled or broadcasted in spring after conventional seedbed cultivations. Broadcasting, surface over sowing and direct drilling (sod seeding) have been used on marginal and rough land in north-east North America (Kunelius *et al*., 1982) and New Zealand (Lowther *et al*., 1996)

Spring is the best moment to sow Lotus species. If they are sown in autumn, sufficient time is needed for plants to develop before the onset of winter cold. Inoculated seeds of *L. corniculatus* and *L. glaber* are sown at 6-10 kg/ha (over seeding 3-5kg/ha) for monocultures and at 3-5 kg/ha when sown with grasses. *L uliginosus* diploid cultivars are sown at 1-3 kg/ha and tetraploids at 1-5 kg/ha. On the other hand, *L. subbiflorus* sowing densities are 3-5 kg/ha (Asuaga, 1994).

Once established, *L. corniculatus* and *L. uliginosus* showed vigour and good growth rhythm and were capable of vigorous growth when well fertilised, but *L. glaber* showed moderate growth. Growth assays carried out in controlled conditions in our laboratory showed that the growth rate of *L. corniculatus* and *L. uliginosus* was higher than that of *L. glaber* and *L. japonicus*. *L. subbiflorus* had the lowest growth rate.

Lotus species grow slowly during early spring and have growth peaks in summer. In South America, they grow during early spring and have the growth peak in late spring (Asuaga, 1994b). *L corniculatus* growth in spring is directly proportional to the build-up of storage carbohydrates in the roots (Alison and Hoveland, 1989).

L. glaber shows a moderate vigour of regrowth after defoliation, but it is more tolerant to intensive grazing than *L. corniculatus*. Regrowth of this last species depends on enough axillary buds and photosynthesising foliage being left after defoliation. This can be done by ensuring that adequate sites for regrowth remain after cutting crops for hay or silage, or by controlling the severity of defoliation when grazing. Prostrate cultivars of *L. corniculatus* cv 'INIA-Draco®' showed ability of more vigorous regrowth than erect types since more of their foliage escape defoliation. *L. corniculatus* growth varies from prostrate to erect with numerous stems arising from a basal, well-developed crown and branches arising from leaf axils. Primary growth originates from the crown but regrowth originates from buds formed at nodes on the stubble left after defoliation. Another Lotus species with good regrowth is *L. subbiflorus* because it has a prostrate habit that avoids an intensive defoliation.

Cultivars of Lotus species with an erect growth habit are the best suited for hay or silage cropping, essentially *L. uliginosus* and *L. corniculatus* cultivars. Two or three cuts of *L. corniculatus* can usually be taken depending on the duration of the growing season and yields are optimal at 6-week cutting intervals (Hoveland and Fales, 1985).

BENEFICIAL PLANT-MICROBE INTERACTIONS

Lotus species are nodulated by both *Mesorhizobium loti* and *Bradyrhizobium* sp. (*Lotus*), usually with different levels of effectiveness. Rhizobial inoculation of seed by effective, specific strains *M. loti* (Díaz et al., 1995; Monza *et al.,* 1997) are needed when sowing, particularly on land where *L. corniculatus* and *L. glaber* has not been cultivated before. A similar management is needed with *L. uliginosus* and *L. subbiflorus*, but inoculating in this case with specific strains of *Bradyrhizobium* sp. (*Lotus*) when native rhizobia species are not present in soils. However, *M. loti* NZP2037 is a particular strain that can form effective root nodules in *L. uliginosus, L. glaber, L. corniculatus* (Vance *et al.,* 1987), and *L. subbiflorus* with different nitrogenase activities (Gonnet and Díaz, 2000). When *L. subbiflorus* was sown in soil with a history of *L. corniculatus,* ineffective root nodules without bacteroids were formed (Figure 3). Flavolans (condensed-tannins) accumulation resulting in red pigmentation of these nodules was observed (Monza *et al.,* 1992). After plant defoliation, root nodule senescence was induced and a new nodule generation was developed during plant regrowth (Vance *et al.*, 1982).

In grass-dominant, grazed tall fescue / *L. glaber* swards on an alkaline hydromorphic soil in the Flooding Pampa area, Argentina, *L. glaber* in symbiosis fixed 27-42 kg N_2/ha annually (assessed by the nitrogenase activity method) while in similar tall fescue / white clover swards, white clover fixed 14-59 kg N_2/ha (Refi and Escuder, 1998). *L. uliginosus* in grass / *L. uliginosus* stand in which the legume constituted 22% of the total, fertilised with lime and phosphate, fixed 35 kg N_2/ha (Laidlaw, 1981). Amounts of 40 to 60 kg N_2/ha were cited for an improved upland pasture in Scotland (Wedderburn, 1980). However, near twice nitrogen fixed was reported for inoculated *L. corniculatus* in North America (Heichel *et al.*, 1985; Farnham and George, 1994; Danso *et al.*, 1991). When grown in association with grass, most of the nitrogen in *L. corniculatus* was derived from nitrogen fixation. In contrast, the grass competitively took up soil nitrogen, and benefited by the transfer of fixed nitrogen by the legume (Farnham and George, 1994).

Roots of Lotus species are infected by arbuscular micorrhizae and several results have shown that non-mycorrhizal Lotus plants required more phosphorous to obtain the same yield than mycorrhizal plants in a phosphorous deficient soil (Mendoza and Pagani, 1997). *L. corniculatus* mycorrhizal plants were more efficient in phosphorous utilization than *L. glaber* mycorrhizal plants and produced larger yields of shoot tissue per unit of phosphorous in a low available phosphorous soil (Mendoza, 2001).

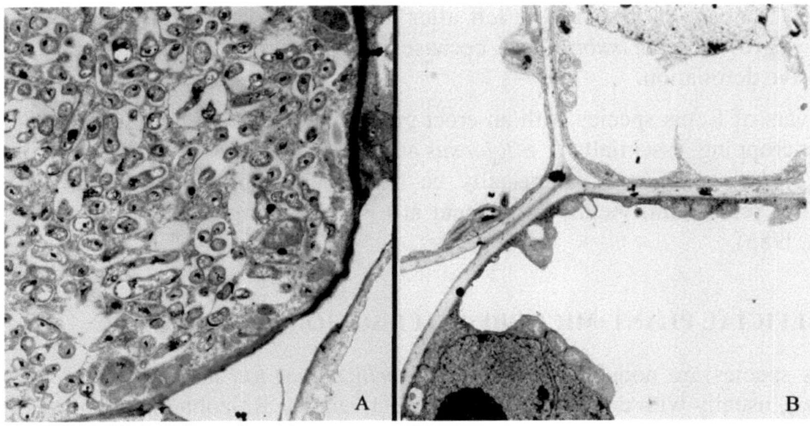

Figure 3. Transmission electron micrograph of nodular cells (3976X). (A) Effective nodule induced by M. loti *strain BM1-5 in the roots of* L. corniculatus. *The cytosolic nodular cell containing the symbiosomes with bacteroids inside can be seen. (B) Ineffective nodule induced by the same strain in the roots of* L. subbiflorus. *A cell with a big vacuole and peripheral mitochondria can be seen (Monza* et al., *1992).*

ANIMAL NUTRITION AND BLOATING

Lotus species are highly acceptable forage either at vegetative stage for grazing or as conserved hay or silage. The feeding value is largely determined by the stage of growth at the time of utilisation since feeding value falls with increasing maturity and associated steminess. In *L. corniculatus* one-third of its dry matter accumulates at the late vegetative stage, and more than half the maximum amount of nutrients accumulates by then. Leaves generally have a higher mineral content than shoots, especially after the start of flowering (McGraw *et al.*, 1986). The rate of forage digestibility decline is greater in late than in early season (Buxton et al., 1985). The overall forage quality under drought conditions is better for Lotus than for lucerne because of its higher leaf:stem ratio, delayed maturity and improved quality of every plant fraction (Peterson *et al.*, 1992). The types and combination of essential amino acids in *L. corniculatus* are optimal for the production of high quality animal products (Waghorn *et al.*, 1990).

One of the most attractive characteristics of Lotus species is the presence of condensed tannins (CT) in the forage that prevents bloating in ruminants. This nutritional advantage is due to the tannins-plant proteins conjugation in the rumen. This reaction protects the protein diet from degradation to ammonium, thus allowing more protein to be utilised in the small intestine (Waghorn *et al.*, 1987; Jean-Blain, 1998). Voluntary intake and rumen digestion are adversely affected if there is a high concentration of CT in the forage fed (Barry and Duncan, 1984). The nutritional value of CT depends upon their concentration in the forage and levels of 20-40 g CT/kg DM are thought to be beneficial (Barry, 1989). A study with *L. uliginosus, L. glaber* and *L. corniculatus* showed that *L. uliginosus* had very high CT levels and concluded that the CT concentrations of this last species were high enough to be detrimental to protein availability and palatability for grazing animals. *L. corniculatus* has moderately low CT, while the CT levels in *L. glaber* may be too low to benefit protein utilisation. (Gebrehiwot *et al.*, 2002). The tannin contents are different in the leaves, flowers, and stem and can vary as consequence of environmental factors, like drought (Carter *et al.*, 1999; Gebrehiwot *et al.*, 2002).

BREEDING OF LOTUS SPECIES

Among the positive attributes of forage Lotus for pasture use we can find: they have high nutritive value but do not induce bloat when direct-grazed, they can persist through reseeding or rhizomes and they tolerate infertile, acidic, drought or wet soil conditions. The improvements of Lotus species intend to make them more persistent, reliable, and productive. A great deal of genetic variation exists in Lotus species for further genetic improvement. Variation for agronomic and herbage quality characteristics including plant decumbency, winter injury, herbage dry mass, and *in vitro* digestible dry matter exists among different accessions.

Improving persistence is a complex task because plant death or loss of stands results from the combined effects of disease susceptibility, adverse environment, low reproduction rate, and limited genetic variation. The persistence of a Lotus stand can vary depending upon the amount of disease, the level of natural reseeding, and the type of cultivar. Characters of specific interest to improve the persistence of Lotus species include rhizomatous growth habit, disease resistance, condensed tannins, and hydrocyanide. The transfer of rhizomatous growth from the putative *L. corniculatus* obtained from Morocco into domesticated *L. corniculatus* with the purpose of increasing long-term persistence has been developed and evaluation studies are carried out in USA in cooperation with INIA-Uruguay.

In Australia, the improvement of *L. uliginosus* pays attention to optimizing the levels of condensed tannins to provide bloat protection. In Argentina, selection among naturalized ecotypes of *L. glaber* has been the basis of cultivar development for alkaline and saline soils of the Pampas.

The breeding of cultivars with improved disease resistance, vigour, ability to reseed or to spread by rhizomes should make these Lotus species more reliable and productive. *L. japonicus* is related taxonomically with agronomical Lotus species and the derived knowledge of this model legume will be transferred to agronomical Lotus. The slow progress observed in Lotus cultivar improvement could be explained by cross pollination of species and the polyploidy nature of some of them. These features are not present *in L. japonicus.*

To determinate the relation between the species and *L. corniculatus* origin will be one of the tasks to attain species improvement. Biochemical and genetic evidences indicate that *L. japonicus, L. glaber,* and *L. uliginosus* together with *L. alpinus* would be the ancestors of *L. corniculatus.*

ACKNOWLEDGEMENTS

We thank P Izaguirre and P Irisarri for their contributions in the correction of the manuscript.

REFERENCES

Alison M, and Hoveland C. (1989) **Birdsfoot trefoil management. 1. Root growth and carbohydrate storage.** *Agricultural Journal* 81, 739-745.

Aoki T, Kamizawa A, and Ayabe S. (2002) **Efficient Agrobacterium-mediated transformation of *Lotus japonicus* with reliable antibiotic selection.** *Plant Cell Reports* 21, 238-243.

Arambarri A. (1999) **Illustrated catalogue of *Lotus* L. Seeds (Fabaceae)** In: *Trefoil: The science and technology of Lotus.* (Beuselinck P, Ed.). American Society of Agronomy, pp. 21-41.

Asuaga A. (1994a) ***Lotus subbiflorus* cv El Rincón, a new alternative for extensive improvements of natural pastures.** In: *Proceedings of the First International Lotus Symposium* (Beuselinck and Roberts, eds.), pp 147-149.

Asuaga A. (1994b) **Use and production of *Lotus corniculatus* in Uruguay.** In: *Proceedings of the First International Lotus Symposium* (Beuselinck and Roberts, Eds.), pp. 134-143.

Banuelos, G, Cardon, G, Mackey, B, Ben-Asher, J, Wu, L, and Beuselinck, P. (1992) **Boron and selenium removal in boron-laden soil by birdsfoot trefoil.** *Lotus Newsletter.* 23, 32-35.

Barry T. (1989) **Condensed tannins: their role in ruminant protein and carbohydrate digestion and possible effects upon the rumen ecology.** In: *The Role of Protozoa and Fungi in Ruminant Digestion* (Nolan JV, Leng RA and Deneyer D.I., Eds.), Pernambul Books, Armidale, Australia, pp. 153-169.

Barry T, and Duncan S. (1984) **The role of condensed tannins in the nutritional value of *Lotus pedunculatus* for sheep.** 1. Voluntary intake. *British Journal of Nutrition* 51, 484-491.

Berreta, E, Risso, D, Montossi, F, and Pigurina, G. (2000) ***Campos* in Uruguay.** In: Grassland Ecophysiology and Grazing Ecology. G. Lemaire, J. Godson, A. de Moraes, C. Nabinger and F. Carvalho, (eds.), CAB International. pp. 377-394.

Beuselinck, P, and Grant, W. (1995) **Birdsfoot trefoil.** In : Barnes, R.F, Miller, D.A, and Nelson, C.J. (eds.) Forages, 5th edn. Vol. 1, An Introduction to Grassland Agriculture. Iowa State University Press, Ames, Iowa, pp. 237-248.

Blumenthal M, and McGraw R. (1999) **Lotus adaptation, use and management.** In: Trefoil: The science and technology of Lotus. Beuselinck P, (ed.) American Society of Agronomy, pp. 97-120.

Buxton D, Hornstein J, Wedin W, and Marten, G. (1985) **Forage quality in stratified canopies of alfalfa, birdsfoot trefoil and red clover.** *Crop Science* 25, 273-279.

Carter E, Theodorou M, and Morris, P. (1999) **Responses of *Lotus corniculatus* to environmental change. 2. Effect of elevated CO_2, temperature and drought on tissue digestion in relation to condensed tannin and carbohydrate accumulation** *Journal of the Science of Food and Agriculture* 79, 1431-1440.

Danso S, Curbelo S, Labandera C, and Pastorini, D. (1991) **Herbage yield and nitrogen fixation in a triple species mixed sward of white clover, lotus and fescue.** *Soil Biology and Biochemistry* 23, 65-70.

Díaz P, Borsani O, and Monza. (1995) **Effect of inoculation and nitrate on nitrate reductase activity and acetylene reduction activity in *Lotus* sp-*Rhizobium loti* symbiosis.** Symbiosis 19:53-63.

Farnham D and George J. (1994) **Harvest management effects on productivity, dinitrogen fixation, and nitrogen transfer in birdsfoot trefoil-orchard grass communities.** *Crop Science* 34, 1650-1653.

Gadgil R, Charlton J, Sandford A, and Allen P. (1986) **Relative growth and persistence of planted legumes in a mid-rotation radiata pine plantation.** *Forestry Ecology and Management* 14, 113-124.

Gebrehiwot L, Beuselinck P, and Roberts C (2002) **Seasonal Variations in Condensed Tannin Concentration of Three Lotus Species.** *Agricultural Journal* 94, 1059-1065.

Gonnet S, and Díaz P. (2000) **Glutamine synthetase and glutamate synthase activities in relation to nitrogen fixation in Lotus spp**. *Brazilian Journal of Plant Physiology* 12, 195-202.

Handberg K, and Stougaard, J. (1992) ***Lotus japonicus*, an autogamous, diploid legume species for classical and molecular genetics.** *The Plant Journal* 2, 487-496.

Heichel G, Vance C, Barnes D, and Henjum K. (1985) **Dinitrogen fixation and N and DM distribution during four-year stands of birdsfoot trefoil and red clover**. *Crop Science* 25, 101-105.

Hoveland, C, and Fales, S. (1985) **Mediterranean germplasm trefoils in the south-eastern USA, Piedmont**. In: *Proceedings of the XV International Grassland Congress*, Kyoto, Japan. The Science Council of Japan and the Japanese Society of Grassland Science, Tochigi-Kem, pp. 126-128.

Izaguirre P, and Beyhaut R. (1998) ***Loteae***. In: *Las leguminosas en Uruguay y regiones vecinas*. Editorial Agropecuaria Hemisferio Sur SRL, pp. 314-327.

Jean-Blain C. (1998) **Nutritional and toxicological aspects of tannins**. *Revista Médica Veterinaria* 149, 911-920.

Jiang G, and Gresshoff P. (1997) **Classical and molecular genetic of the model legume *Lotus japonicus***. *Molecular Plant-Microbe Interaction* 10, 59-68.

Kunelius H, Campbell A, McCrea K, and Ivany J. (1982) **Effects of vegetation suppression and drilling techniques on the establishment and growth of sod-seeded alfalfa and birdsfoot trefoil in grass dominant swards**. *Canadian Journal of Plant Science* 62, 667-675.

Laidlaw A. (1981) **Establishment, persistence and nitrogen fixation of white clover and marsh trefoil on blanket peat**. *Grass Forage Science* 36, 227-230.

Lowther W, Horrell R, Fraser W, Trainor K, and Johnstone P. (1996) **Effectiveness of a strip seeder direct drill for pasture establishment**. *Soil and Tillage Research* 38, 161-174.

McGraw R, Russelle M, and Grava J. (1986) **Accumulation and distribution of dry mass and nutrients in birdsfoot trefoil**. *Agronomy Journal* 78: 124-131.

Monza J, De Felipe R, and Bedmar E. (1992) **Nódulos Fix$^-$ formados por rhizobios de crecimiento rápido en *Lotus subbiflorus***. *XVI Reunión Latinoamericana de Rhizobiología*. La Pampa Argentina. p. 42.

Monza J, Díaz P, Borsani O, Ruiz-Argüeso T, and Palacios J. (1997) **Evaluation and improvement of the energy efficiency of nitrogen fixation in *Lotus corniculatus* nodules induced by *Rhizobium loti* strains indigenous to Uruguay**. *World Journal of Microbiology and Biotechnology* 13, 565-571.

Mendoza R. (2001) **Phosphorus nutrition and mycorrhizal growth response of broadleaf and narrowleaf birdsfoot trefoils**. *Journal of Plant Nutrition* 24, 203-214.

Mendoza R, and Pagani E. (1997) **Influence of phosphorus nutrition on mycorrhizal growth response and morphology of mycorrhizae in *Lotus tenuis***. *Journal of Plant Nutrition* 20, 625-639.

Orea A, Pajuelo P, Pajuelo E, Quidiello C, Romero J, and Márquez A. (2002) **Isolation of photorespiratory mutants from *Lotus japonicus* deficient in glutamine synthetase**. *Physiologia Plantarum* 114, 352-361.

Papadopoulos Y, and Kelman W. (1999) **Traditional breeding of Lotus species. In:** *Trefoil: The science and technology of Lotus* (Beuselinck P, Ed.) American Society of Agronomy pp. 187-198.

Parniske M. (2001) **Euroconference on Molecular Genetics of Model Legumes. Impact for Legume Biology and Breeding**. John Innes Centre, Norwich

Peterson P, Sheaffer C, and Hall M. (1992) **Drought effects on perennial forage legume yields and quality**. *Agronomy Journal* 84, 774-779.

Refi R, and Escuder C. (1998) **Nitrogen fixation by *Trifolium repens* and *Lotus tenuis*-based pastures in the Flooding Pampa, Argentina**. *Agronomie Paris* 18, 285-297.

Russelle M, McGraw R, and Leep R. (1991) **Birdsfoot trefoil response to phosphorous and potassium**. *Journal of Production Agriculture* 4, 114-120.

Shiferaw W, Shelton H, and So H. (1992) **Tolerance of some subtropical pasture legumes to waterlogging**. *Tropical Grasses* 26, 187-195.

Udvardi M. (2001) **Euroconference on Molecular Genetics of Model Legumes. Impact for Legume Biology and Breeding**. Max Planck Institute. Golm.

Vance, C, Johnson, L, Stade, S, and Groat, R. (1982) **Birdsfoot trefoil (*Lotus corniculatus*) root nodules: morphogenesis and its effect for forage harvest on struture and function**. *Canadian Journal of Botany* 60, 505-518.

Vance C, Reibach P, and Pankhurt C. (1987) **Symbiotic properties of *Lotus pedunculatus* root nodules induced by *Rhizobium loti* and *Bradyrhizobium* sp (Lotus)** *Physiologia Plantarum* 69, 435-442.

Vignolio O, Fernández O, and Maceira O. (1994) **Response of *Lotus tenuis* and *L. corniculatus* to flooding in seedling stage**. In: *Proceedings of the First International Lotus Symposium* (Beuselinck and Roberts , Eds.), pp. 160-163.

Vignolio O, Fernández O, and Maceira N. (1999) **Flooding tolerance in five populations of *Lotus glaber* Mill. (syn. *Lotus tenuis* Waldst. et. Kit.)** *Australian Journal of Agricultural Research* 50, 555-559.

Waghorn G, Jones W, Shelton I, and McNabb W. (1990) **Condensed tannins and the nutritive value of pasture**. *Procceedings of New Zealand Grasses Associations* 51, 171-176.

Waghorn G, Ulyatt M, John A, and Fisher M. (1987) **The effect of condensed tannins on the site of digestion of amino acids and other nutrients in sheep fed on *Lotus corniculatus***. *British Journal of Nutrition* 57, 115-126.

Wedderburn M. (1980) **Investigations into Lotus species**. Ph.D. Thesis, University of Glasgow.

Weeb, K, Jones S, Robbins M, and Minchin F. (1990) **Characterization of transgenic root culture of *Trifolium repens, T. pratense* and *Lotus corniculatus* and transgenic *L. corniculatus***. *Plant Science* 70, 243-254.

Wheeler D, and Dodd M. (1995) **The effect of aluminum on the growth of a range of temperate legume species and cultivars: a summary of results. Plant-soil interactions at low pH: principles and management**. *Proceedings of the Third International Symposium, Brisbane, Queensland, Australia, 12-16 September 1993. p.439-445. Developments in Plant and Soil Sciences Vol. 64.* (Date RA, Grundon NJ, Rayment GE, Probert ME, Eds.) Kluwer Academic Publishers Dordrecht, Netherlands.

Wheeler D, Edmeades D, Christie R, and Gardner R. (1992) **Effect of aluminium on the growth of 34 plant species: a summary of results obtained in low ionic strength solution culture**. *Plant Soil* 146, 61-66.

Chapter 1.3

DROUGHT AND SALINE STRESS

Pedro Díaz[1,2], Jorge Monza[1], and Antonio Márquez[2*]

[1]Laboratorio de Bioquímica. Departamento de Biología Vegetal. Facultad de Agronomía. Montevideo, URUGUAY. [2]Departamento de Bioquímica Vegetal y Biología Molecular, Facultad de Química Universidad de Sevilla. Sevilla, SPAIN. [*]Corresponding author.

Email: cabeza@us.es

Phone: +34 954 557 145 Fax: +34 954 626 853

Keywords: Proline, oxidative stress, salt stress, drought stress, TBARS

This chapter examines the response of L. japonicus *plants to different types of osmotic stress in the context of the changes most commonly noticed in other plant species. Up to 12-fold proline accumulation was detected in* L. japonicus *plants in relation to drought and salt stress treatments. In parallel to the accumulation of proline as an osmolyte, the above mentioned stress conditions produced in* L. japonicus *an increased level of oxidative damage as evaluated through measurement of thio barbituric acid reactive substances (TBARS). Characterisation of the* L. japonicus *response to osmotic stress is a prerequisite for the use of this plant as a model legume to assist in applied projects aimed to improve salt and drought tolerance of other closely related Lotus species. Different Lotus species are currently used to improve both pastures and hay quality wherein other forage legume species are not suitable (see chapter 1.2). However, many of these Lotus species are not well adapted to high temperature, salt, or drought.*

INTRODUCTION

Drought and salinity are the major limitations on crop productivity world-wide. One third of the earth surface is classified as arid or semiarid, and in most of the humid region, wherein much of world food is produced, the crops are subjected to periods of severe drought. Moreover, nearly 40% of the world land surface can be categorised as having potential salinity problems (Boyer, 1982). Other abiotic stresses such as atmospheric pollutants, presence of heavy metals, low temperature, or soil acidity can also affect plant productivity and quality (Figure 1). Drought, high salinity, and low temperature induce common metabolic effects because all of them decrease water activity inside the cell. Once stress is set, highly complex biochemical and physiological mechanisms are switched on in order to protect major

A.J. Márquez (ed.), Lotus japonicus Handbook, 39-50.

process such as cell respiration, photosynthetic activity and nutrient transport. Also, stress tolerance responses are induced to preserve cell organelles and tissue structures such as meristems (Kramer, 1983).

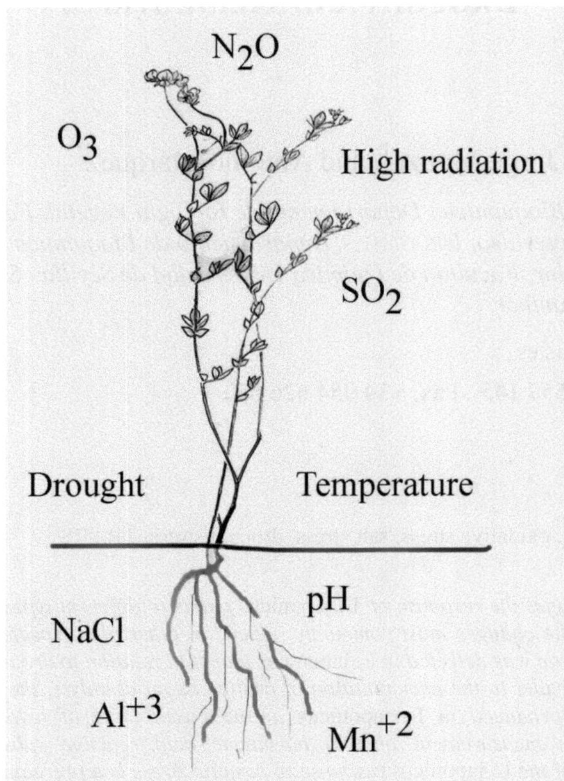

Figure 1. Main types of plant abiotic stresses that may affect growth and productivity of Lotus plants.

Animal production in the temperate region is based principally on pastures composed by grasses and legumes. These plants species are subjected to different abiotic stresses that joined with biotic stress decrease growth rate and dry matter production affecting dramatically the plant survival in the field.

Legumes are one of the most important components of the pasture because they are the main crude protein source in the animal food. Amongst the genera used, the most frequently found are *Medicago, Trifolium,* and *Lotus.* The last of these species is currently used to improve both pastures and hay quality wherein other forage legume species are not suitable (Papadopulous and Kelman, 1999). Actually, a number of *Lotus* species and cultivars exhibit a wide and different degree of tolerance to a compromised edaphic condition such as low pH, salinity, and high aluminium soil content. Moreover, considering these conditions, almost all Lotus species are more tolerant than lucerne or trefoils (Wheeler *et al.*, 1992). However, unlike lucerne, Lotus species are not well adapted to high temperature, salinity, or drought (Blumenthal and McGraw, 1999).

Recently, the use of molecular techniques, such as genetic markers and other biotechnological approaches are providing new tools to traditional plant breeding programs. In the near future, these approaches jointly with traditional breeding could make considerable progress in obtaining new improved cultivars with higher tolerance to water and saline stresses.

The knowledge of the biochemical and molecular basis of plant responses to stress are required to successfully generate improved stress-tolerance forage legumes. Among all forage species, Lotus is a privileged candidate to be improved for higher persistence and productivity in regions where other legumes are no suitable to be sown.

L. japonicus is closely related taxonomically to *L corniculatus* (the most agronomically important *Lotus* species), and offer a number of advantages for genetic studies. This specie can be seen as an excellent model to understand and improve stress tolerance in *Lotus* spp. In this chapter we make a little background of the main plant responses to drought and salt stress in order to analyse afterwards some of these responses in the genus *Lotus* and, in particular, in the model legume *L. japonicus.*

PLANT RESPONSES TO DROUGHT STRESS

In broad terms, drought can be permanent, seasonal, or random (Kramer, 1983). Permanent drought is present in desert areas wherein lack of rainfall is very common; seasonal drought occurs when either winter or summer seasons are regularly dry. Finally, random drought occurs in humid regions subjected to occasional variation in amount and distribution of rainfall. In addition, another type of drought to be considered is non-apparent drought. It is observed on hot summer days when high temperature or wind conditions induce an increase in transpiration rate that exceeds water root absorption rate. Consequently, despite adequate soil water content, plants can show drought symptoms (Sánchez-Díaz y Aguirreolea, 1993).

Plants have different strategies to avoid the deleterious effects of drought and these can be classified as follows:

1. Drought escapes i.e. short-life annual plants that develop their life cycle exclusively during the year period when soil water is available.
2. Drought tolerance i.e. plants cannot avoid the water stress; consequently they develop a number of adaptation mechanisms aimed to increase drought tolerance.

Both anatomical and morphological strategies are developed in plants to resist dehydration, such as deep and branched root systems or presence of thick cuticle in leaves and stems. All the strategies play an important role in drought tolerance by either increasing the water input by absorption or by reducing water output by transpiration (Kramer, 1983). In addition to these strategies, plant tolerance to drought is also mediated by metabolic responses that will be mentioned later in this chapter.

Lotus species in the geographical zones where they are sowed could be exposed to random or non-apparent drought and they are adapted to temperate and humid environments, wherein these types of drought are frequent. Therefore, they show mechanisms to tolerate the drought and not to avoid it.

PLANT RESPONSES TO SALT STRESS

Salt stress caused by high salt levels in the external solution of plant cells produces several deleterious consequences, such as ionic imbalance (Niu *et al.*, 1995; Zhu *et al.*, 1997). Also, when salinity results from an excess of NaCl, by far the most common salt stress, the increased intracellular Na^+ and Cl^- levels became deleterious to cellular systems (Serrano *et al.*, 1999). In addition, not only Na^+ and Cl^- homeostasis is disturbed, but also K^+ and Ca^{2+} (Hasegawa *et al.*, 2000, Rodríguez-Navarro, 2000, Serrano *et al.*, 1999). As a result, plant survival and growth are dependent on adaptations that re-establish ionic homeostasis, thereby minimising the duration of cellular exposure to ion disequilibrium. Moreover, high concentrations of salt impose a hyperosmotic shock by decreasing the chemical activity of water and causing loss of cell turgor. This negative effect in the plant cell is thought to be similar to the effects caused by drought.

Plants respond to salt stress, as well as drought, at three different levels, i.e., cellular, tissue and whole plant level. Cell-based mechanisms of ion homeostasis and the synthesis of osmoprotectants are essential determinants for salt tolerance.

Salt tolerance mechanisms strategies can be classified as those that function to minimise osmotic stress or ion unbalance, or alleviate the consequent secondary effects caused by the stresses. Some of these mechanisms are similar to drought mechanisms. Other strategies are different, such as ion homeostasis and ion transport systems. The cytotoxic ions in saline environments, typically sodium and chloride, are compartmentalised into the vacuole and used as osmotic solutes (Borsani *et al.*, 2003).

Halophytes have a particular ability to grow under high concentrations of NaCl, unlike halophytes most crop species are sensitive or hypersensitive to salt presence in the growth medium. It seems clear that the NaCl tolerance of halophytes is not the result of adaptive mechanisms or metabolic processes that are unique to these plants (Glenn *et al.*, 1999). On the contrary, it seems that the biochemical mechanisms leading to salt tolerance in halophytes plants, are regulated in such way that allow a more successful response to salt stress than in other plants (Hasegawa *et al.*, 2000). The question to be addressed now is whether the same mechanisms employed by halophytes in salt stress tolerance could be employed in glycophytes without a loss in productivity.

BIOCHEMICAL RESPONSES TO OSMOTIC STRESS

Salinity and drought induce osmotic stress and have similar biochemical responses, that include the accumulation of compatibles osmolytes and increase of reactive oxygen species (ROS) levels (Zhu *et al.*, 1997). According to this, and from a practical point of view, salt stress can be imposed in laboratory settings more easily and precisely by applying controlled osmotic stress.

For several years there has been known that plants respond to drought and salt stress by undergoing biochemical adaptive processes such as ion transport and by accumulating different compounds, named compatible solutes (Hanson and Hitz, 1982; Smirnoff and Stewart, 1985). Solutes known that accumulate under osmotic stress include carbohydrates, organic acids, polyamines, and amino acids. These molecules can accumulate to high concentration without interfering with cell metabolism (Bray, 1993).

Stressful conditions such as drought or salt stress, associated with high light intensity are able to promote ROS production and oxidative stress. Water and salt stress induced physiological changes such as decrease of the cell water potential and stomata closing. These events, in turn, result in lower CO_2 availability to the green tissues favouring the Mehler reaction that result in superoxide production. Moreover, during drought, plant glycolate oxidase activity is induced generating hydrogen peroxide. The increased activity of antioxidant enzymes such as superoxide dismutase (SOD), peroxidase, and ascorbate peroxidase during stress has been reported in several organisms. Oxidative stress is associated with osmotic stress damage to membrane lipids and nucleic acids. The levels of thio barbituric acid reactive substances (TBARS) can be used to evaluate this damage. Another ROS that are produced in stress conditions are the hydroxyl radical and singlet oxygen. These molecules are inactivated by non-enzymatic systems. Osmolytes such as proline and poliols must act as scavengers (Smirnoff and Cumbes, 1989; Alia *et al.* 2001. Hong *et al.*, 1999) and could play a protective role from oxidative damage to macromolecules such as proteins.

L. JAPONICUS RESPONSE TO DROUGHT STRESS

Few reports are available concerning the biochemistry and molecular responses to osmotic stress (drought and salt stress) in *Lotus* spp. While most of reports referred to *L. corniculatus* response to drought (Olsson *et al.*, 1996; Carter *et al.*, 1997; Borsani *et al.*, 1999; and Borsani *et al.*, 2001), none study has been carried out in the biochemistry of *L. japonicus* response to different stress conditions.

The drought imposition form determines the types of biochemical response that the plant will induce. In plants the drought can be imposed in two ways: a) fast drought that can mimic the named non-apparent drought wherein evaporation rate is higher than the water absorption, b) slow drought similar to seasonal and random drought wherein the water loss is slowly imposed. Experimentally the form of drought imposition will determine the biochemical response to be studied.

L. japonicus plants show a well-developed growth in hydroponic assemblies (Borsani *et al.*, 1999; Orea *et al.*, chapter 7.3, this book) fed with Hornum nutrient solution (Handberg and Stougaard, 1992). When these plants were subjected to fast drought for 12 h, withholding the nutrient solution, the relative water content (RWC) decreased from 90 to 60%. However, when *L. japonicus* plants were grown in plastic pots filled with a soil-type support (vermiculite or montmorillonite-based terragreen^{R}) the RWC dropped to 50% when they were subjected to slow drought for 4 days withholding the irrigation.

Working with 30 days-old *L. japonicus* plants, fast and slow drought periods should not exceed 12 hours and 4 days, respectively, since for longer drought periods the plants were unable to rehydrate irreversible drought. In these experimental conditions, for both fast and slow drought assays, *L. japonicus* plants recovered 90% of RWC at 6-8 h after re-establishing the water supply.

During the above mentioned drought treatments, the plants showed slight wilt symptoms but no change in pigment contents was observed (Table 1). Therefore, the photosynthetic apparatus appears not to have been seriously damaged by water stress. For this reason, changes observed in other metabolic pathways in response to water stress should not be associated with photosynthesis alterations.

	Control	**Drought** (12h)	**Drought** (4d)	**Salt stress** (4d)
Hydric deficit (%)	14.3 ± 2.5	30.1 ± 6.5	49.5 ± 7.0	15.3 ± 2.2
chlorophyll a (mg. g^{-1} DW)	6.54 ± 0.58	6.05 ± 1.64	5.68 ± 1.02	5.82 ± 0.97
chlorophyll b (mg. g^{-1} DW)	3.14 ± 0.18	2.70 ± 0.74	2.78 ± 0.43	2.87 ± 0.41
a/b ratio	2.14 ± 0.24	2.22 ± 0.11	2.08 ± 0.10	2.12 ± 0.22
Proteins (mg. g^{-1} DW)	115 ± 25	99 ± 18	103 ± 17	111 ± 9
Amino acids (µmol. g^{-1} DW)	*250 ± 62*	*263 ± 42*	*233 ± 18*	*234 ± 76*

Table 1. Hydric deficit, chlorophyll a/b ratio, proteins, and amino acids in wild type leaves of L. japonicus *plants subjected to drought and salt stress. Values are the mean of four replicates ± standard deviations. All parameters were determined according to Borsani* et al. *(1999).*

We have studied the ability of *L. japonicus* plants to accumulate proline as osmolyte in response to drought. Three to 12-fold proline accumulation was detected, respectively, in leaves from plants subjected to fast or slow drought (Figure 2), in spite of the fact that leaves in both treatments began with similar RWC. A concomitant increase in proline could be detected during the slow drought process. In fact, a linear relationship could be observed between the proline content and the hydric deficit produced on the plants at different times of this process (Figure 3).

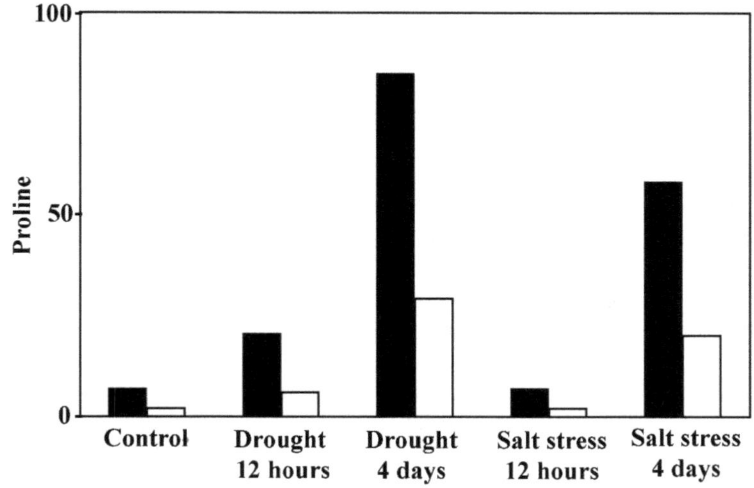

Figure 2. Proline content in L. japonicus *leaves from plants subjected to different drought and salt stress treatments. Black bars: absolute content of proline (μmol per g dry weight). White bars: relative content of proline (% of total amino acid content). Plants were grown as described by Orea* et al. *(Chapter 7.1, this book). Proline was determined according to Borsani* et al. *(1999). Other experimental details are given in the text.*

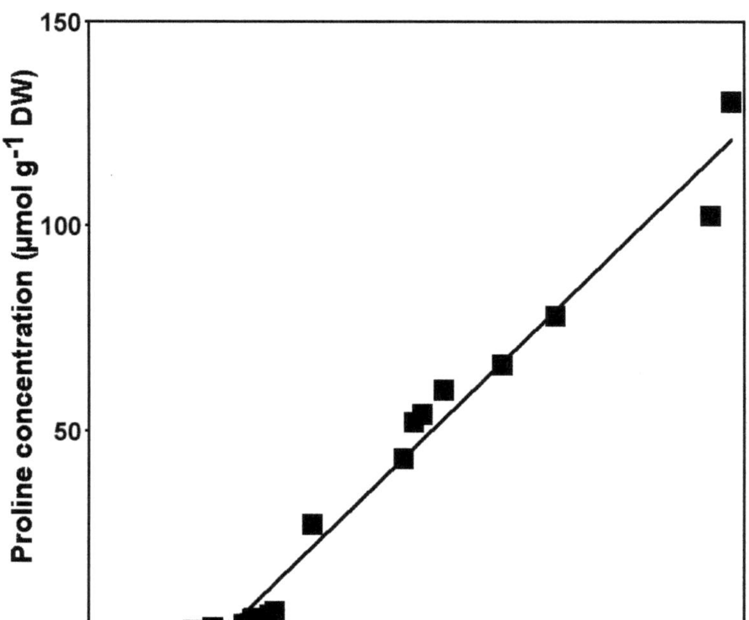

Figure 3. Relationship between Hydric Deficit (HD) and proline content in leaves of L. japonicus *subjected to different times of slow drought applied by removal of the irrigation to the plants.*

Although the amount of proline produced was lower in the fast drought compared to the slow drought, it is important to say that the actual rate of proline accumulation was always higher in the fast drought than in the slow drought if considering the length of time applied in both processes. This result is easy to explain considering that proline accumulation may result from both induction of proline biosynthesis and/or inhibition of its oxidation (Hong *et al.*, 2000). Moreover, there are previous reports that activation of pyrroline-5-carboxylate synthetase (an enzyme from proline biosynthesis) as well as inhibition of proline dehydrogenase (an enzyme from proline degradation), are both faster in fast drought than in slow drought. An additional explanation considered the oxidation of proline to glutamate through the proline dehydrogenase pathway during the dark period of slow drought treatment (Sanada *el al.*, 1995). Leaf proline accumulation in response to drought seems a peculiar characteristic of *Lotus* since it has been reported in other in other species such as *L. corniculatus* cv. San Gabriel (Borsani *et al.* 1999) and *L. uliginosus, L. glaber, and L. subbiflorus.* In addition, proline content was higher in ammonium than nitrate fertilized *Lotus* plants suggesting that nitrate could be used as osmolyte instead of proline under water stress conditions in the last case (Diaz *et al.*, manuscript in preparation).

The level of oxidative stress associated with drought in *L. japonicus* has been also examined. Figure 4 shows that the level of TBARS increased up to 80% during drought, being much higher during the slow than in the fast drought.

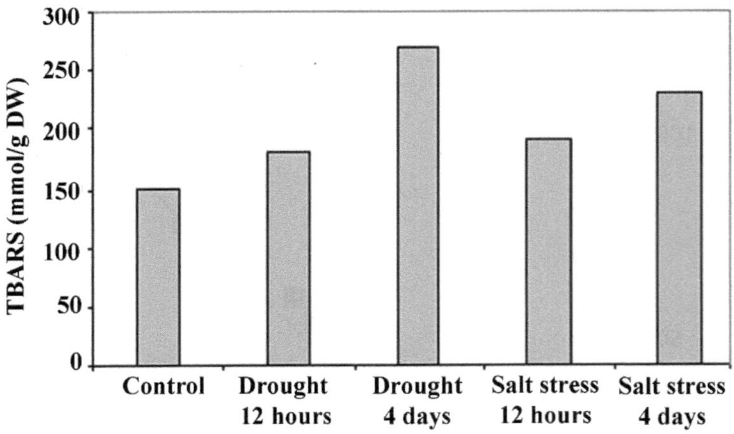

Figure 4. TBARS content in L. japonicus *leaves from plants subjected to different drought and salt stress treatments. TBARS were measured as described by Borsani* et al. *(2001).*

This result indicates that water loss is able to generate an important extent of oxidative damage in *L. japonicus* leaves. Similarly, in *L. corniculatus* leaves it was also shown a drought increased oxidative stress and total SOD activity. Additionally, accumulation of Cu-ZnSOD transcripts as well as the presence of a new Cu-ZnSOD protein were detected in water stressed *L. corniculatus* leaves (Borsani et al., 2001). This enzymatic response may constitute in plants an advantageous adaptation to water stress by protecting the plant cell form oxidative reactions (Bowler et al., 1992). For example, glutamine synthetase purified from *L. corniculatus* leaves was effectively protected against inactivation from in vitro hydroxyl production by addition of proline, sorbitol, and mannitol. According to the scavenger index in decreasing order, the ranking was sorbitol, mannitol, and proline (unpublished data). Molecules as polyamines and nitric oxide could act as non-enzymatic defences scavenging superoxide radical (Bors et al., 1989; Radi et al., 2001).

THE *L. JAPONICUS* RESPONSE TO SALT STRESS

L japonicus salt stress studies were carried out in similar conditions than for previously described drought studies. Thirty day-old *L. japonicus* plants were subjected to salt stress by irrigating with Hornum NaCl supplemented solution for 4 days until reaching final concentration of 250 mM. Neither wilt symptoms nor changes in pigment contents were observed in this stress conditions. Whether short time salt tolerance of *L. japonicus* plants is due to salt exclusion or compartmentation in vacuole is unknown. However, when plants were subjected to salt stress, but in a lower relative humidity (RH) around 50%, the plants showed a dehydrated phenotype. This fact may indicate that low RH increases transpiration rates and consequently rises in the water and Na^+ input in *L. japonicus* tissues.

We have studied the extent of accumulation of proline because of application to *L. japonicus* plants of different concentrations of NaCl and for different lengths of time. Figure 5A shows that concentrations higher than 100 mM of NaCl were required to produce significant accumulation of proline after 4 days of salt treatment. Using 250 mM NaCl, this increase in proline was observed at 3-4 days of treatment (Figure 5B). In these conditions the maximum level of proline increased by 7-fold as a result of the salt treatment. This increase in proline happened both in absolute terms, reaching values around 50 μmol per g DW, as well as in its relative proportion with regard to total amino acid content (Figure 2). However, these increases were lower that those observed in *L. japonicus* plants subjected to drought at the same time (4 days, Figure 2). A similar behaviour in proline accumulation was observed in *Triticum durum* subjected to salt stress and drought (Mattioni *et al.*, 1997).

A 50% increase in TBARS species was detected in *L. japonicus* plants subjected to 250 mM NaCl treatment for 4 days (Figure 4). This result also indicates that the salt treatment was sufficient to generate an important but still reversible oxidative damage in *L. japonicus* leaves.

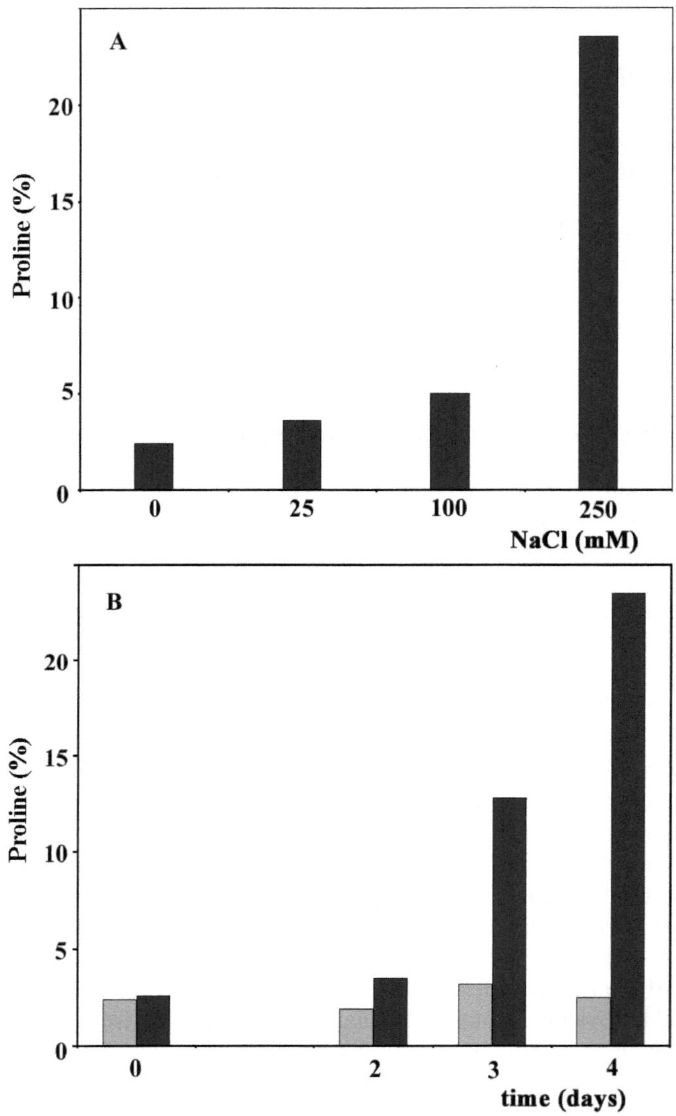

Figure 5. Proline accumulation during salt stress in L. japonicus *leaves. A: Effect of different NaCl concentrations for 4 days of salt treatment. B: Proline accumulation at different times of treatment with 250 mM NaCl (black bars) and control without NaCl (grey bars). Other details as described in previous figures and in the text.*

ACKNOWLEDGEMENTS

Authors wish to thank funding from joint research projects from MEC-AECI and BMC2001-3162 from MCYT and BFU2004-02753 from MEC (Spain), and HPRN-CT2000-00086 and MRTN-CT-2003-505227 and INCO FP6-2003-INCO-DEV-2-

517617 from the European Union, as well as support given by Junta de Andalucía (Spain) to group CVI-163. Technical assistance of MJ Cubas is also gratefully acknowledged. Also, we would like to thank to Dr Omar Borsani for critical review of the manuscript and Dr Clara Pritch for English correction.

REFERENCES

Alia A, Mohanty P, and Matysk J. (2001). **Effect of proline in the production of synglet oxygen.** *Amino acid* 21, 195-200.

Blumenthal M and McGraw R. (1999). **Lotus adaptation, use and management**. In: *Trefoil: The science and technology of Lotus.* (Beuselinck P , Ed.). American Society of Agronomy, pp. 97-120.

Bors W, Langebartels C, Michel C, Sandermann JH. (1989). **Polyamines as radicals scavengers and protectants against ozone damage.** *Phytochemistry* 28, 1589-1595.

Borsani O, Díaz P, Agius F, Valpuesta V, and Monza J. (2001). **Water stress generates an oxidative streess through the induction of a specific Cu/Zn superoxide dismutase in Lotus corniculatus leaves.** *Plant Science* 161, 757-763.

Borsani O, Díaz P, and Monza J. (1999). **Proline is involved in water stress responses of Lotus corniculatus nitrogen fixing and nitrate fed plants.** *Journal of Plant Physiology* 155, 269-273.

Borsani O, Valpuesta V, and Botella J. (2003). **Developing salt tolerant plants in a new century: a molecular biology approach.** *Plant Cell Tissue and Organ Culture* 73, 101-115.

Bowler C, Van Montagnou M and Inzé D. (1992). **Superoxide dismutase and stress tolerance.** *Annual Review of Plant Physiology* 43, 83-116.

Boyer J. (1982). **Plant productivity and environment.** *Science* 218, 443-448.

Bray E. (1993). **Molecular responses to water deficit.** *Plant Physiology* 103, 1035-1040.

Carter E, Theodorus M, and Morris P. (1997). **Responses of Lotus corniculatus to environmental change I. Effect of elevated CO_2, temperature and drought on growth and plant development.** *New Phytologist* 136, 245-253.

Glenn E, Brown J, and Blumwald E. (1999). **Salt tolerance and crop potential of halophytes.** *Critical Reviews in Plant Science* 18, 227-255.

Handberg K, and Stougaard J. (1992). ***Lotus japonicus*, an autogamous, diploid legume species for classical and molecular genetics.** *The Plant Journal* 2, 487-496.

Hanson A, and Hitz W. (1982). **Metabolics responses of mesophytes to plant water deficits.** *Annual Review of Plant Physiology* 33, 163-203.

Hasegawa M, Bressan R, Zhu J-K, and Bhonert H. (2000). **Plant Cellular and Molecular Responses to High Salinity.** *Annual Reiew of Plant Physiology* 51, 493-499.

Hong Z, Lakkineni K, Zhang Z, and Verma D. (2000). **Removal of feedback inhibition of pyrroline 5 carboxylate reductase results in increased proline accumulation and protection of plants from osmotic stress.** *Plant Physiology* 122, 1129-1136.

Kramer P. (1983). *Water relations of plants.* Academic Press, 489 pp.

Mattioni, C, Lacerenza, N. Troccoli, A, De Leonardis, A, and Di Fonzo, N. (1997). **Water and salt stress-induced alterations in proline metabolism of *Triticum durum* seedlings.** Physiol. Plant. 101:787-792.

Niu X, Bressan R, Hasegawa P, and Pardo J. (1995). **Ion homoestasis in NaCl stress environments.** *Plant Physiology* 109, 735-742.

Olsson M, Nilsson K, Liljenberg C, and Hendry G. (1996). **Drought stress in seedlings: lipid metabolism and lipid peroxidation during recovery from drought in *Lotus corniculatus* and *Cerastium fontanum*.** *Physiologia Plantarum* 96, 577-584.

Papadopoulos Y and Kelman W. (1999). **Traditional breeding of Lotus species**. In: *Trefoil: The science and technology of Lotus* (Beuselinck P, Ed.). American Society of Agronomy pp187-198.

Radi, R, Peluffo, G, Alvarez, M, Naviliat, M, and Cayota, M. (2001). **Unravelling peroxynitrite formation in biological systems**. *Free Radical Biology & Medicine* 230, 463-488.

Rodríguez-Navarro A. (2000). **Potassium transport in fungi and plants**. Biochimica and Biophysica Acta 1469, 1-30.

Sanada Y, Ueda H, Kuribayashi K, Andoh T, Hayashi F, Tamai N, and Wada K. (1995). **Novel light-dark change of proline levels in halophythe (*Mesembryanthemum crystallinum L.*) and glycophytes (*Hordeum vulgare L* and *Triticum aestivum L.*): Leaves and roots under salt stress**. *Plant Cell Physiology* 36, 965-970.

Sánchez-Díaz M y Aguirreolea J. (1993). ***Efectos fisiológicos que causa la falta persistente de agua en los cultivos***. *Phytoma España* 51, 26-36.

Serrano R, Mulet JM, Ríos G, Márquez JA, de Larriona IF, Leube MP, Mendizabal I, Pascual-Ahuir A, Proft, MRR, and Montesinos C. (1999). **A glimpse of the mechanism of ion homeostasis during salt stress**. *Journal of Experimental Botany* 50, 1023-1036.

Smirnof N, and Cumbes Q. (1989). **Hydroxyl radical scavenging activity of compatible solutes**. *Phytochemsitry* 28, 1057-1060.

Smirnoff N, and Stewart G. (1985). **Nitrate assimilation and traslocation by higher plants: Comparative physiology and ecological consequences**. *Physiologia Plantarum* 64, 133-140.

Wheeler D, Edmeades D, Christie R, and Gardner R. (1992). **Effect of aluminium on the growth of 34 plant species: a summary of results obtained in low ionic strength solution culture**. *Plant Soil* 146, 61-66.

Zhu JK, Hasegawa PM, and Bressan RA. (1997). **Molecular aspects of osmotic stress**. *Critical Reviews in Plant Science* 16, 253-277.

SECTION 2

SYMBIOTIC PROCESSES

Chapter 2.1

METHODS FOR STUDYING NODULE DEVELOPMENT AND FUNCTION

Makoto Hayashi[1], Myra L Tansengco[1], Norio Suganuma[2], Krzysztof Szczyglowski[3], Lene Krusell[4], Thomas Ott[4], and Michael Udvardi[4*]

[1]Osaka University, Graduate School of Engineering, Department of Biotechnology, Yamadaoka 2-1, Suita, Osaka 565-0871, JAPAN; [2]Department of Life Science, Aichi University of Education, Kariya Aichi, 448-8542, JAPAN; [3]Agriculture and Agri-Food Canada, Southern Crop Protection and Food Research Centre, 1391 Sandford Street, London, Ontario N5V 4T3, CANADA; [4]Max Planck Institute of Molecular Plant Physiology, Am Mühlenberg 1, 14476 Golm, GERMANY; [*]Corresponding author.

Email: udvardi@mpimp-golm.mpg.de

Phone: +49 331 567 8149 Fax: +49 331 567 8134

Keywords: Plant growth systems, microscopy, lacZ-expressing rhizobia, acetylene reduction assay, molecular markers, real-time RT-PCR, immunoblotting

Interaction between Lotus japonicus *and* Mesorhizobium loti *results in the development of a specialised organ: the root nodule. Lotus root nodules develop from de-differentiated root cells, which form a meristem that undergoes a limited number of cell divisions. The result is a determinate, roughly spherical organ. Invading rhizobia colonise cells in the nodule cortex, each of which ultimately accommodates many thousands of nitrogen-fixing bacteria called bacteroids. Differentiation of both plant and bacterial cells is crucial for symbiotic nitrogen fixation (SNF), and genetic defects in either partner can compromise SNF. Identification of Lotus and* M. loti *genes that are necessary for nodule development and function is a major focus of current research. This chapter briefly describes the major steps in Lotus nodule development and differentiation, before presenting methodologies that are used routinely to characterise these processes in wild type and mutant interactions.*

LOTUS NODULE DEVELOPMENT AND DIFFERENTIATION

Lotus japonicus is a diploid, perennial legume with a natural habitat in the Far East (Handberg et al., 1992). As most leguminous species, Lotus interacts symbiotically with the beneficial soil bacteria to form root-derived organs, the nitrogen fixing nodules. Fast growing strains of *Mezorhizobium loti* (e.g. NZP2235, JRL501), the

A.J. Márquez (ed.), Lotus japonicus Handbook, 53-82.

broad host-range *Rhizobium* strain NGR234, as well as some slow growing *Bradyrhizobium spp.* have been reported to nodulate Lotus. However, *M. loti* is the most commonly used species since it is an effective microsymbiont, and the complete nucleotide sequence of the *M. loti* strain JRL501 (MAFf303099) genome has recently become available (Kaneko et al., 2000).

Like soybean, Lotus forms spherical root nodule structures of the determinate type (Figure 1). These nodules are characterized by lack of persistent meristematic activity, a developmental feature that distinguishes them from the elongated, indeterminate root nodules of temperate legumes such as pea and alfalfa. However, the organogenesis of Lotus nodules has been shown to combine some features of indeterminate and determinate nodule development (van Sprousen et al., 2001).

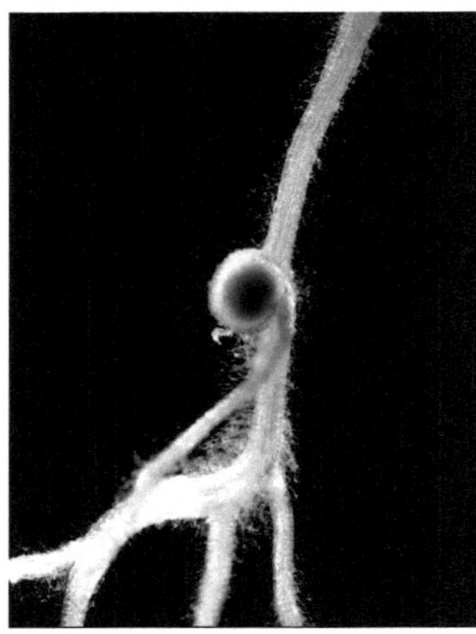

Figure 1. Fully mature nitrogen-fixing nodule formed on L. japonicus *roots. See colour plate 2 (G) for this figure in colour.*

Lotus japonicus responds to *M. loti* by initiating a complex developmental program for nodule organogenesis within the susceptible zone of the root system. The morphogenic signals of bacterial origin, called nodulation (Nod) factors, have been characterized from several different *M. loti* strains (Lopez-Lara et al., 1995; Olsthoorm et al., 1998; Niwa et al., 2001). They were shown to consist of a mixture of lipochitin oligosaccharide (LCO) molecules that are synthesized and secreted from the bacteria in response to an as yet uncharacterized inducer molecule(s), likely to represent isoflavonoids, derived from Lotus roots. The major component of the LCO mixture was found to be N-acetylglucosamine pentasaccharide in which the non-reducing residue is N-acylated with a C18:1 acyl moiety, N-methylated, and carries a carbamoyl group, while the reducing end is substituted with 4-*0*-acetylfucose (Niwa et al., 2001). Ectopic application of the LCO mixture on Lotus roots incites various cellular and molecular responses that are reminiscent of the early responses to Nod factors and/or rhizobial infection in other legume species (Lopez-Lara et al., 1995; Niwa et al., 2001).

The earliest responses to Nod factors described in Lotus include membrane depolarization, extra-cellular alkalinization and calcium spiking in root hair cells (Harris et al., 2003; Radutoiu et al., 2003). These physiological reactions occur within a few minutes upon application of purified Nod factors and/or living bacteria to the root system and were shown to depend on the presence of a functional root perception apparatus that involves NFR1/NFR5 LysM receptor-like kinases (Madsen et al., 2003; Radutoiu et al. 2003).

On the cytological level, cells of the epidermis, cortex and root pericycle respond to the presence of *M. loti* and/or application of Nod factors by initiating various growth patterns. In the epidermis, root hair cells deform by tip swelling, branching and curling, leading to the formation of the typical "shepherd's crook" structures (Szczyglowski et al., 1998, Niwa et al., 2001). These structures entrap the bacteria and serve as a starting point for initiation and growth of the infection thread (IT). Detailed microscopic examination of infected root hairs revealed the presence of bona fide ITs originating at the curled tip and extending the full length of the root hair (Figure 2; Szczyglowski et al., 1998).

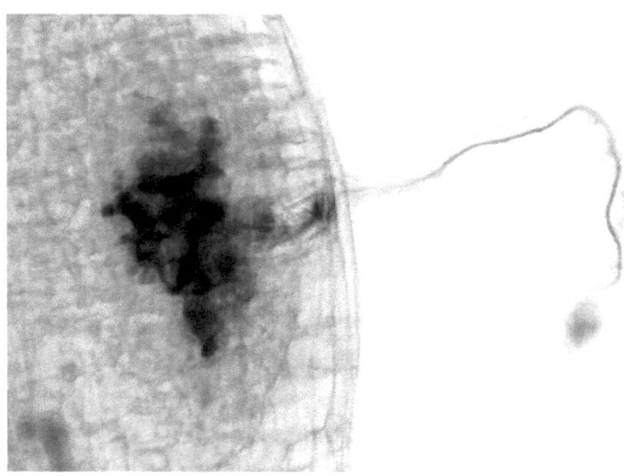

Figure 2. Brightfield micrograph of L. japonicus *nodule 10 days after inoculation with* M. loti *strain NZP2253 carrying a hemA:LacZ reporter gene fusion. Roots were stain for β-galactosidase activity, cleared, and examined using brightfield microscopy. See colour plate 2 (E) for this figure in colour.*

The Lotus ITs were shown to be tubular structures consisting of a multilayered fibrillar cell wall, an overlying enclosure membrane continuous with the cytoplasmic membrane of the infected host cell, and a lumen containing vegetative bacteria embedded in an amorphous matrix. Unlike many legumes, Lotus forms very broad infection threads (Figure 3) that extend from infected root hairs through polarized cortical cells of the outer cortex to the meristematic cells of the nodule primordium. Thick and prominent infection threads, usually observed in association with indeterminate-type nodulation, spread between and inside Lotus cortical root cells forming multiple, narrow lateral branches that initiate host cell invasion (Szczyglowski et al., 1998, van Spronsen et al., 2001).

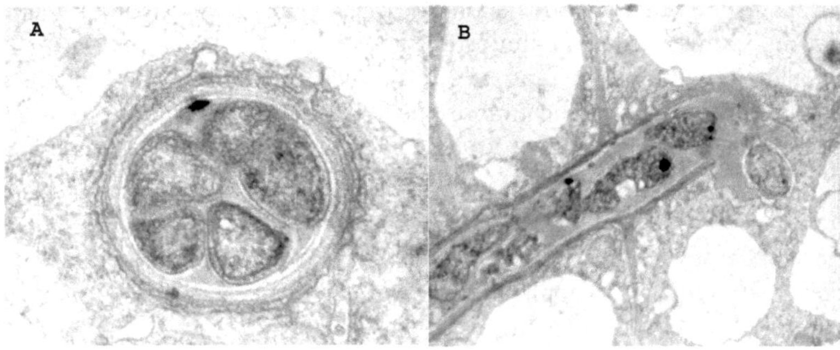

Figure 3. Transmission electron micrographs of a cross-section (A) and longitudinal section (B) of infection threads containing bacteria in L. japonicus *roots inoculated with* M. loti *strain NZP2235.*

In the root cortex, the cells assume an interesting position dependent pattern of dedifferentiation. The outermost cortical cells positioned immediately underneath the infected root hairs either do not divide or make a single anticlinal division. They undergo significant swelling with a concomitant polarization of the cytoplasm in a radial direction to gives rise to so called cytoplasmic bridges. (Niwa et al., 2001; von Spronsen, 2001). The formation of cytoplasmic bridges has been observed in the first, second and only occasionally in the third cortical root cell layer (van Spronsen et al., 2001). This type of cell polarization is assumed to guide the progression of the infection thread towards the underlying nodule primordium (Szczyglowski and Amyot, 2003 and references therein).

The third cortical cell layer, as opposed to the outer cortical layer typically involved in the determinate nodule initiation, is proposed to give rise to the first cell divisions leading to the formation of nodule primordia in inoculated Lotus roots (van Spronsen et al., 2001). The fully activated cortical cells of the inner cortex are characterized by enlarged, centrally located, nuclei (Figure 4), and contain trans-vacuolar cytoplasmic strands radiating from the nucleus to the cell wall. These cells undergo successive divisions, which during further development spread to surrounding cell layers, and eventually, give rise to a clearly defined nodule meristem. Incoming ITs ramify within the meristematic region of the nodule primordia where endocytosis of the bacteria is initiated. The release of the vegetative bacteria into the cytosol of the host plant cell has been shown to occur through small erosion zones in the ITs. Individual bacterial cells migrate from within the IT to infection droplets and to the cytosol of the host cells (Figure 5; Szczyglowski et al., 1998). They remain separated from the host cytoplasm by the peribacteroid (symbiosome) membrane. In wild type Lotus nodules, only a subtle morphological transition from the vegetative to bacteroid states occurs, such that the *M. loti* bacteroids undergo a slight increase in size while retaining the same straight rod shape as vegetative cells (Szczyglowski et al., 1998, Suganuma et al., 2003; Figure 6). However, branched and pleomorphic-shaped bacteroids were occasionally observed (Szczyglowski et al., 1998).

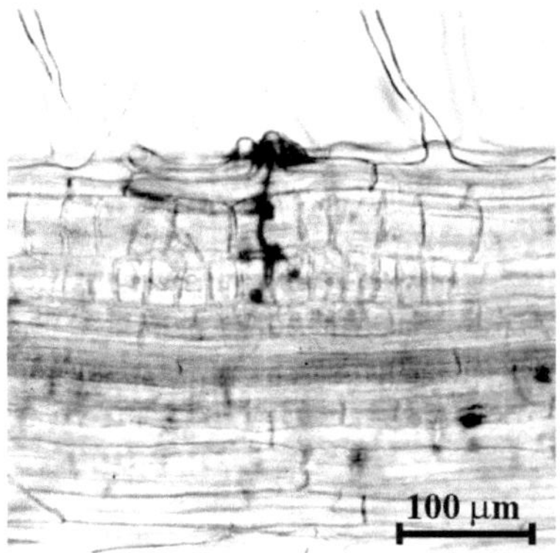

Figure 4. Brightfield micrograph of L. japonicus *roots showing the progression of the IT towards the developing nodule primordium.* L. japonicus *roots were inoculated with* M. loti *strain NZP2253 carrying a hemA:LacZ reporter gene fusion, stained for β-galactosidase activity, and examined using brightfield microscopy. See colour plate 2 (F) for this figure in colour.*

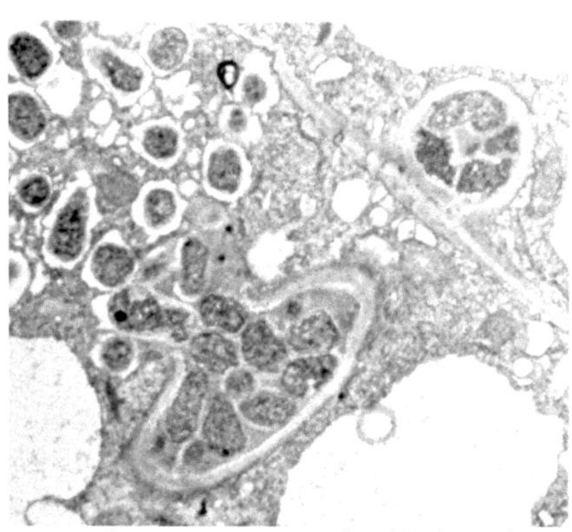

Figure 5. Transmission electron micrograph of L. japonicus *cell invasion by* M. loti *strain NZP2235. The cell wall has eroded at the side of an infection thread and individual bacteria have migrated from the IT into the infection droplets (see also Figure 3B).*

Lotus nodules initiate nitrogen fixation approximately 10 days after inoculation with *M. loti,* coinciding with the emergence of young nodules from the root cortex. Twenty one days after inoculation, the mature, spherical, nodules display a characteristic histology consisting of a large central nitrogen fixing area, primarily composed of the bacteroids-containing cells, surrounded by a concentric layer of uninfected nodule cortex. Together with peripherally located vascular elements and the apparent lack of a discernible meristem, these cytological features typify a determinate type of nodule (Szczyglowski et al., 1998).

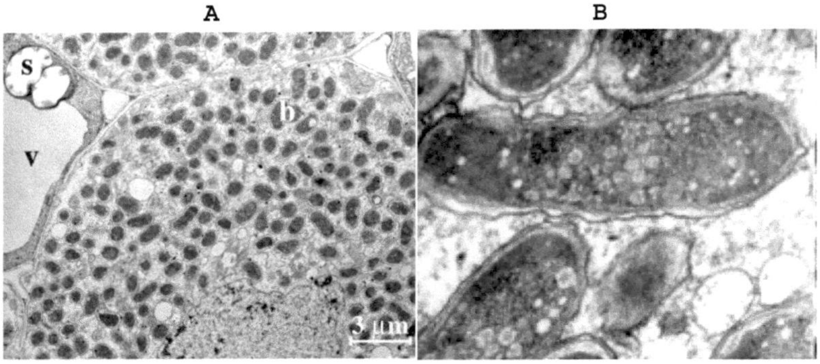

Figure 6. Transmission electron micrographs of L. japonicus *nodule infected cell containing many symbiosomes (A), and a close up of* M. loti *bacteroids surrounded by peribacteroid membranes (B).*

Although certainly variable and dependent on particular growth conditions used, Lotus establishes a homeostasis of symbiotic root development by limiting the extent of nodulation events, such that only a defined number of nitrogen fixing nodules is formed (Wopereis et al., 2001). Three weeks after inoculation with *M. loti*, 10-15 mature nitrogen fixing nodules can be found on the upper portion of the wild type Lotus root grown under nitrogen limiting conditions (Figure 7). The mechanism involved in this interesting homeostatic regulation is currently being characterized on the molecular level and at least with respect to one aspect of the mechanism involved, a strict requirement for a shoot-derived *Har1* receptor kinase function was recently reported (Wopereis et al., 2001; Krussel et al. 2002; Nishimura et al. 2001).

PLANT GROWTH SYSTEMS FOR NODULATION STUDIES

A variety of substrates and growth conditions are used to culture inoculated Lotus plants for studies of nodulation and symbiotic nitrogen fixation. Growth substrates include course sand, clay beads, and solidified agarose, some of which are described elsewhere in this book. In this chapter, a glass-slide system that facilitates studies of the initial stages of infection, and a plastic pillow system for the production of mature nodulated plants are described.

Glass slide culture of seedlings for microscopy

Liquid culture of seedlings on a slide glass is useful for observation of roots under a microscope without damaging root hairs. This technique can be adopted for root hair deformation assay by application of LCOs (Niwa et al., 2001) or inoculation of *M. loti*, or pharmaceutical assays. This technique can be adopted for root hair deformation assay using LCOs (Niwa et al., 2001) or rhizobium application, or pharmacological assays. This system is useful for studying early infection and nodulation events, but is not appropriate for later stages when plants become too large.

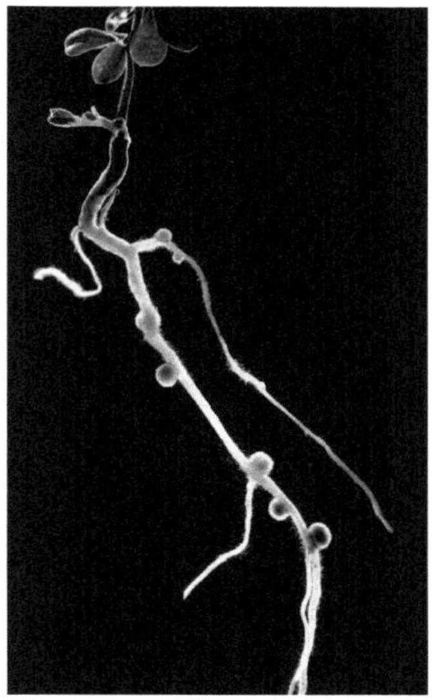

Figure 7. L. japonicus *nodulation phenotype 3 weeks after inoculation with* M. loti *strain NZP2235. See cover for this figure in colour.*

Equipment and reagents

- Glass slides and cover slips
- Silicone caulk
- Staining jars
- Forceps
- B & D medium

Method

1. Preparation of Fåhraeus slides
    a. Put roughly 10 μl of silicone caulk at each corner of a cover slip.
    b. Set the cover slip on a glass slide in a way keeping 1 mm space between them, and 1 cm above from the bottom of the glass slide (Figure 8).

Figure 8. Fåhraeus glass slide with cover slip.

c. Let them still for a day in order to solidify silicone.

d. Sterilize by autoclave in a glass jar filled with distilled water.

2. Transfer of seedlings to the slides

   a. Put 50 ml of B & D medium in a sterilized staining jar.

   b. Using forceps, transfer seedlings to the slide one by one, without damaging root hairs (Figure 9). Select only those with straight roots (Miyakojima, 2 day old seedlings; Gifu, 3 day old seedlings as in the section of spot inoculation).

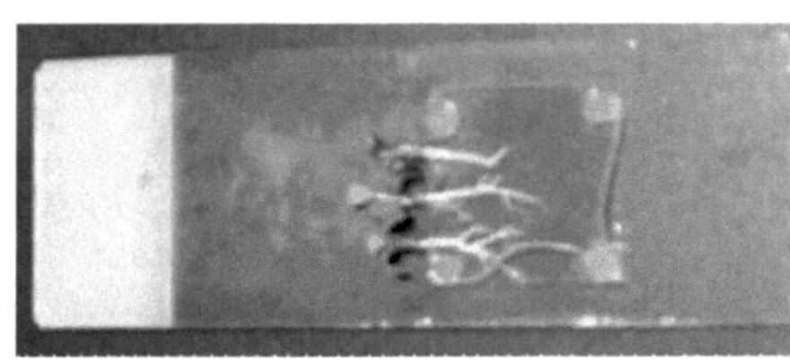

Figure 9. Seedlings in the slide.

   c. Insert slides with seedlings to the staining jar (Figure 10).

   d. Cover the bottom half of the jar with black paper and aluminium foil to avoid light to roots (Figure 11), culture them at 26°C, 16h light/23°C, 8h dark regime.

Note: We set seedlings four days before applying Nod factors or other chemicals in order to avoid potential wounding responses.

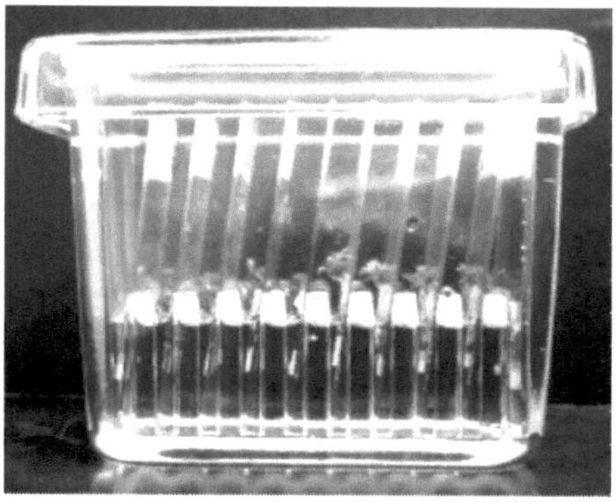

Figure 10. Slides set in the staining jar.

Figure 11. The jar covered by black paper and aluminium foil.

The pillow growth system

The pillow system allows effective nodulation analysis of a large plant population with a relatively small area needed for growing the samples. As several plants can be grown simultaneously, the time-consuming cleaning of root tissue is reduced substantially.

To observe and screen plants for nodule phenotype, it is necessary to germinate and establish the plants on agar media, then transplant them to soil after about a week. Root growth and nodulation kinetics using this method resemble those observed in plants grown in vermiculite/sand mixtures.

I. Sowing seeds in agar medium

Equipment and reagents

- Plastic tray
- 50 ml plastic tube
- Germination medium: 0.8% (w/v) Bacto agar
- Bleach solution: 10% NaClO in sterile distilled water
- Sterile distilled water
- Sterile filter paper
- Sterile Petri dishes or rectangular plastic plates
- Parafilm

Method

1. Prepare the germination medium by autoclaving at 121°C for 15 min and then pour into pre-sterilized Petri dishes.
2. Place the seeds in a 50 ml plastic tube and add 10-15 ml concentrated sulfuric acid. Vortex vigorously for 10-15 min and then pour the solution into a container of tap water. Rinse the seeds at least three times.
3. Put the seeds in a clean plastic tube and add the bleach solution. Shake the container gently for 10 min.
4. Pour out the bleach solution and then wash 4-5 times with sterile water. Keep the seeds in sterile distilled water and shake the container gently for 2-3 hr, then discard the water.
5. Place the seeds in sterile filter paper to dry, and transfer the seeds in germination medium (work in laminar flow hood and use sterile forceps). Seal the Petri plates with Parafilm and incubate in growth chamber under 16-h-light/26°C and 8-h-dark/23°C cycle with 60°C relative humidity.
6. Keep the plates under dark conditions for 2 days and then allow the seeds to germinate for another 7-10 days (i.e. once two to four true leaves have formed).

II. Growing seedlings in the pillow system

Equipment and reagents

- Plastic tray: 30 x 10 x 10 cm
- Pillow bags: polypropylene tea packs (120 x 95 mm) or nylon bags
- Vermiculite:perlite mix (6:1, v/v)
- Broughton and Dilworth (B&D) medium (Table 1)
- 10 μM KNO_3

- Bleach solution: 10% commercial bleach diluted in distilled water
- Saran wrap
- 2 to 3-day-old culture of *Mesorhizobium loti* strain

Method

1. Fill the pillow bags with vermiculite:perlite mix and autoclave at 121°C for 20 min.
2. Soak the pillows in sterile B&D nutrient solution containing minimal nitrate concentration (10 μM KNO_3) until the substrate has absorbed the liquid.
3. Sterilize the plastic tray with bleach solution and wash several times with distilled water. Place 10-15 pillow bags side by side in the plastic tray (Figure 12).

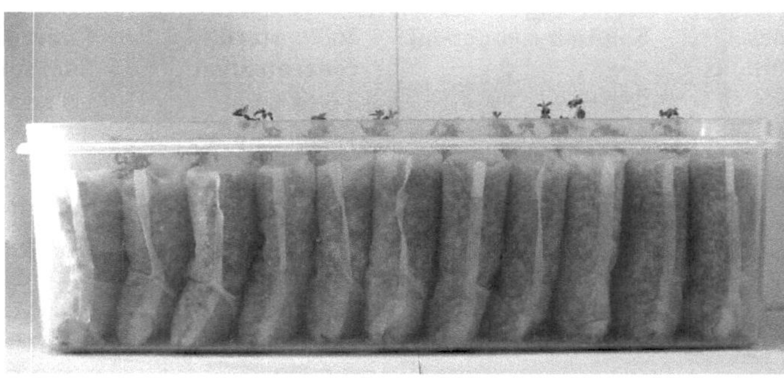

Figure 12. The pillow system for growing L. japonicus *plants. Pillow bags containing a vermiculite-perlite mixture are placed side by side in the narrow tray.*

4. Sow the seedlings in between individual pillows (five plants per row, Figure 13).

Figure 13. L. japonicus *seedlings in between pillows.*

5. Incubate the trays in growth chamber. Cover the trays with Saran wrap for a few days to maintain high humidity.

6. After 2 days in the pillow system, inoculate roots with *M loti* culture at a cell density of 10^8 cells/ml.
7. Water the plants with B&D nutrient solution (without nitrate) for at least 2 weeks and then with distilled water for the succeeding weeks. Avoid having the growth condition overly wet, or overly dry.
8. For nodule examination, carefully remove the pillows individually taking care not to damage or cut the nodulated root parts.
9. Place the plants in a clean dish with water and examine using stereomicroscopy.

Stock solutions	**Solution component**	**2000x stock concentration**	**Concentration in final solution**
A	$CaCl_2$	2.0 M	1.0 mM
B	KH_2PO_4	1.0 M	0.5 mM
C	Fe-citrate[b]	0.02 M	10 μM
D	$MgSO_4$	0.5 M	0.25 mM
E	K_2SO_4	0.5 M	0.25 mM
F	$MnSO_4$	2 mM	1.0 μM
G	H_3BO_3	4 mM	2.0 μM
H	$ZnSO_4$	1 mM	0.5 μM
I	$CuSO_4$	0.4 mM	0.2 μM
J	$CoSO_4$	0.2 mM	0.1 μM
K	Na_2MoO_4	0.2 mM	0.1 μM

Table 1. Broughton and Dilworth (1971) nutrient solution. [a]Adjust the pH to 6.8 with KOH. [b]Dissolve Fe-citrate with heating and keep in dark bottle

MICROSCOPY METHODS

Light-microscope analysis of bacterial infection and colonization of root cells using lacZ-expressing rhizobia and X-Gal staining

Investigation of infection process is important for analysing physiologic and genetic aspects of nodulation. Ethylene and nitrogen compounds like nitrate are known to inhibit nodulation, and there are several mutants both in plants and rhizobia that show defect in infection. The method shown here is conventional but powerful tool to see colonization of rhizobium in shepherd crook and infection thread development.

Equipment and Reagents

- 15 ml or 50 ml plastic tubes
- Incubator at 30°C
- 1XPBS, pH7.0
- 2% Glutaraldehyde in 1XPBS
- 20 mg/ml X-Gal in N-Dimethylformamide (keep at -20°C)
- 50 mM Potassium hexacyanoferrate (III) (4°C)
- 50 mM Potassium hexacyanoferrate (II) trihydrate (4°C)

Method

1. Prepare inoculated plants with *Mesorhizobium loti* harbouring constitutively expressing lacZ (Tansengco et al., 2003). Infection threads can be observed after three days of inoculation, and nodules are visible after seven days. If plants are grown in vermiculite/perlite mixture, remove them carefully in PBS by forceps. Vermiculite is easy to remove after soaking plants in PBS for several minutes. Similarly, if plants are grown on agar plate, remove agar.
2. Add glutaraldehyde solution to a plastic tube, put samples in, and then vacuum for 20-30 minutes. Subsequently, let samples keep one hour at room temperature.
3. Wash samples with PBS for three times.
4. Prepare staining solution (0.8 mg/ml of X-Gal, 2.5 mM of potassium ferricyanide in 0.2x PBS) and add the solution to the samples.
5. Vacuum 3-5 minutes (longer vacuum is necessary for staining inside the nodule). Normally, when the samples are wilted, vacuum is enough.
6. Keep samples in staining solution for overnight at room temperature.
7. Transfer samples to PBS.
8. Root tissue can be cleared by chloral hydrate solution mentioned below.
9. Observe samples under stereomicroscope or inverted light microscope.

Note: For observation of infection threads, a shorter period (90 min to 3 hrs) of staining is better, because endogenous activity of lacZ in root stele interferes with the observation of stained infection threads.

Clearing tissues with chloral hydrate for light microscopy

Tissue clearing is useful for the analysis of internal structures of whole mount plant tissue. Presented here is a simple clearing technique that can then be observed under DIC (differential interference contrast) microscopy. This is an efficient way to examine root and nodule development, as well as other non-ovule materials, without sectioning. The protocol can also be used for clearing *LacZ*- or GUS-stained materials, after chlorophyll is removed by 70% ethanol.

Equipment and reagents

- Clearing solution: chloral hydrate:glycerol:water, 8:1:2 (dissolve by stirring for 1 hr)
- Fixative: ethanol:acetic acid, 9:1 (v/v)
- Sterile distilled water
- Ethanol/water solution: 90%, 70%, 50%, 30% (v/v)
- Glass vials
- Light microscope

Method

1. Place the tissues in a glass vial and immerse in ethanol:acetic acid solution.
2. Fix the samples under vacuum for 10-15 min and then overnight at room temperature.
3. Discard the fixing solution and rinse briefly with sterile water.
4. Immerse the tissues in ethanol/water series in the following steps: 90%, 70%, 50%, and 30% (v/v) for 30 min each.
5. Rinse the samples briefly with sterile water and place in clearing solution for several hours or overnight depending on the specimen. Tissues can be stored in clearing solution at room temperature for several weeks.
6. Observe the samples under a light microscope (Figure 14).

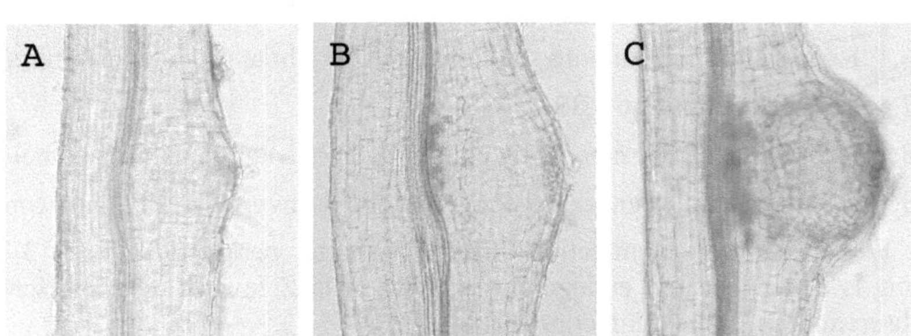

Figure 14. Cleared tissues showing successive stages of nodule development. A: Cortical cell division, B: Bump, C: Developing nodule. See colour plate 2 (A, B, C) for this figure in colour.

Glycol methacrylate (Technovit) embedding of plant tissues for fluorescence microscopy

Thin sectioning using glycol methacrylate (GMA)-based embedding medium (commercially available as Technovit 7100) is especially suited for the analysis of many biological samples. The polymerization of GMA is initiated by means of a barbituric acid derivative in combination with chloride-ions and benzoyl peroxide. Its low viscosity for complete and rapid infiltration and the hardness of the resulting polymer allow thinner sections to be cut providing clear images of cellular detail. The protocol is simple giving a minimum processing time of only a few days. Moreover, staining and enzymatic reactions using standard methods are possible without removing the resin.

I. Fixation and embedding of plant tissues

Equipment and reagents

- Glass vials
- Vacuum dessicator
- Rotator (4 rpm)
- Sample embedding molds
- Fixative: 4% (w/v) paraformaldehyde in 20 mM sodium cacodylate buffer pH 7
- Ethanol: 30%, 50%, 70%, 80%, 90%, 95%, 99%, 100% (v/v)
- Technovit 7100 embedding kit (Heraeaus Kulzer, Wehrheim, Germany)
- Technovit 3020 kit (Heraeaus Kulzer)

Method

1. Preparation of Fåhraeus slides
    a. Put roughly 10 μl of silicone caulk at each corner of a cover slip.
    b. Set the cover slip on a glass slide in a way keeping 1 mm space between them, and 1 cm above from the bottom of the glass slide (Figure 8).
    c. Let them still for a day in order to solidify silicone.
    d. Sterilize by autoclave in a glass jar filled with distilled water.

2. Fixation
    a. Cut small pieces of tissue (<5 mm). Try to keep the sample size small to facilitate processing, i.e. fixation, infiltration, and sectioning.
    b. Prepare fresh 4% (w/v) paraformaldehyde solution by dissolving the powder to the buffer with heating in fume hood and add 1-2 drops of 1N NaOH solution. Adjust pH after cooling down. Fix samples for 30 min under vacuum.
    c. Change to fresh fixative and fix overnight at 4°C.
3. Dehydration
    a. Dehydrate tissue in an ethanol/water series at RT in the following steps: twice in 30%, 50%, 70%, 80% (v/v) for 15 min each, twice in 90%, 95%, 99% for 5 min each, then in 100% for 1 hr.
    b. Use a rotator (4 rpm) or a rocking platform for agitation.
4. Infiltration
    a. Prepare infiltration solution consisting of 100 ml Technovit resin (2-hydroxyethyl methacrylate) and 1 g Hardener I (benzoyl peroxide, as accelerator).
    b. Mix until the powder is completely dissolved and store solution in dark bottle. At 4°C, the solution remains stable for approximately 4 weeks (avoid air).
    c. Immerse the tissue in fresh 1 ml absolute ethanol and add 250 μl infiltration solution every 15 min.
    d. Mix the sample after adding the infiltration solution and keep the tissue vials in rotator.
    e. Remove half volume of the solution every hour (for a total of 3 hr) making the solution 50% infiltration solution after 1 hr, 75% after 2 hr, and 82.5% after 3 hr.
    f. Leave the sample in 100% infiltration solution overnight in a rotator.
    g. Change to fresh infiltration solution and leave for 2-3 hr with agitation.
5. Embedding
    a. Prepare embedding solution consisting of 15 parts of infiltration solution and 1 part of Hardener II.
    b. Mix the solution and immediately pour in a suitable mold (we use a plastic film case) and add the samples. Arrange the specimen appropriately and allow polymerizing for at least several hours.
6. Mounting
    a. Mix Technovit 3040 in a volume ratio of 2 parts powder to 1 part liquid.

b. Mix vigorously and immediately pour on top of the embedded sample.

c. Remove the block from the mold after about 10 min. If the sample remains soft due to high temperature and humidity, store the samples in auto-dry desiccators for one to several days.

II. Sectioning and staining

Equipment and reagents

- Fret saw
- Epoxy glue
- Plastic or metal stubs
- Razorblade
- Ultramicrotome
- Sterile distilled water
- Glass slides and cover slips
- Forceps
- Hotplate
- 0.05% Toluidine Blue O in 1x PBS (137 mM NaCl, 8.10 mM $Na_2HPO_4 \cdot 12H_2O$, 2.68 mM KCl, 1.47 mM KH_2PO_4, pH 7.4)
- DAPI (4',6-diamidino-2-phenylindole) staining solution: 10 µg/ml in vectashield antifade mounting medium (Vector Laboratories, Inc.) with 5 µg/ml fluorescent brightener 28 (Sigma-Aldrich)

Method

1. Separate embedded tissues in plastic blocks by sawing the block into small pieces using a fret saw.
2. Mount the blocks onto metal or plastic stubs using epoxy glue.
3. Trim the edges of the block to a rectangular face using a razor blade.
4. Prepare triangular glass knives using a knife maker machine.
5. Set the block and glass knife in the ultramicrotome such that the cutting face of the block is parallel with the knife-edge.
6. Cut sections at 1-2 µm thick and float individual sections one at a time on distilled water on a cover slip by forceps.
7. Handle each section with a forceps and let go sections just before it touches the water.
8. Place cover slips at 50°C on a hot plate and let the water evaporate.
9. Stain with toluidine blue solution and observe under brightfield light microscopy (Figure 15).

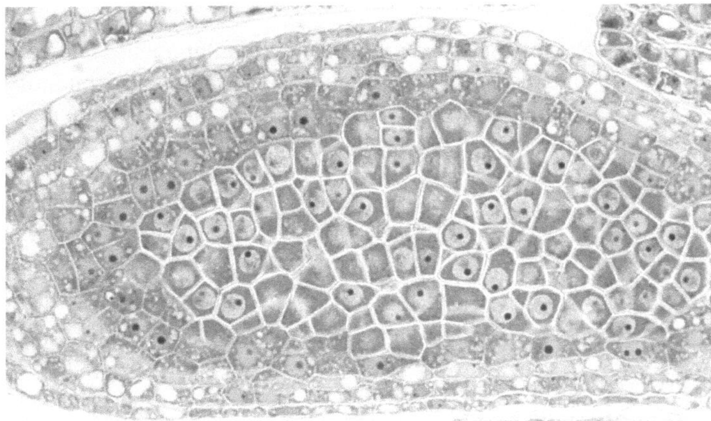

Figure 15. Longitudinal section of L. japonicus *anther showing pollen mother cells. TBO staining.*

10. Alternatively, for fluorescent microscopy, stain with DAPI solution containing fluorescent brightener to stain the nuclei and cell wall, respectively (Figure 16)

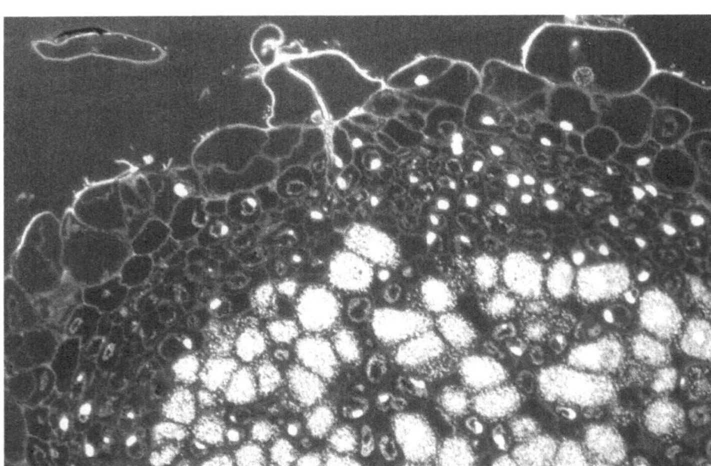

Figure 16. Lotus japonicus *nodule 2 weeks after infection with* Mesorhizobium loti. *DAPI staining. See colour plate 2 (D) for this figure in colour.*

Spot inoculation

To investigate the initial events of nodule development, spot inoculation is a powerful method in terms of temporal observation because it is easy to identify the focus of nodule initiation. In contrast to the slide inoculation method described below, it is not easy to evaluate the quantity of root hair deformation using this method, but the slide inoculation hardly induces nodules and they are not spatially fixed. For inducing nodule primordia, one can apply rhizobium (Imaizumi-Anraku et al., 2000) as well as LCOs, but the latter fail to induce complete (empty) nodules (Niwa et al., 2001). In general, MG-20 Miyakojima forms bigger nodule primordia than B-129 Gifu accession by LCOs spot. For inoculation, we use agarose for matrix to spot rhizobium or LCOs, other application method using quartz sands (van Spronsen et al., 2001) however would be also possible.

Equipment and reagents

1. Square Petri dish (either 8 cm x 12 cm or 8 cm square)
2. Low melting agarose (SeaPlaque GTG agarose, BMA)
3. TY medium
4. B & D medium
5. LCOs mixture from *M. loti* JRL501 harbouring pMP2112 (Niwa et al., 2001)

Method

1. Preparation of seedlings for spot inoculation
   a. Treat seeds with sulfuric acid (conc) for 5 to 8 min (depending on the seed condition) followed by extensive washing by tap water. Scarification by a piece of sandpaper is the alternative.
   b. Sterilize seeds with sodium hypochlorite solution (1% active chlorine) for less than 10 min, and rinse by sterilized water for three times.
   c. Imbibe seeds in sterile water for a few hours at room temperature, with gentle shaking.
   d. Transfer seeds to Petri dish containing 0.8% Bacto Agar (BD). Cover the dish with aluminium foil. Incubate at 26°C-16h-light/23°C-8h-dark regime. Remove foil after one day (Miyakojima) or two days (Gifu), and incubate one extra day.
   e. Transfer seedlings to the square Petri dish containing B&D medium solidified with 1% Bacto Agar.
   f. Cover the dish with black paper and aluminium foil (leave top 1 cm open), put the dish vertically, and incubate for 2 days (Figure 17). Addition of AVG is of choice, use 10 μM in case.

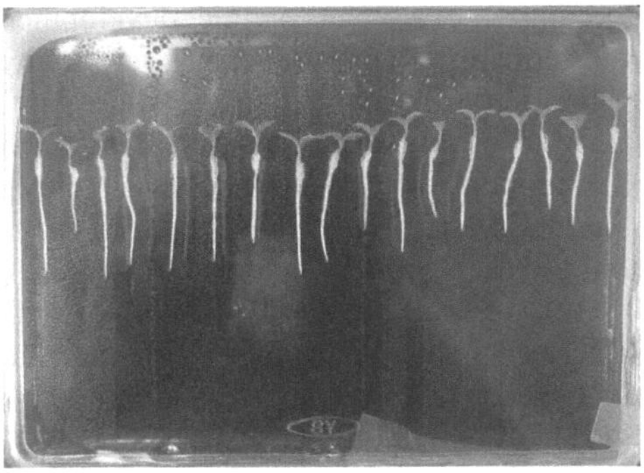

Figure 17. Miyakojima seedlings transferred to the square dish set vertically.

2. Preparation of spot solution, Rhizobium
    a. Inoculate *M. loti* from plate to 3 ml of TY liquid medium, culture at 28°C for 2 days.
    b. Spin down culture by centrifuge at 8000 rpm, 2 min, 4°C.
    c. Wash pellet with saline (0.8% NaCl) 2 times, adjust the concentration to $O.D._{600}$=0.1.
    d. Mix 25µl of the suspension with 50 µl of sterilized low melting agarose (1%).
3. Preparation of spot solution, LCOs:
    a. Mix 25µl of LCOs (10^{-5} to 10^{-3}M) with 50 µl of sterilized low melting agarose (1%).
4. Spot inoculation
    a. Pipette out 10µl of the solution above, and spot 1 µl of it to the root tip, using micropipette.
    b. Wrap the rid of the dish with Parafilm. Use surgical tape (Micropore: 3M) at the top to avoid excess humidity.
    c. Set black paper and aluminium foil again as above, and incubate.
    d. After one week, developing nodule can be observed by naked eye (rhizobium inoculation). In case of LCOs, 10 days are enough to recognize the nodule under stereomicroscope.

Notes: Higher concentrations of LCOs bring better nodulation, but not 100%. Normally more than 50% of the spotted seedlings show nodulation. Nodules can be observed 0 to 2 cm below the spot point. In the original papers, the authors recommended adding black ink so that the spot point can be easily recognized. In our experience, however, black ink inhibits root elongation, so we do not add it. Holding the dish up to the light can identify the spot point.

ACETYLENE REDUCTION ASSAY FOR NITROGENASE ACTIVITY

Biological nitrogen fixation is assessed using several methods, such as Kjeldahl analysis, ^{15}N-incorporation assays, acetylene reduction assays, and H_2-evolution assays. Of these methods, the closed acetylene reduction assay is simple, rapid, and sensitive (Hardy et al. 1968; Vessey 1994), and has been widely used for estimating the nitrogen-fixing activity of legumes. This assay takes advantage of nitrogenase's high affinity for acetylene, and in the presence of acetylene total electron flow through nitrogenase is virtually used to reduce acetylene to ethylene. The ethylene produced can be detected by hydrogen flame ionization after gas chromatography (Figure 18).

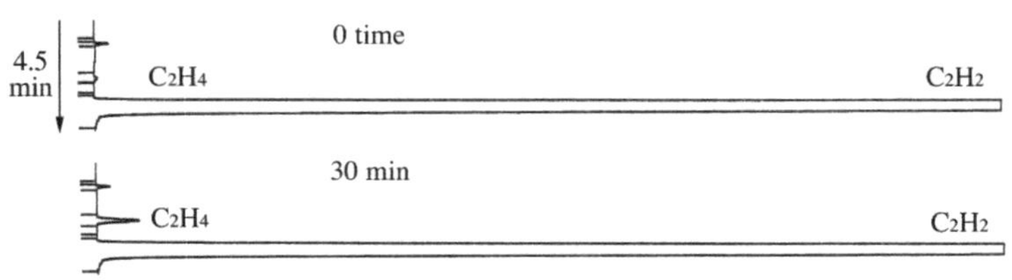

Figure 18. Nitrogenase acetylene reduction assay. Typical GC chromatograms with separate peaks for acetylene and ethylene are shown. A small amount of ethylene is detected just after injecting acetylene into a vial (0 time). After incubating the vial for 30 min, the peak of ethylene increases by reduction of acetylene (30 min).

Reagents and apparatuses

- Acetylene
- Ethylene (for standard)
- 35-mL vial (Any vials with different volumes are available if the vial can be sealed with a cap, from which a gas sample can be withdrawn with a syringe.)
- Serum cap
- 1-mL disposable syringe with a needle (25G x 1'')
- Glass column (2 mm i.d. x 2 m) packed with Porapak N (Waters, Milford, MA, USA)
- Gas chromatograph equipped with a hydrogen flame-ionization detector (FID)

- Integrator (for determination of the peak area separated by gas chromatography).

Procedures

- Detach a nodulated root from a freshly harvested plant.
- Place the nodulated root in a 35-mL vial and close with a serum cap. A single nodulated root is enough for determination of activity using the 35-mL vial.
- Withdraw 3.5 mL of air from the closed vial and inject the same volume of acetylene gas to be 0.1 atmosphere of acetylene in the vial.
- Immediately withdraw 0.5 mL of gas sample from the closed vial and measure the concentration of ethylene contaminated in the 0.1 atmosphere of acetylene injected in the vial with gas chromatograph.
- Incubate the closed vial containing the nodulated root in darkness at 25°C.
- After 30 min, withdraw 0.5 mL of gas sample and measure the concentration of ethylene.
- Based on standard ethylene, calculate the ethylene produced during the incubation after subtracting the ethylene contaminated in acetylene gas.

Notes

- Acetylene cylinder contains a small amount of ethylene and the ethylene impurity should be determined daily since it varies inversely with the pressure in the cylinder.
- Preparations of tissue seriously affect acetylene reduction activity (Vessey 1994). The least disturbed, most intact tissue should be used in acetylene reduction assay.
- Rates of acetylene reduction are linear up to 60 minutes, but it is recommended that the assay time should be kept to within 30 minutes (Hardy et al. 1968).
- Nodulated roots evolve endogenous ethylene (Suganuma et al. 1995a), though the amounts of evolution are thought to be negligible compared with acetylene reduction activity determined within 30 min.
- Replacement of air (N_2) with $Ar:O_2$ slightly increases acetylene reduction activity (Hardy et al. 1968). If required, air in each vial should be replaced by introducing a mixture of $Ar:0_2$ (0.8:0.2 atmosphere) gas before adding acetylene gas.
- Greater information concerning nitrogenase assay is referred to 'Methods for Evaluating Biological Nitrogen Fixation' edited by FJ Bergersen (1980) John Wiley & Sons, Inc, New York, NY.

MOLECULAR MARKERS AND MEASURING METHODS

Nodule development and differentiation are orchestrated by changes in gene expression that occur in a precise temporal sequence. Nodule-specific plant genes that are induced within hours or days after contact with rhizobia are called early nodulin genes, while those that are induced a week or so later, just prior to or following the start of nitrogen fixation are called late nodulin genes. Early nodulins are believed to play important roles during bacterial colonisation or nodule development, while late nodulins presumably fulfil structural or functional roles during differentiation and/or nitrogen fixation. Sequential changes in rhizobial gene expression also accompany nodule development and differentiation. Because of their well-defined temporal expression patterns, many plant and bacterial nodule-induced genes, or the proteins that they encode, have been used as molecular markers for different stages of nodule development. Such markers are especially useful for studies with plant or bacterial mutants that are defective in nodulation or nitrogen fixation, as they help to identify the stage at which nodule development or differentiation is impaired. Expression of marker genes in wild type and ineffective nodules has been examined by Northern analysis (Purdom and Trese 1995; Kato et al. 2002), in situ hybridization (Banba et al. 2001; Kawashima et al. 2001), immunoblotting (Egli et al. 1991; Romanov et al. 1995; Suganuma et al. 2003) and use of bacterial reporter gene fusions (Voroshilova et al. 2001). The following two sub-sections describe the methods of real-time RT-PCR and immunoblotting for measuring gene transcript and protein levels, respectively.

Transcript profiling using real-time RT-PCR

Real-time reverse transcription-polymerase chain reaction (RT-PCR) is an extremely sensitive and quantitative method of measuring gene transcript levels that can be multiplexed to handle transcripts of many different genes in parallel. However, care must be taken to remove all genomic DNA from isolated RNA before RT-PCR measurements (Czechowski et al., 2004).

Materials and Methods

Harvested plant material is typically frozen immediately in liquid nitrogen and stored at -70 °C, prior to use. Total RNA is extracted from frozen organs of *Lotus* plants using an RNeasy kit (Qiagen, Hilden, Germany). Before cDNA synthesis, 1-2 µg of total RNA is treated with RNase-free DNaseI (Sigma, St Louis, MI, USA), to destroy genomic DNA. DNAseI-treated RNA is subsequently phenol-chloroform extracted, precipitated, and resuspended in 10 µL water. PCR, using gene-specific primers, is performed on the resulting RNA to confirm the absence of DNA. Reverse transcription is then performed with SuperScript™ III reverse transcriptase (Invitrogen, Carlsbad, CA, USA) usinng an oligo(dT) primer. The efficiency of cDNA synthesis is assessed by real-time PCR amplification of ubiquitin (see Table 2), and only those reactions that exhibit similar efficiencies (similar C_T values for ubiquitin) are analysed further.

PCR reactions can be performed in a 96-well plate with an Applied Biosystems ABI Prism 5700 Sequence Detection System (Foster City, CA, USA), using SYBR Green to monitor dsDNA synthesis. A standard PCR reaction contains approximately 4 ng cDNA template, 10 µL of SYBR Green Mix (Applied Biosystems, Foster City CA, USA), 5 pmol forward and reverse primers, and water to a final volume of 20 µL. The following standard thermal profile is used for all PCR reactions: 50°C for 2 min; 95°C for 10 min; 40 cycles of 95°C for 15 sec and 60°C for 1 min. Data is analysed using SDS 2.0 software (Applied Biosystems, Foster city, CA, USA). For quality control, PCR products are run on a 3 % agarose gel to confirm the presence of a single amplicon. Gene-specific primers can be designed with PrimerExpress software (Applied Biosystems, Foster city, CA, USA), and typically have melting temperatures of 60 +/- 2 and GC content of between 45-55%. Primers are designed to produce short amplicons, typically between 60-150 bp to ensure high PCR efficiencies.

Threshold cycle (C_T) values for each gene transcript are normalized to those of polyubiquitin (GenBank AW720576; see Table 2) before making comparisons between samples (Colebatch et al. 2002). Gene expression ratios are calculated using the following equations:

$$nC_T = C_T - C_TLj_{Ubi}$$

$$\Delta C_T = nC_{Tx} - nC_{Ty}$$

$$nRatio = 2^{-\Delta CT}$$

Where C_T is the PCR cycle number at which a set threshold value (usually 0.1) of SYBR Green fluorescence is reached, nC_T is the normalised C_T value, ΔC_T is the difference in normalised C_T values in sample x and y, and $Ratio_n$ is the ratio of gene transcript level in sample x compared to y. The last equation assumes a PCR efficiency (E) of 100%. Generally, this is not the case, and several methods are available to estimate PCR efficiency. The first method of measuring PCR efficiency uses template dilutions and the equation $(1+E)=10^{(-1/slope)}$, as described previously (Pfaffl *et al.*, 2001). The second method uses data obtained from the exponential phase of each individual amplification plot and the equation $(1+E)=10^{slope}$ (Ramakers *et al.,* 2003). If E is known, the gene expression ratio can be calculated using the equation:

$$nRatio = (1+E)^{-\Delta CT}$$

Gene	Primer sequences (F, forward; R, reverse)	Comment
LjUbi	F: 5'- TTCACCTTGTGCTCCGTCTTC-3' R: 5'- AACAACAGCACACACAGACAATCC-3'	Constitutive
LjSbtS	F: 5'- ATGAAGACTGAGAAGAATCCC-3' R: 5'-GTAATTATATTTGGTCCCACGGAAG-3'	Induced 2-5 hpi
LjNin	F: 5'- AATGCTCTTGATCAGGCTG-3' R: 5'- AGGAGCCCAAGTGAGTGCTA-3'	Induced 5 hpi
Leghaemo-globin	F: 5' –AAAGACATGTTCTCCTTTCT- 3' R: 5' – CATTGCCTTCTTAATTGCAG - 3'	Induced 7 dpi
LjEnod2	F: 5'- CAGGAAAAACCACCACCTGT-3' R: 5'- ATGGAGGCGAATACACTGGTG-3'	Induced12 dpi
LjKUP	F: 5'-CGTTGATTCTCGCCTACAGGAC-3' R: 5'-TAAAGCGGCGATGTCACGA-3'	Induced 14dpi

Table 2: Marker genes and corresponding primers for real-time RT-PCR. The polyubiquitin gene, LjUbi is a constitutively expressed gene used for normalization of data (Colebatch et al., 2002). Other genes are used as molecular markers for early or late stages of nodule development and differentiation. The symbiosis-induced subtilase gene, LjSbtS (Martin Parniske, personal communication), and the nodule inception gene, LjNin (Schauser et al. 1999), are induced within a few hours post-inoculation (hpi). The other genes, including LjKUP (Desbrosses et al., 2004), are induced several days post-inoculation (dpi).

Measuring changes in protein abundance using immunoblotting

When the appropriate antibody is available, expression of marker genes can be easily and sensitively analysed using immunoblotting. In immunoblotting, methods for detection of target proteins are variable. As an example, visualizations of immunoreactive leghaemoglobin, a representative late nodulin, and nitrogenase in bacteroids with a ProtoBlot immunoblotting system (Promega, Madison, WI, USA) are described (Figure 19).

Reagents and Apparatuses

- Grinding medium (50mM Tris-HCl pH7.4, 0.3M sucrose, 1mM EDTA, 5mM DDT, 1mM PMSF)
- Polyclar AT
- PVDF membrane (Immobilon-P membrane; Millipore, Bedford, MA, USA)
- TBS (20mM Tris-HCl pH 7.5, 150mM NaCl)
- TBST (20mM Tris-HCl pH 7.5, 150mM NaCl, 0.05% Tween 20)
- Blocking solution (1% (w/v) BSA in TBST)

- Alkaline phosphate buffer (100mM Tris-HCl pH 9.5, 100mM NaCl, 5mM $MgCl_2$)
- Colour development solution (BCIP/NBT in alkaline phosphate buffer)
- Primary antibody
- Anti-IgG secondary antibody conjugated to alkaline phosphatase
- Sonicator
- Electrophoresis system
- Blotting apparatus

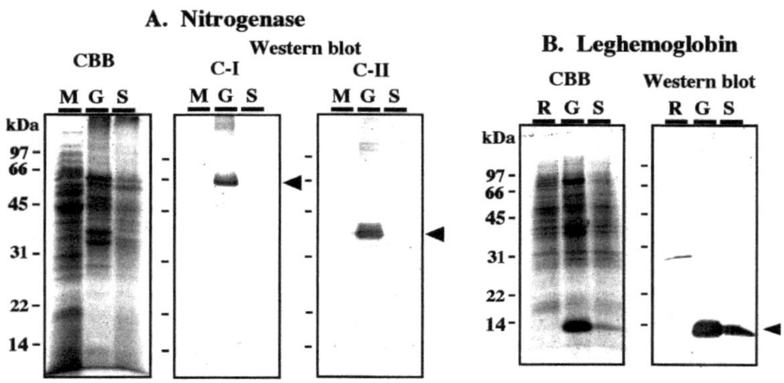

Figure 19. Examples of immunoblotting analysis of nitrogenase component I (C-I) (A) and component II (C-II) in a crude bacteroids, and of leghaemoglobin (B) in a host plant fraction isolated from nodules on wild type Gifu (G) and Lotus japonicus *sen1 (Ljsym75) mutant (S) plants. For reference, free-living* Mesorhizobium loti *(M) and root extracts from Gifu plants (R) were also tested, respectively.*

Procedures

Preparation of extracts for crude bacteroids and for plant cytosol

- Homogenize nodules with a mortar and pestle in grinding medium with Polyclar AT powder in a ratio of 0.3 g per g fresh weight of nodules.
- Squeeze the homogenate through four layers of gauze and centrifuge the filtrate at 200 x g for 10 min at 4°C.
- Centrifuge the supernatant at 5,000 x g for 10 min at 4°C and wash the pellet with grinding medium three times.
- Sonicate the final suspension ten times for 10 sec each time at 100W and centrifuge at 16,000 x g for 30 min at 4°C. The resulting supernatant is used the as the bacteroid fraction in immunoblotting.

- Centrifuge the supernatant, obtained by the centrifugation at 5,000 x g for 10 min, at 16,000 x g for 30 min at 4°C. The resulting supernatant is used as the plant cytosol fraction in immunoblotting.

Electrophoresis and transfer to membrane

- Separate the polypeptides by SDS-PAGE (polyacrylamide gel electrophoresis) on a 12% (w/v) polyacrylamide gel. Bacteroid fraction is used for detection of nitrogenase and plant cytosol fraction is for detection of leghaemoglobin, respectively.
- Transfer the polypeptides separated by SDS-PAGE to PVDF membranes electrophoretically (Towbin et al. 1979).

Immunodetection

- Incubate the membrane in blocking solution for 60 min to block nonspecific protein binding.
- Incubate the membrane in TBST containing the appropriate dilution of primary antibody for 30 min with gentle agitation. Antibodies against nitrogenase components I and II isolated from *Rhizobium leguminosarum* bv viciae bacteroids (Bisseling et al 1980) were diluted to 1:10,000 and that against leghaemoglobin isolated from soybean nodules (Suganuma et al. 1995b) was diluted to 1:10,000.
- Wash the membrane three times in TBST for 10 min each.
- Incubate the membrane in TBST containing the appropriate dilution of anti-IgG alkaline phosphatase conjugate for 30 min with gentle agitation.
- Wash the membrane three times in TBST for 10 min each wash.
- Rinse the membrane briefly in TBS two times.
- Incubate the membrane in colour development solution.
- Stop the reaction by washing the membrane in deionized water for 15 min when the colour has developed to the desired intensity.

Notes

- PVDF membrane, a hydrophobic membrane, must be pre-wet first in 100% methanol or ethanol. After pre-wetting, submerge the membrane to blotting solution until use.
- Do not dry out the membrane during any steps of the procedures after pre-wetting.
- You can store the membrane covered with a plastic sheet at 4°C after blotting. For the dried PVDF membrane, first rewet in methanol or ethanol followed by floating in TBS.

- Detailed procedures for electrophoresis and electrophoretic transfer are usually included with commercial devices.

REFERENCES

Banba M, Siddique ABM, Kouchi H, Izui K, Hata S (2001) ***Lotus japonicus* forms early senescent root nodules with *Rhizobium etli***. *Molecular Plant-Microbe Interactions* 14, 173-180.

Berleth T, Jurgens G (1993) **The role of the *monopteros* gene in organizing the basal body region of the Arabidopsis embryo**. *Development* 118, 575-587.

Bisseling T, Moen AA, van den Bos RC, van Kammen A (1980) **The sequence of appearance of leghaemoglobin and nitrogenase components I and II in root nodules of *Pisum sativum* L**. *Journal of General Microbiology* 118, 377-381.

Broughton.Wj, and Dilworth MJ. (1971). **Control of Leghaemoglobin Synthesis in Snake Beans**. *Biochemical Journal* 125, 1075-1080.

Colebatch G, Kloska S, Trevaskis B, Freund S, Altmann T, and Udvardi MK. (2002b). **Novel aspects of symbiotic nitrogen fixation uncovered by transcript profiling with cDNA arrays**. *Molecular Plant-Microbe Interactions* 15, 411-420.

Czechowski T, Bari RP, Stitt M, Scheible W-R, and Udvardi MK (2004) **Real-time RT-PCR profiling of over 1,400 *Arabidopsis* transcription factors: unprecedented sensitivity reveals novel root- and shoot-specific genes**. *The Plant Journal* 38, 366-79.

Desbrosses G, Kopka C, Ott T, and Udvardi MK. (2004) ***Lotus japonicus LjKUP* encodes a potassium transporter of the plasma membrane and is induced significantly during nodule organogenesis**. *Molecular Plant-Microbe Interactions* 17, 789-797.

Egli MA, Larson RJ, Hruschka WR, and Vance CP (1991) **Synthesis of nodulins and nodule-enhanced polypeptides by plant gene-controlled ineffective alfalfa nodules**. *Journal of Experimental Botany* 42, 969-977.

Gerrits PO, and Smid L (1983) **A new, less toxic polymerization system for the embedding of soft tissues in glycol methacrylate and subsequent preparing of serial sections**. *Journal of Microscopy* 132, 81-85.

Handberg K, and Stougaard J. (1992*)* ***Lotus japonicus*, an autogamous, diploid legume species for classical and molecular genetics**. *The Plant Journal* 2, 487-496.

Hardy RWF, Holsten RD, Jackson EK, and Burns RC (1968) **The acetylene – ethylene assay for N_2 fixation: Laboratory and field evaluation**. *Plant Physiology* 43, 1185-1207.

Harris JM, Wais R, and Long SR. (2003) **Rhizobium-induced calcium spiking in *Lotus japonicus***. *Molecular Plant-Microbe Interactions* 16, 335-341.

Imaizumi-Anraku H, Kouchi H, Syono K, Akao S, and Kawaguchi M (2000) **Analysis of *ENOD40* expression in alb1, a symbiotic mutant of *Lotus japonicus* that forms empty nodules with incompletely developed nodule vascular bundles**. *Molecular and General Genetics* 264, 402-410.

Kaneko T, Nakamura Y, Sato S, Asamizu E, Kato T, Sasamoto S, Watanabe A, Idesawa K, Ishikawa A, Kawashima K, Kimura T, Kishida Y, Kiyokawa C, Kohara M, Matsumoto M, Matsuno A, Mochizuki Y, Nakayama S, Nakazaki N, Shimpo S, Sugimoto M, Takeuchi C, Yamada M, and Tabata S (2000) **Complete genome structure of the nitrogen-fixing symbiotic bacterium *Mezorhizobium loti***. *DNA Research* 7, 331-338.

Kato T, Kawashima K, Miwa M, Mimura Y, Tamaoki M, Kouchi H, and Suganuma N (2002) **Expression of genes encoding late nodulins characterized by a putative signal peptide**

and conserved cysteine residues is reduced in ineffective pea nodules. *Molecular Plant-Microbe Interactions* 15, 129-137.

Kawashima K, Suganuma N, Tamaoki M, and Kouchi H (2001) **Two types of pea leghaemoglobin genes showing different O_2-binding affinities and distinct patterns of spatial expression in nodules.** *Plant Physiology* 125, 641-651.

Krussel L, Madsen LH, Sato S, Aubert G, Genua A, Szczyglowski K, Duc G, Tabata S, de Bruijn F, Pajuelo E, Sandal N, and Stougaard J. (2002) **Shoot control of nodule organogenesis and root development is mediated by a serine/threonine receptor kinase.** *Nature* 420, 422-426.

Lopez-Lara I, van den Berg JDJ, Thomas-Oates JE, Glushka J, Lugtenberg BJJ, and Spaink, HP. (1995) **Structural identification of the lipo-chitin oligosaccharide nodulation signals of *Rhizobium loti*.** *Molecular. Microbiology* 15, 627-638.

Madsen EB, Madsen LH, Radutoiu S, Olbryt M, Rakwalska M, Szczyglowski K, Sato S, Kaneko T, Tabata S, Sandal N, and Stougaard J. (2003) **A receptor kinase gene of the LysM type is involved in legume perception of rhizobial signals.** *Nature* 425, 637-640.

Nishimura R, Hayashi M, Wu GJ, Kouchi H, Imaizumi-Anraku H, Murakami Y, Kawasaki S, Akao S, Ohmori M, Nagasawa M, Harada K, and Kawaguchi M (2002) **HAR1 mediates systemic regulation of symbiotic organ development.** *Nature* 420, 426-429.

Nishimura R, Ohmori M, and Kawaguchi M (2002) **The novel symbiotic phenotype of enhanced-nodulating mutant of *Lotus japonicus*: astray mutant is an early nodulating mutant with wider nodulation zone.** *Plant & Cell Physiology* 43, 853-859.

Niwa S, Kawaguchi M, Imaizumi-Anraku H, Chechetka SA, Ishizaka M, Ikuta A, and Kouchi H. (2001) **Responses of a model legume *Lotus japonicus* to lipochitin oligosaccharide nodulation factors purified from *Mesorhizobium loti* JRL501.** *Molecular Plant-Microbe Interactions* 14, 848-856.

Olsthoorn MMA, Lopez-Lara IM, Peterson BO, Bock K, Haverkamp J, Spaink HP, and Thomas-Oates JE. (1998) **Novel branched Nod factor structure results from alpha-(1->3) fucosyl transfer activity: The major lipo-chitin oligosaccharides from *Mesorhizobium loti* strain NZP2213 bear alpha-(1->3) fucosyl substituent on a nonterminal backbone residue.** *Biochemistry* 37, 9024-9032.

Purdom D, Trese AT (1995) **Morphological and molecular characteristics of host-conditioned ineffective root nodules in cowpea.** *Plant Physiology* 109, 239-244.

Rodutoiu, S, Madsen, LH, Madsen, EB, Felle, HH, Umehara, Y, Gronlund, M, Sato, S, Nakamura, Y, Tabata, S, Sandal, N, and Stougaard, J. (2003) **Plant recognition of symbiotic bacteria requires two LysM receptor-like kinases.** *Nature* 425, 585-592.

Romanov VI, Gordon AJ, Minchin FR, Witty JF, Skot L, James CL, Borisov AY, and Tikhonovich IA (1995) **Anatomy, physiology and biochemistry of root nodules of Sprint-2 Fix$^-$, a symbiotically defective mutant of pea (*Pisum sativum* L).** *Journal of Experimental Botany* 46, 1809-1816.

Ruzin SE (1999) **Infiltration and embedding tissues.** In: *Plant Microtechnique and Microscopy*. Oxford University Press, NY, pp. 61-72.

Schauser L, Roussis A, Stiller J, and Stougaard J. (1999). **A plant regulator controlling development of symbiotic root nodules.** *Nature* 402, 191-195.

van Spronsen PC, Gronlund M, Bras CP, Spaink H, and Kijne JW. (2001) **Cell biological changes of outer cortical root cells in early determinate nodulation.** *Molecular Plant-Microbe Interactions* 14, 839-847.

Suganuma, N, Nakamura, Y, Yamamoto, M, Ohta, T, Koiwa, H, Akao, S, and Kawaguchi, M. (2003) **The *Lotus japonicus Sen1* gene controls rhizobial differentiation into nitrogen-fixing bacteroids in nodules**. *Molecular Genetics and Genomics* 269, 312-320.

Suganuma N, Yamauchi H, and Yamamoto K (1995a) **Enhanced production of ethylene by soybean roots after inoculation with *Bradyrhizobium japonicum*.** *Plant Science* 111, 163-168.

Suganuma N, Tamaoki M, and Kouchi H (1995b) **Expression of nodulin genes in plant-determined ineffective nodules of pea**. *Plant Molecular Biology* 28:1027-1038.

Szczyglowski K, and Amyot L. **Symbiosis, Inventiveness by Recruitment?** (2003) *Plant Physiology* 131, 935-940.

Szczyglowski K, Shaw S.R, Wopereis J, Hamburger D, Copeland S, Dazzo F.B, and de Bruijn FJ. (1998) **Nodule organogenesis and symbiotic mutants of the model legume *Lotus japonicus.*** *Molecular Plant-Microbe Interactions* 11, 684-697.

Tansengco ML, Hayashi M, Kawaguchi M, Imaizumi-Anraku H, and Murooka Y. (2003) ***crinkle*, a novel symbiotic mutant that affects the infection thread growth and alters the root hair, trichome, and seed development in *Lotus japonicus*.** *Plant Physiology* 131,1054-1063.

Towbin H, Staehelin T, and Gordon J (1979) **Electrophoretic transfer of proteins from polyacrylamide gels to nitrocellulose sheets: procedure and some applications**. *Proceedings of the National Academy of Sciences USA* 76, 4350-4353.

Vessey JK. (1994) **Measurement of nitrogenase activity in legume root nodules: In defence of the acetylene reduction assay**. *Plant and Soil* 158, 151-162.

Voroshilova VA, Boesten B, Tsyganov VE, Borisov AY, Tikhonovich IA, and Priefer UB. (2001) **Effect of mutations in *Pisum sativum* L. genes blocking different stages of nodule development on the expression of late symbiotic genes in *Rhizobium leguminosarum* bv. viciae**. *Molecular Plant-Microbe Interactions* 14, 471-476.

Wopereis J, Pajuelo E, Gresshoff PM, Dazzo FB, de Bruijn FJ, Stougaard J, and Szczyglowski K. (2000) **Short root mutant of *Lotus japonicus* with a dramatically altered symbiotic phenotype**. *The Plant Journal* 23, 97-114.

Chapter 2.2

A PROCEDURE FOR *IN VITRO* NODULATION STUDIES

Ani Barbulova and Maurizio Chiurazzi*

*Institute of Genetics and Biophysics – A Buzzati Traverso. Via Marconi 12, 80125, Napoli, ITALY; *Corresponding author.*

Email: chiurazz@igb.cnr.it

Phone: +39 081 725 7256 Fax: +39 081 593 6123

Keywords: Rhizobium, nodules, infection

Standardised in vitro *conditions yielding successful, synchronised, reproducible cycles of symbiotic nitrogen fixation are essential experimental tools for investigating the different stages of the Rhizobium-legume interaction. A standard* in vitro *procedure can facilitate analysis of the kinetics of phenotype at the molecular level occurring after Rhizobium infection. This is essential when comparing, for instance, the expression pattern of nodulin or marker genes or the phenotypes of wild type and mutant lines. We present here a feasible, reliable procedure for* in vitro *nodulation studies of* L. japonicus. *All the different steps of the symbiotic interaction have been analysed and different parameters of plant growth followed to demonstrate the reproducibility of the procedure.*

MATERIALS AND METHODS

Media

Bacterial medium (TyR)

- Bacto triptone 5 g/L
- Yeast extract 5 g/L

After sterilization, add $CaCl_2$ to a final concentration of 6 mM

Plant germination and growth medium (Jensen's)

To prepare 1 litre of 0.1% Jensen medium, use 1 ml of each of the below solutions:

- $CaHPO_4$ 40 g/400 ml
- K_2HPO_4 8 g/400 ml

A.J. Márquez (ed.), Lotus japonicus Handbook, 83-86.

- $MgSO_4.7H_2O$ 8 g/400 ml
- NaCl 8 g/400 ml
- Fe citrate* 4 g/400 ml

*Fe citrate can be added in appropriate amounts directly into the medium before autoclaving, without preparing a stock solution.

Add 2.5 µl of the below per 1 litre 0.1% Jensen medium:

- $Na_2MoO_4.4H_2O$ 16 g

Add 2.5 µl of the below per 1 litre 0.1% Jensen medium:

- $CuSO_4.5H_2O$ 1.4 g/l
- $MnSO_4.4H_2O$ 24.4 g/l
- $ZnSO_4.7H_20$ 3.88 g/l

Add 25 µl of the below per 1 litre 0.1% Jensen medium:

- H_3BO_3 (10 x stock) 5.08 g

To solidify Jensen's medium, add 1% plant agar (Duchefa). Both TyR and Jensen's media are sterilized for 20 min. at 120°C.

Instrumentation

Preparation of Petri dishes

The Petri dishes used are 140x10 mm. After sterilization, 85 ml of 0.1% Jensen medium is poured into each Petri dish in a laminar flow cabinet. After drying for 20 min., one semicircle of filter paper is added in each Petri dish. Petri dishes, closed by Parafilm could be stored for 1-2 months at 4°C.

Filter paper preparation

Semicircles of filter paper with average diameter 5 cm are cut following the border of the base of the Petri dishes and wrapped in aluminium foil and sterilized for 30 min. at 120°C. Under sterile conditions in a laminar flow cabinet, each paper is put on the top of the already poured 0.1% Jensen's medium using forceps.

Growth conditions

Germination of the plants and their further growth after infection takes place at 23°C, with 16 /8 hours light/dark period and light intensity of 246 µE s^{-1} m^{-2}.

Procedure

Seed sterilization and germination

Lotus japonicus B-129 GIFU ecotype seeds are transferred to Eppendorf tubes (up to 100/Eppendorf tube), surface sterilized for 20 min in 25% commercial bleach (1%

hypochlorite) and 0,1% Triton, washed 6 times in sterile H_2O and kept over night in water at 4°C. After 20-24 hours, the seeds are transferred to Petri dishes (140X10 mm), containing 0.1% solidified Jensen medium with a semicircles of sterilized filter paper placed on the top of the medium as to cover half of the Petri dish. Seeds are placed on the border of the filter paper (ten seeds for each Petri dish), so that the emerging roots will stay attached to it. The seeds are left overnight in the dark at 4°C lid side down before Petri dishes are moved into the culture cabinet. Here they are arranged in vertical position and kept for 24 hrs in the dark. Then the seeds are exposed to the light. At this step, it is important to maintain the young emerging roots in contact with the filter paper.

Preparation of the bacterial suspension

A single bacterial colony of *Rhizobium loti* R7A strain is grown overnight at 30°C in TyR medium, containing 6mM $CaCl_2$ and 100μg/ml Rifampicin. The day of inoculation, bacterial cells are collected by centrifugation at 4500 rpm for 10 min and rinsed twice with 1XPBS solution. The OD then is measured at λ=600 and 20 μl of bacterial suspension with 5.10^7 cells are used for inoculation of each primary root meristem.

Infection

The inoculation is performed 4-5 days after germination of the seeds (primary root length about 1 cm). It is important to discard the unsynchronised seedling. Each primary root meristem is inoculated with 20 μl of the freshly prepared bacterial suspension and Petri dishes are sealed with Parafilm. The filter paper support is crucial to obtain a complete and uniform diffusion of the drop of bacteria over the root meristems. Four days after infection, the filter paper is removed (to avoid drying) and the plants are left for further growth on the same Petri dishes. This is done simply by taking the plants out of the Petri dish, removing the filter paper using forceps, and putting back the plants into the same Petri dish, paying attention to provide a direct contact of the roots with the medium. Aluminium foil is wrapped around the lower part of the Petri dishes to keep roots in the dark.

CONCLUSIONS

We screened the plants for different parameters of growth and symbiotic phenotypes five weeks after inoculation with *M loti* (Figure 1). The pattern of nodulation is uniform and the plants show conserved root and shoot phenotypes (Table 1). The basic experimental procedure reported here could be further improved by addition, into the Jensen medium, of factors improving the efficiency of nodulation as the ethylene production inhibitor, L-α-(2-aminoethoxyvinil) glycine (AVG) which is known to stimulate the nodulation of *L japonicus* (Pacios Bras et al., 2000, MPMI Vol.13: 475-479).

Figure 1. L japonicus *plants five weeks after inoculation with* M loti *on 0.1% Jensen's medium.*

However, the results in Table 1 show a detailed analysis and standard deviation of different parameters of nodule, root and shoot development that indicate a high reproducibility of the infection process. Such a reproducible experimental tool gives the opportunity for precise screening and detection of weak phenotypes.

Parameters	**Average mean (±SD)**
Number of nodules	8 ± 2
Primary root length	7.2 ± 0.4
Number of secondary roots	5.2 ± 0.6
Shoot length	8.5 ± 1.2
Number of trifoliate leaves	6± 1.3

Table 1. Average mean of different growth parameters calculated 5 weeks after inoculation with M. loti *on 0.1% Jensen medium. The results summarize the data from the screening of 900 plants.*

REFERENCES

Pacios Bras C, Jorda MA, Wijfjes AH, Harteveld M, Stuurman NThomas-Oates JE, and Spaink HP. (2000) **A *Lotus japonicus* nodulation system based on heterologous expression of the fucosyl transferase NodZ and the acetyl transferase NoIL in *Rhizobium leguminosarum*.** *Molecular Plant-Microbe Interactions.* 13, 475-479.

Chapter 2.3

ARBUSCULAR MYCORRHIZA

Sonja Kosuta, Thilo Winzer, and Martin Parniske*

*Sainsbury Laboratory, John Innes Centre, Norwich Research Park, Colney, Norwich NR4 7UH, UNITED KINGDOM; *Corresponding author.*

Email: martin.parniske@lrz.uni-muenchen.de

Phone: +44 1 603 450 249 Fax: +44 1 603 450 011

Keywords: Glomeromycota, *Glomus intraradices*, fungal inoculum, nurse-culture, ink staining

Symbiotic root colonisation by arbuscular mycorrhizal (AM) fungi plays a critical role in the capture of nutrients such as phosphorous from the soil. The roots of almost all vascular plants can interact with AM fungi with the notable exception of many brassicas, including the model plant Arabidopsis thaliana. *Legumes, which also form symbiotic nitrogen-fixing nodules, have been intensively studied over the past decade, revealing a surprising genetic overlap between the early stages of nodulation and AM formation. For this reason, legumes have become useful model systems for AM research. Here we present some basic techniques for rapid and consistent AM infection and for the visualisation of AM structures in plant roots. These techniques are useful for molecular and cellular studies and have been adapted to the specific requirements of* Lotus japonicus.

INTRODUCTION

Arbuscular mycorrhiza

Arbuscular mycorrhiza (AM) is an ecologically important and widespread symbiotic association occurring between roughly 200 species of microscopic soil fungi, the *Glomeromycota* (Schüssler et al., 2001), and over 90 % of vascular plant species (Smith and Read, 1997). Both partners benefit from this association; the plant supplies the fungal partner with hexose and the fungus delivers phosphates and other mineral nutrients to the plant. Benefits to plant growth, disease resistance, and survival, as well as soil particulate structure and microbial nutrient cycling make AM of fundamental importance to most ecosystems. AM may play a particularly significant role when nutrient availability is low due to rapid nutrient cycling and leaching in tropical soils (Bowman and Panton, 1993) or after soil disturbance and depletion in agricultural soils (Gianinazzi et al., 1995). It is also critical under

A.J. Márquez (ed.), Lotus japonicus Handbook, 87-95.

conditions where plant survival is restricted due to climate or contamination as in arctic (Dalpé and Aiken, 1998), saline (Al-Karaki and Hammad, 2001), desert (Bohrer et al., 2003), and polluted soils (Vosatka and Dodd, 1998).

Despite the ecological and potential agronomic importance of AM, study of the symbiosis at the molecular and genetic level has been hampered by technical problems in manipulating the AM fungal partner. First, AM fungi cannot complete their life cycle in the absence of a plant host and are therefore difficult to grow in axenic culture, a prerequisite for the production of standardised AM inoculum. Second, AM infection events are relatively rare and non-synchronous as the infection process occurs over several days with new infections occurring as new roots grow. Once arbuscules and vesicles have formed, all stages of infection can be present at once making gene expression studies based on total RNA extractions difficult to interpret. Finally, although transient marker gene expression using a biolistic approach has been successful in an AM fungus, the multinucleate spores and coenocytic hyphal growth of AM fungi make establishment of a robust, stable transformation system for the fungus unlikely in the near future. Full genome sequencing of *Glomus intraradices* will soon be in progress however, a suitable transformation system is a requirement for further studies in functional genomics.

On the plant side, AM research has benefited over the past decade from the discovery that some plant genes required for the formation of symbiotic nodules on legume roots by nitrogen-fixing Rhizobium bacteria are also required for AM infection (Duc et al., 1989). The mutants defective for both nodulation (Nod^-) and AM infection (Myc^-) that have been identified in *L. japonicus* indicate that at least six independent genes are common to the two root symbioses and are referred to as the common *SYM* genes (Kistner and Parniske, 2002). Nodulation and mycorrhization are very different processes, involving unrelated microsymbionts and giving rise to completely different symbiotic structures in the host root. However, the striking genetic overlap between these root symbioses in legumes and the fact that AM predates nodulation by roughly 300 M years, suggest that some plant functions from the ancient AM symbiosis may have been recruited in the evolution of root nodulation (LaRue and Weeden, 1994).

Stages of AM infection

AM-fungi are obligate symbionts that are present in the soil as multinucleate spores surrounded by thick cell walls. When temperature and humidity are favourable, spores germinate, forming hyphae that grow through the soil and begin to ramify. Hyphal growth is further stimulated by CO_2 and compounds secreted by the roots of potential host plants (Buée et al,. 2000). In turn, plant roots respond to diffusible signals from the AM fungus with changes to gene expression (Kosuta et al. 2003). At the surface of host roots, hyphae attach themselves and form slight swellings called appressoria, often in the groove between adjacent epidermal cells. Under the appressorium, fungal hyphae penetrate the epidermal cell, a process associated with the highly localised expression of specific plant genes (Blilou et al., 2000; Chabaud et al., 2002). Hyphae then exit the epidermal cell and grow intracellularly to the inner cortex. Here, the fungus forms long intercellular runner hyphae, from which hyphal branches penetrate cortical cells to form tree-like intracellular structures called arbuscules.

Even while inside the plant cell, the fungus does not penetrate the cytoplasm. In the arbuscule-containing cell, a plant-derived membrane is formed that surrounds the arbuscule. The fungal cell wall becomes progressively thinner as the arbuscule develops, leading to the establishment of an extensive intracellular interface, which is where the phosphate and carbon are thought to be transferred between the plant and the fungus. In the intercellular spaces in the inner cortex, many *Glomus sp.* fungi also form vesicles: thin-walled, lipid-filled structures thought to be connected to energy storage.

TECHNIQUES FOR THE STUDY OF ARBUSCULAR MYCORRHIZA IN LOTUS

Glomus intraradices infection using a chive nurse pot system

The method described below can be used for both propagating *G. intraradices* and for infecting *L. japonicus* or any other susceptible plant of choice. The system delivers relatively synchronous infection as well as a high infection pressure.

Initially, when setting up the system, fungal inoculum (dried AM-infected substrate or surface-sterilised spores) is required to infect a young chive plant grown from seed. Subsequently, the infected chive plant is transferred into a small pot (10 cm^3) containing substrate composed of a 1:1 v/v mixture of Terragreen (Oil-Dri Ltd., Wisbech Cambs, UK) and sand (Double Arches 16/30 Sand, Hepworth Minerals and Chemicals, Cheshire, UK). The infected chive serves as nurse plant, allowing a dense AM hyphal network to be established throughout the entire substrate, which usually takes six to eight weeks. More chive plants are then infected by sowing chive seeds into the substrate around the initial nurse plant. Five to six weeks after germination, once a significant level of AM infection has been confirmed by staining some of the young chive roots, these young chive plants are individually transferred into fresh substrate in smaller pots (8.5 cm^3). To minimise the risk of transferring an un-infected plant, two young chives may be transferred to each smaller pot. After four to six weeks, these new chive nurse pots are in turn ready to be used for co-cultivation with *L. japonicus*. Pre-germinated seedlings are carefully transferred into the chive nurse pots. To minimize damage to the fragile hyphal network in the substrate, seedlings are placed into carefully dug, small holes not deeper than 2 cm. Scarified, non-germinated seeds can be sown directly into the chive nurse pots. It is advisable not to use chive nurse pots much older than six weeks for co-cultivation, as nutrient competition between the growing chive plants and the young seedlings may limit growth of the latter. After four to six weeks, smaller-than-average chive plants may not be sufficiently infected.

The maintenance and propagation of the chive nurse pots should be kept separate from *L. japonicus* plants in order to avoid contamination with *Mesorhizobium loti* (ideally in different glasshouses, growth chambers etc.). Moreover, older chive nurse pots should be discarded regularly in order to avoid potential accumulation of pathogens within the substrate.

Chive nurse pot nutrient solution
(Thilo Winzer, unpublished)

Macro and micro elements	Final concentration (mg/L)
$MgSO_4.7H_2O$	24.65
KNO_3	5.05
KH_2PO_4	0.115
K_2HPO_4	0.725
$CaCl_2.2H_2O$	18.55
H_3BO_3	0.0715
$MnSO_4.4H_2O$	0.051
$ZnSO_4.7H_2O$	0.011
$CuSO_4.5H_2O$	0.004
$Na_2MoO_4.5H_2O$	0.0025
$CoCl_2.4H_2O$	0.005
Fe-EDTA-di-hydroxyphenylacetate	0.835

Soak chive pots with this nutrient solution instead of water once a week.

Glomus intraradices infection using an axenic nurse culture system

The use of an axenic system is advantageous for gene expression studies and a prerequisite for detailed cell biology. However, possibly due to reduced plant vigour, AM infection can be slower and less reliable in the Petri dish than in the greenhouse. None the less, consistent AM infection of *L. japonicus* can be achieved under sterile conditions, with high infection pressure and relatively synchronous infection, using the nurse culture system based on Ri T-DNA transformed roots described below.

Since Mosse and Hepper (1975) first established monoxenic mycorrhiza with *Glomus mosseae*, at least 27 AM fungal species have been successfully cultured monoxenically in association with *Agrobacterium rhizogenes* T-DNA transformed or non-transformed excised host roots of many plant species (Fortin et al. 2002). The most widely-used plant-fungus combination is carrot (*Daucus carota*) and *G. intraradices* (Bécard and Piche, 1992) because of the vigour of this system and the ease with which it can be propagated. Sterile starter cultures of several AM fungal species to be used for research purposes can be obtained from the Glomales *in vitro* Collection (GINCO) (see www.mbla.ucl.ac.be/ginco-bel or http://res2.agr.ca/ecorc/ginco-can/ for a list of available species and ordering instructions) and various protocols for *A. rhizogenes* transformation have been published (Stougaard et al., 1987; Hansen et al., 1989; Boisson-Dernier et al., 2001).

To infect Ri T-DNA transformed roots with AM fungi, place sterile spores within 1 cm of roots, close enough so that AM-stimulatory compounds produced by roots will reach the fungus. While infection can be achieved on moist sterile filter paper, long-term cultures require a nutrient-rich culture medium. To maximise infection, the medium described by Bécard and Fortin (1988) contains minimal phosphorus and has low pH (5.5). Furthermore, the use of gellan gum (Phytagel, Sigma, St Louis, USA) as a gelling agent provides a completely transparent medium, which allows growth of the microscopic AM fungal hyphae to be followed and photographed with ease. Infection can take 2-3 weeks to become established, after which time the extraradical hyphae will begin to grow throughout the Petri dish, forming a dense hyphal mat throughout the medium over a period of 4-6 weeks. The negative geotropism of Ri T-DNA transformed carrot roots can be further exploited by turning the Petri dish on its side so that roots remain confined to the upper third of the Petri dish. AM co-cultures grow well in the dark at 22-28°C, in Petri dishes sealed with parafilm to conserve moisture. The direct transfer of a portion of the *G. intraradices*-infected carrot roots to a new Petri dish containing fresh medium will propagate the co-culture.

M medium (pH 5.5)
(Bécard and Fortin 1988)

Macro and micro elements	**Final concentration (mg/L)**
$MgSO_47H_2O$	731
KNO_3	80
KCl	65
KH_2PO_4	4.8
$Ca(NO_3)_24H_2O$	288
NaFeEDTA	8
KI	0.75
$MnCl_24H_2O$	6
$ZnCl_27H_2O$	2.65
H_3BO_3	1.5
$CuSO_44H_2O$	0.13
$Na_2MoO_45H_2O$	0.0024
Vitamins	
Glycine	3
Thiamine HCl	0.1
Pyridoxine HCl	0.1
Nicotinic acid	0.5
Myo inositol	50
Sucrose	10,000
Phytagel	
Horizontal cultures	4,000
Vertical cultures (Petri dishes turned on their side)	5,000

To infect *L. japonicus*, pre-germinated surface-sterilised seeds are placed directly into the Petri dish on a root-free part of the hyphal mat and grown in a well-lighted growth chamber (16 hr photoperiod, 3200 cd m^{-2} light intensity) at 22-28°C. Petri dishes containing seedlings should be sealed with breathable medical tape (Micropore, 3M Health Care, St Paul, MN), propped at an angle of 70° and the bottom portion of the Petri dish covered with black plastic to protect the *L. japonicus* seedling roots from the light. The first signs of AM infection (appressorium formation) will be visible on *L. japonicus* roots after 5-10 days, and infection sites with arbuscules and vesicles should be present by 15 days after planting. Seedlings can be grown under these conditions for 4-6 weeks before nutrients in the medium become limiting.

Staining protocol to visualise AM fungi in and on roots

This procedure is a modification of common mycological staining techniques, in which roots are cleared by heating in potassium hydroxide, then acidified and stained. Protocols for AM staining with a variety of colorants, including the most commonly used Chlorazol Black E, Trypan Blue, and Acid Fuschin have been described in detail (Brundrett et al., 1994; Brundrett et al. 1996). Ink staining is preferable because of the rapidity of this procedure and the relative non-toxicity of ink compared to some of the other colorants.

Rapid ink staining (adapted from Vierheilig et al., 1998 by M Chabaud, INRA Toulouse)

1. Place root fragments (1-5 cm long) in an Eppendorf tube. Make a hole in the tube cover with a syringe to avoid the tube bursting open during heating. Separate large root samples into several smaller batches in several tubes. Packing large root samples into a single tube will result in inefficient clearing and staining.
2. Cover root fragments with 10% KOH (1 mL) and heat 7-10 minutes at 95°C using a block heater or water bath. Older roots are generally thicker, tougher and more highly-pigmented and thus require longer clearing with KOH. To determine the clearing time required, replace the discoloured KOH in the tube after 10 min with fresh KOH and heat for 10 min more. Repeat until KOH remains colourless after heating. Gloves, safety goggles, and extreme care are necessary when handling hot KOH.
3. Remove KOH and rinse roots twice with 5% acetic acid. Avoid touching roots as they are very fragile after heating in KOH and will be damaged by rough treatment.
4. Cover root fragments with 5% black ink (Sheaffer Manufacturing Co., Ft. Madison, IA) in 5% acetic acid and heat 5 min at 95°C.
5. Remove ink solution and rinse roots twice with 5% acetic acid. Ink solution can be filtered and re-used.
6. De-stain roots by soaking in 5% acetic acid for 20 min or longer, depending on the intensity of colour desired. Stained roots can be left in water for several days, but for a longer term should be stored in 1:1 glycerol:water.

Adaptations for high throughput staining

A large volume of independent samples can be stained simultaneously using a multi-well support and a series of clearing, rinse, and staining baths. An appropriately sized support for root systems of individual 2-8 week old plants is a 0.8 mL, 96-well storage plate (ABgene House, Surrey, UK). A 1 mm diameter hole drilled into the bottom of each well allows it to drain and fill with various solutions while retaining root fragments. KOH and stain baths can be heated in any chemical- and heat-resistant plastic or glass container, such as a clear 174x115x60mm lunch box (Watkins & Doncaster the Naturalists, Kent, UK). These particular boxes are heat resistant only up to 90°C. The staining protocol is very similar to that described above, but the following adaptations have been made to accommodate the larger volume of samples.

1. Preheat KOH and ink stain baths to 90°C in a large water bath placed in the fume hood.
2. Place root samples/fragments into individual wells, curling the root pieces and adjusting them with forceps to ensure that they are wedged at the bottom of the well.
3. Lower the root support containing root samples into the hot KOH bath. The KOH level should be just high enough to ensure that roots are submerged. Cover the water bath and heat at least 30 min at 90°C.
4. Carefully remove root support from KOH bath using long forceps and allow KOH several minutes to drain away.
5. Rinse root samples in 5% acetic acid by lowering the root support into rinse baths of similar volume to the KOH bath, ensuring that roots are submerged each time, and allowing several minutes draining between rinses.
6. After draining away the final rinse, lower the root support into the hot ink stain bath, ensuring that roots are completely submerged in the stain. Cover the water bath and heat at least 15 min at 90°C.
7. Remove root support from ink bath using long forceps, allow the ink to drain, then rinse and de-stain 20 min in 5% acetic acid baths. Root samples can be stored in the support in a water bath for several days.

CONCLUSION

This review briefly describes some of the basic techniques for AM culture and study that have proven to be effective in *L. japonicus*. Further information about AM fungi, including helpful culture and staining protocols, can be obtained from the following organisations/sites:

- The Mycorrhiza Information Exchange http://mycorrhiza.ag.utk.edu/
- The International Culture Collection of Vesicular Arbuscular Mycorrhizal Fungi (INVAM) http://invam.caf.wvu.edu/
- The International Bank for the Glomeromycota (BEG) www.kent.ac.uk/bio/beg/

- The Glomales In Vitro Collection (GINCO) www.mbla.ucl.ac.be/ginco-bel or www.agr.gc.ca/science/ecorc/ginco-can/index_e.htm

REFERENCES

Al-Karaki GN, and Hammad R. (2001) **Mycorrhizal influence on fruit yield and mineral content of tomato grown under salt stress**. *Journal of Plant Nutrition* 24, 1311-1323.

Bécard G, and Fortin A. (1988) **Early events of vesicular-arbuscular mycorrhiza formation on Ri T-DNA transformed roots**. *New Phytologist* 108, 211-218.

Bécard G, and Piche Y. (1992) **Establishment of vesicular-arbuscular mycorrhiza in root organ culture: review and proposed methodology.** In: *Methods in Microbiology* (Norris JR, Read DJ, Varma AK, Eds.), Academic Press, London 24, 89-108.

Blilou I, Ocampo JA, and Garcia-Garrido JM. (2000) **Induction of *Ltp* (lipid transfer protein) and *Pal* (phenylalanine ammonia-lyase) gene expression in rice roots colonised by the arbuscular mycorrhizal fungus *Glomus moseae*.** *Journal of Experimental Botany* 51, 1969-1977.

Bohrer G, Kagan-Zur V, Roth-Bejerano N, Ward D, Beck G, and Bonifacio E. (2003) **Effects of different Kalahari-desert VA mycorrhizal communities on mineral acquisition and depletion from the soil by host plants**. *Journal of Arid Environment* 55,193-208.

Boisson-Dernier A, Chabaud M, Garcia F, Bécard G, Rosenberg C, and Barker DG. (2001) ***Agrobacterium rhizogenes*-transformed roots of *Medicago truncatula* for the study of nitrogen-fixing and endomycorrhizal symbiotic associations**. *Molecular Plant-Microbe Interactions* 14, 695-700.

Bowman DMJS, and Panton WJ. (1993) **Factors that control monsoon-rain-forest seedling establishment and growth in North Australian eucalyptus savanna**. *Journal of Ecology* 81, 297-304.

Brundrett M, Melville L, and Peterson L. (1994) **Practical Methods in Mycorrhizal Research**. Mycologue Publications, Waterloo.

Brundrett M, Bougher N, Dell B, Grove T, and Malajczuk N. (1996) **Working with Mycorrhizas in Forestry and Agriculture**. Australian Centre for International Agricultural Research Monograph 32, Canberra. Available online: http:// mycorrhiza.ag.utk.edu /mstain. htm.

Buée M, Rossignol M, Jauneau A, Ranjeva R, and Bécard G. (2000) **The presymbiotic growth of arbuscular mycorrhizal fungi is induced by a branching factor partially purified from plant root exudates**. *Molecular Plant Microbe Interactions* 13, 693-698.

Chabaud M, Vernard C, Defaux-Petras A, Bécard G, and Barker DG. (2002) **Targeted inoculation of *Medicago truncatula* in vitro root cultures reveals *MtENOD11* expression during early stages of infection by arbuscular mycorrhizal fungi**. *New Phytologist* 156, 265-273.

Dalpe Y, and Aiken SG. (1998) **Arbuscular mycorrhizal fungi associated with Festuca species in the Canadian high arctic**. *Canadian Journal of Botany* 76, 1930-1938.

Duc G, Trouvelot A, Gianinazzi-Pearson V, Gianinazzi S 1989 **First report of non-mycorrhizal plant mutants (Myc$^-$) obtained in pea (*Pisum sativum* L.) and faba bean (*Vicia faba* L.)** Plant Sci 60:215-222.

Fortin JA, Bécard G, Declerck S, Dalpe Y, St-Arnaud M, Coughlan AP, and Piche Y. (2002) **Arbuscular mycorrhiza on root-organ cultures**. *Canadian Journal of Botany* 80, 1-20.

Gianinazzi S, Trouvelot A, Lovato P, VanTuinen D, Franken P, and Gianinazzi-Pearson V. (1995) **Arbuscular mycorrhizal fungi in plant-production of tremperate agroecosystems**. *Critical Reviews in Biotechnology* 15, 305-311.

Hansen J, Jorgensen J-E, Stougaard J, and Marcker KA. (1989) **Hairy Roots-a short cut to transgenic root nodules**. *Plant Cell Reports* 8, 12-15.

Kistner C, and Parniske M. (2002) **Evolution of signal transduction in intracellular symbiosis.** *Trends in Plant Science* 7, 511-518.

Kosuta S, Chabaud M, Lougnon G, Gough C, Dénarié J, Barker DG, and Bécard G. (2003) **A diffusible factor from arbuscular mycorrhizal fungi induces symbiosis-specific *MtENOD11* expression in roots of *Medicago truncatula*.** *Plant Physiology* 131, 952-962.

LaRue TA, and Weeden NF. (1994) **The symbiosis genes of the host**. In: *Proceedings of the First European Nitrogen Fixation Conference* (GB Kiss, G Endre, Eds.) Aug 28-Sept 2, Officina Press, Szeged, Hungary. pp. 147-151.

Mosse B, and Hepper CM (1975) **Vesicular arbuscular infections in root organ cultures**. *Physiological Plant Pathology* 5, 215-223.

Schüssler A, Schwarzott D, and Walker C. (2001) **A new fungal phylum, the *Glomeromycota*: phylogeny and evolution**. *Mycological Research* 105, 1413-1421.

Smith SE, and Read DJ. (1997) **Mycorrhizal Symbiosis**, 2nd Edition. Academic Press, London.

Stougaard J, Abildsten D, and Marcker KA. (1987) **The Agrobacterium rhizogenes pRi TL-DNA segment as a gene vector system for transformation of plants**. *Molecular General Genetics* 207, 251-235.

Vierheilig H, Coughlan AP, Wyss U, Piche Y. (1998) **Ink and vinegar, a simple staining technique for arbuscular-mycorrhizal fungi**. *Applied Environmental Microbiology* 64, 5004-5007.

Vosatka M and Dodd JC. (1998) **The role of different arbuscular mycorrhizal fungi in the growth of *Calamagrostis villosa* and *Deschampsia flexuosa*, in experiments with simulated acid rain**. *Plant & Soil* 200, 251-263.

SECTION 3

CELL AND MOLECULAR BIOLOGY PROTOCOLS

Chapter 3.1

CONCURRENT VISUALIZATION OF *GUSA* AND *LACZ* REPORTER GENE EXPRESSION

Clara L Díaz, Helmi RM Schlaman, and Herman P Spaink*

*Institute of Biology, Clusius Laboratory, Wassenaaarseweg 64 2333 AL Leiden, THE NETHERLANDS, *Corresponding author*

Email: spaink@rulbim.leidenuniv.nl

Phone: +31 71 527 5063 Fax: +31 71 527 5088

Keywords: Reporter gene staining, nodulation, roots, rhizobia

*Here we describe conditions under which two reporter genes, encoding β-glucuronidase (*gusA*) and β-galactosidase (*lacZ*), can be used to simultaneously distinguish between the sites of expression of plant and bacterial genes, respectively, in symbiotic tissues. These conditions involve the fixation of cytoplasmic proteins with cold methanol before detecting *gusA *expression, and the addition of detergents to the incubation solution in order to inactivate endogenous plant β-galactosidases.*

INTRODUCTION

The nodulation of the roots of leguminous plants by rhizobia is the result of concerted gene expression of both symbiotic partners. In plant cells, the timing and localization of genes involved in this process is usually visualized with promoter-*gus*A (Jefferson et al, 1987) fusions. Concurrent with plant gene expression, subsequent staining of rhizobia carrying constitutively expressed *lacZ*-encoding β-galactosidase (*β-gal*) as a reporter gene with a contrasting substrate has been used to localize rhizobia at different nodulation steps (Pichon et al, 1994a).

Here we describe the procedures we follow to obtain double staining of reporter gene activities during nodulation of *Lotus japonicus* roots. We adapted the protocol applied by Pichon et al. (1994a) on nodulated *Medicago truncatula* roots because Lotus roots seem to present a barrier for efficient inactivation of endogenous β-galactosidases and rapid penetration of enzyme substrates. We carried out precipitation of cytoplasmic proteins in cold methanol before the detection of plant promoter *gus*A fusions (Wijers et al, 2001). In addition, we added detergents to the

A.J. Márquez (ed.), Lotus japonicus Handbook, 99-109.

solutions used for the inactivation of endogenous β-galactosidases and for the detection of activities of reporter genes. These modifications gave a reduction of background staining originated by endogenous root β-galactosidases and a shortening of the reaction time of the promoter-reporter gene fusions.

To set up this protocol, we obtained hairy roots of *L. japonicus* by transformation with *Agrobacterium rhizogenes* LBA 1334 containing pPB12B, a derivative of the binary vector pLP100 carrying the *Medicago sativa ENOD12B* promoter fused to *gus*A. This construct presented a nodulation-independent expression in lateral roots, as well as induced expression upon inoculation with *Sinorhizoium meliloti* in transgenic *M truncatula* plants (Trinh et al, 1998; see Chapter 6.2 for a protocol for hairy roots induction). We took advantage of the expression of *MsENOD12B-gusA* during nodulation of Lotus roots by *Mesorhizobium loti* R7A containing pXLGD4, that carries the constitutively expressed *hemA* promoter fused to *lacZ* (Leong et al, 1985), for the improvement of concurrent visualization of plant and bacterial reporter genes. A detailed analysis of the expression of *MsENOD12B-gus*A is beyond the scope of this work.

MATERIALS

Stock solutions

Stock	Volume prepared	Procedure	Storage
90% (v/v) Methanol	100 mL	Mix 90 mL methanol with 10 mL water[1]	-20°C
0.2 M Na_2HPO_4-NaH_2PO4 pH 7	1000 mL	Dissolve 22.2 g $Na_2HPO_4 \cdot 2H_2O$ and 10.4 g $NaH_2PO_4 \cdot H_2O$ in water, check and correct pH.	100 mL aliquots, sterilized solution at RT
0.2 M Na_2HPO_4-NaH_2PO4 pH 7.4	1000 mL	Dissolve 29.2 g $Na_2HPO_4 \cdot 2H_2O$ and 4.9 $NaH_2PO_4 \cdot H_2O$ in water, check and correct pH	100 mL aliquots, sterilized solution at RT
1% (v/v) Triton X-100	100 mL	Add 1 mL detergent to 99 mL water	25 mL aliquots at -20°C, and liquid at 4°C
1% (w/v) N-laurylsarcosine	100 mL	Dissolve 1 g in water	25 mL aliquots at -20°C, and liquid at 4°C
0.5 M NaEDTA, pH 8	200 ml		
0.2 M K_2HPO_4-KH_2PO_4, pH 7	1000 mL	Dissolve 21.6 g K_2HPO_4 and 10.34 g $KH_2PO_4 \cdot 2H_2O$ in water, check and correct pH	100 mL aliquots, sterilized solution at RT

Stock	Volume prepared	Procedure	Storage
0.1 M $MgSO_4$	100 mL	Dissolve 2.465 g $MgSO_4 \cdot 7\ H_2O$ in water	25 mL aliquots, sterilized solution at RT
1 M KCl	100 mL	Dissolve 7.46 g KCl in water	25 mL aliquots, sterilized solution at RT
25 mg mL^{-1} X-gluc (5-Bromo-4-chloro-3-indolyl--β-D-glucoronic acid, cyclohexylammonium salt	5 mL	Dissolve 125 mg X-gluc in 5 mL dimethyl-formamide (DMFO) in a dark glass vial	Keeps for 4-6 months at -20°C
40 mg mL^{-1} Magenta-gal (M-gal, 5-Bromo-6-Chloro-3-indolyl-β-D-galactopyranoside)[3].	2.5 mL	Dissolve 100 mg M-gal in 2.5 mL dimethyl-formamide (DMFO) in a dark glass vial	Keeps for 4-6 months at -20°C

Deionized, sterilized water is used to prepare all solutions. To avoid bacterial contamination, solutions are sterilized for 20 min at 120°C then stored at room temperature (RT). Open bottles are stored at 4°C. M-gal can be ordered from AppliChem www.applichem.de or from www.biotium.com.

Solutions for detecting plant promoter-*gus*A fusions

GUS fixative

- 90% methanol

Pi-Ferri buffer

0.1 M Na_2HPO_4-NaH_2PO_4 pH 7 containing 1 mM $K_3Fe(CN)_6$. Use within 2 weeks; store in the dark at 4°C.

To prepare	200 mL
0.2 M Na_2HPO_4- NaH_2PO_4 pH 7	100 mL
0.1 M $K_3Fe(CN)_6$	2 mL
water	98 mL

GUS reaction buffer

0.1 M Na_2HPO_4-NaH_2PO_4 pH 7 containing 1 mM $K_3Fe(CN)_6$, 1 mM $K_4Fe(CN)_6$, 10mM EDTA, 0.1% Triton X-100, 0.1% N-laurylsarcosine and 0.5 mg mL^{-1} X-gluc.

To prepare, just before use	10 mL	50 mL
0.2 M Na_2HPO_4-NaH_2PO4 pH7	5 mL	25 mL

0.1 M $K_3Fe(CN)_6$	0.1 mL	0.5 mL
0.1 M $K_4Fe(CN)_6$	0.1 mL	0.5 mL
0.5 M NaEDTA	0.1 mL	0.5 mL
1% Triton X-100	1 mL	5 mL
1% N-laurylsarcosine	1 mL	5 mL
25 mg mL^{-1} X-gluc	0.2 mL	1 mL
Water	2.5 mL	12.5 mL

Washing buffer

0.1 M Na_2HPO_4-NaH_2PO_4 pH 7.4

To prepare	**200 mL**
0.2 M Na_2HPO_4-NaH_2PO_4 pH7.4	100 mL
Water	100 mL

Solutions for the detection of bacterial promoter-*lacZ* fusions

Endogenous β-gal inactivation buffer

0.1 M Na_2HPO_4-NaH_2PO_4 pH 7.4 containing 1.25% glutaraldehyde. (0.1% Triton X-100, 0.1% N-laurylsarcosine can be added when nodulated roots are being assayed 4 weeks after inoculation). Use electron microscopy grade glutaraldehyde, store bottle at 4°C in the dark. Take volume needed with a needle and syringe.

To prepare, just before use	**10 mL**	**50 mL**
0.2 M Na_2HPO_4-NaH_2PO_4, pH7.4	5 mL	25 mL
Glutaraldehyde 25%	0.5 mL	2.5 mL
Water	4.5 mL	22.5 mL

Z buffer

0.1 M K_2HPO_4-KH_2PO_4 pH 7 containing 1 mM $MgSO_4$, 10mM KCl, 0.1% Triton X-100 and 0.1% N-laurylsarcosine.

To prepare	**200 mL**
0.2 M K_2HPO_4-KH_2PO_4 pH 7	100 mL
0.1 M $MgSO_4$,	2 mL
1 M KCl	2 mL
1% Triton X-100	2 mL
1% N-laurylsarcosine	2 mL
Water	92 mL

β-gal reaction buffer

Z buffer containing 5 mM $K_3Fe(CN)_6$, 5 mM $K_4Fe(CN)_6$, 0.1% Triton X-100, 0.1% N-laurylsarcosine and 0.8 mg mL^{-1} M-gal.

To prepare, just before use	**10 mL**	**25 mL**
0.2 M K_2HPO_4-KH_2PO_4 pH 7	5 mL	12.5 mL
0.1 M $MgSO_4$,	0.1 mL	0.25 mL
1 M KCl	0.1 mL	0.25 mL
1% Triton X-100	1 mL	2.50 mL
1% N-laurylsarcosine	1 mL	2.50 mL
0.1 M $K_3Fe(CN)_6$	0.5 mL	1.25 mL
0.1 M $K_4Fe(CN)_6$	0.5 mL	1.25 mL
40 mg mL^{-1} M-gal	0.2 mL	0.50 mL
Water	1.6 mL	4 mL

PROCEDURES

GUS fixation

Submerge freshly cut roots in cold GUS fixative. Do not let methanol to warm up. Place roots immediately at -20°C for 2 h. Alternatively, roots may be kept in water during sample collection. Dry them gently with a tissue paper before submerging in fixative.

Washes

Discard fixative and submerge roots in a generous volume of Pi-Ferri buffer. Incubate by gently agitation for 10 min. Discard solution and add fresh buffer, incubate as described. Repeat washing for a third time. These steps are carried out at room temperature.

GUS reaction

Submerge roots in GUS reaction buffer. Infiltration in vacuum (20-30 min) will speed up the reaction by forcing penetration of the substrate, but will not have an appreciable effect for incubations longer that 4 h at 37°C. The reaction should be monitored to avoid excess of deposition of ClBr-indigo. A too dark blue appearance of plant cells interferes with the observation of the red precipitate product of M-gal.

Ending GUS reaction

Submerge roots in washing buffer (0.1 M Na_2HPO_4-NaH_2PO_4 pH7.4), proceeding as described in the washing step. At this stage, roots can be examined and pieces of interest for staining of β-gal activity can be sectioned.

Inactivation of endogenous plant β-galactosidases

Place roots in endogenous β-gal inactivation buffer, infiltrate in vacuum for 30 min and let stand for 2 to 4 h at room temperature.

Equilibration in Z buffer

Roots are incubated as described above in Z buffer.

β-gal reaction

Roots are submerged in freshly prepared β-gal reaction buffer and infiltrated for 30 min in vacuum. The vials are placed at 28°C for the reaction to take place. The reaction time should be carefully monitored, to avoid over staining.

Ending β-gal reaction

Discard reaction buffer and submerge roots in deionised water, for at least 1 h. Storing the roots overnight at 4°C will enhance deposition of product.

Long-term preservation of double-stained tissue

Root pieces are post-fixed in 2.5% glutaraldehyde in 0.1 M Na_2HPO_4-NaH_2PO_4 pH 7.4 for 4 h at room temperature or overnight at 4°C. Root pieces are rinsed in deionized water for 30 min and dehydrated by subsequent incubations (20 min per step) in 20, 50, and 70% ethanol. Roots can be stored for many months in 70% ethanol at room temperature or at 4°C.

RESULTS

Observations appearing in Figure 1 were made using a Leica MZ12 stereoscopic microscope; pictures were taken with a Sony DKC-5000 digital camera. Roots have been cleared by submersion in chlorallactophenol (Beekman and Engler, 1994), with exception of the root appearing in panel D. In panels A, B, and D, rhizobia are easily detected in infection threads in root hairs. It seems that infection threads did not (yet) penetrate the root cortex. MsENOD12B is expressed in cell layers close to the pericycle in a rather long lateral root (panel A) and in an emerging lateral root, where is also strongly expressed at the tip of the root (panel B). Similar observations have been reported for MsENOD12A expression in roots of transgenic *M. sativa* ssp varia (Bauer et al, 1996) and MsENOD12B in transgenic *M. truncatula* roots (Trinh et al, 1998). In panel C, MsENOD12B seems to be induced in internal cortical cells below an extensive zone of infection thread formation, which have already penetrated external cortex cells. Cells expressing MsENOD12B cells correspond to rapidly dividing internal cortical cells that lead to the formation of nodule primordial in *L. japonicus* (Spronsen et al, 2001) Similarly, induction of

MsENOD12A in dividing cortical cells triggered by inoculation of *M. melitoti* Nod factors and cytokinins has been observed in transgenic *M. sativa* ssp varia roots (Bauer et al, 1996). Note that the gene is expressed in the cortex only at loci of infection, as reported for pea (Scheres et al, 1990) and *M. truncatula* ENOD12 genes (Pichon et al, 1994b).

In panel D, MsENOD12B is strongly expressed in a young, emerging nodule in *L. japonicus* roots, as has been observed for *M. truncatula* roots (Trinh et al. 1998).

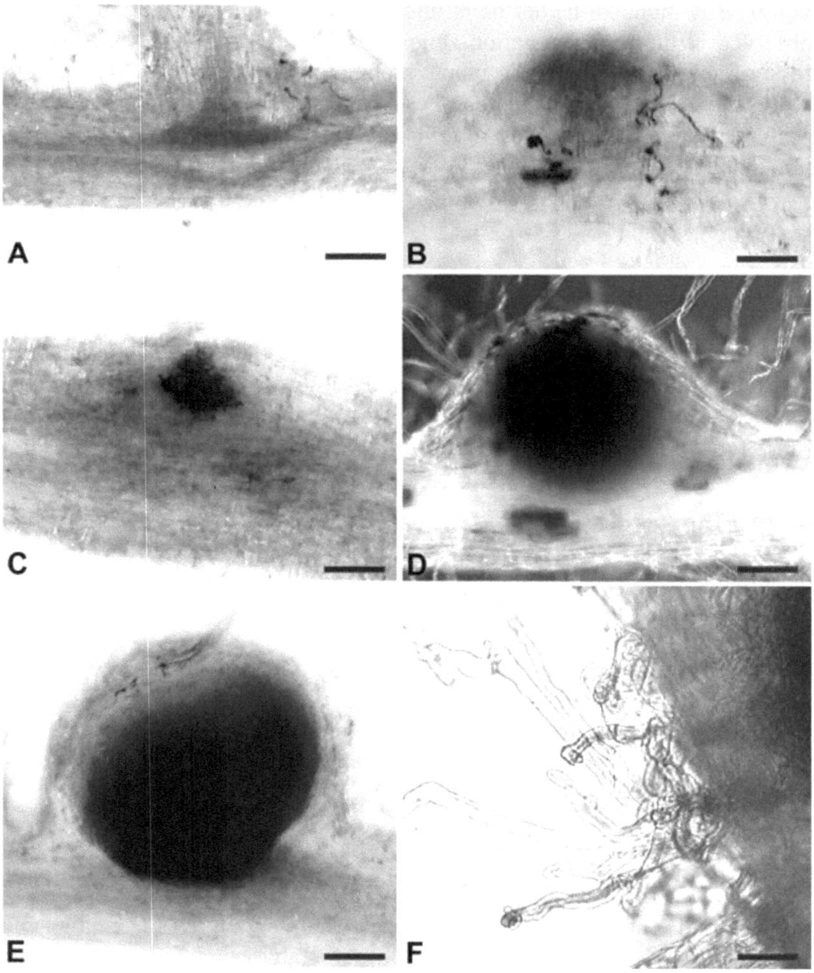

Figure 1. Double staining of L. japonicus *hairy roots expressing MsENOD12B-gusA (X-gluc as substrate, blue precipitate) during nodulation induced by* M. loti *R7A constitutively expressing β-gal (M-gal as substrate, red precipitate). Bar in panels A, B, D, and E=125 μm. Bar in panel C=95 μm. Bar in panel F=30 μm. See colour plate 3 (A-F) for this figure in colour.*

Bacteria in infection threads in the epidermis can be seen by the red M-gal precipitate.

In panel E, MsENOD12B expression can be clearly observed at the base of a young nodule. Sectioning of embedded nodules is necessary for the precise localization of GUS expression and rhizobia inside nodules such as the ones presented in panels D and E. In panel F, a close-up of a root hair carrying an infection thread and deformed roots hairs observed on the surface of a mature nodule, show that *MsENOD12B* is induced in root hairs interacting with rhizobia. GUS activity was also observed in the root elongation zone (close to root tips) and the growing root hair zone of some lateral roots in plants inoculated by with *M. loti,* including root hairs (not shown). Taken together, these observations suggest that *MsENO12B* share similar patters of expression in *L. japonicus* roots as the other *ENOD12* genes of plants that have an indeterminate type of nodulation.

COMMENTS

Solutions used for the detection of promoter-reporter gene activity

Freshly prepared solutions should be used to avoid bacterial contamination that could give rise to artefacts of staining. The pH of these solutions has to be checked before use, as the pH tends to fluctuate by salt precipitation after sterilization and long time storage of stock solutions.

GUS-staining

Cytoplasmic protein precipitation in cold methanol is a very convenient way to fix *gus*A-expressing cells: we have not observed leakage of product even after long incubation time (overnight) of strongly expressed promoters. GUS expression of methanol-fixed roots compared well with that of paraformaldehyde-fixed roots. We have also observed that methanol fixation increases the permeability of *L. japonicus* root cells to glutaraldehyde and enzyme substrates, even if detergents are omitted from the incubation solutions used for endogenous β-galactosidases inactivation and reporter gene detection.

Endogenous β-galactosidases inactivation

We often found out that blocking of background staining of endogenous β-galactosidases in *L. japonicus* roots was not predictable, even if incubation time in inactivation buffer was extended overnight. This was particularly true when roots of old plants (4 to 8 weeks) were used for staining. The combination of methanol fixation and addition of detergents to the inactivation buffer resulted in complete blocking of background staining in the roots of plants up to 4 weeks old, and in considerable background reduction in the roots of older plants. In general, inactivation of endogenous β-galactosidases in older roots and mature nodules requires longer incubation times.

Staining for β-galactosidase activity in rhizobia

We have observed that storage of roots after inactivation of endogenous β-galactosidases for 48 h at 4°C does not seem to affect β-galactosidase activity encoded by *lacZ* pXLGD4. The stability of this reporter gene allows time for careful examination of GUS-stained roots before proceeding with the detection of bacterial β-galactosidase activity. During staining of the second reporter gene, roots can be inspected, pictures made, and the reaction continued by placing the roots back in the reaction buffer and incubating at 28°C. Alternatively, the reaction can be slowed down by storage at 4°C overnight, and continued the following day if necessary.

Long-term storage of double stained tissue

As a standard, we perform post-fixation and store *L. japonicus* roots in 70% ethanol. Apart for ensuring aseptic conditions of storage, ethanol extracts many colouring substances from Lotus roots. Taking pictures and picking of root pieces and nodules for embedding can be made in 70% ethanol.

Photography

Although many observations can be made directly on roots kept in 70% ethanol (see fig. 1D), detailed observations like bacteria in infections threads is greatly facilitated in cleared tissue. We have applied different recipes for the clearing of *L. japonicus* roots and we have found out that only submersion in chlorallactophenol (chloralhydrate: lactic acid: phenol 2:1:1, by weight, Beeckman and Engler, 1994)) results in totally cleared roots. This solution is a fixative by itself; however, it tends to dissolve the red M-gal precipitate of freshly stained roots. Therefore, we take post-fixed roots that have been kept in 70% ethanol to 50 and 30% ethanol (20 min each) before incubation in three changes of water (at least 1 h each change) to ensure that ethanol residues have been washed away.

Roots are completely cleared after overnight incubation in clearing solution at room temperature; they can be stored for long periods in this solution in a tightly closed glass vial. Because chlorallactophenol is highly toxic and produces skin burns, only roots that have been carefully examined to address a particular detail are cleared. For mounting cleared roots, we use microscope slides that are prepared by fixing two cover slides of 20x20 mm on both ends of a slide with a contact epoxy resin, to prepare a chamber where one or two root pieces can be contained. The prepared slide is placed as a bridge on top of two slides positioned about 1.5 cm apart on the interior surface of a lid of a plastic Petri dish (diameter 9.5 cm) to its facilitate picking up. Root pieces are transferred on a drop of clearing solution to the middle of the chamber and the Petri dish is closed.

Roots can be screened in closed Petri dishes using a stereoscopic microscope. A cover slide is positioned on top of the roots before final observation and photographing; mounted slides can be kept for a few days at 4°C in Petri dishes closed with Parafilm. The mounted glass slide can be easily picked up from the glass bridge and placed on a clean glass or plastic surface that can protect the microscope stage from contamination with clearing solution. We recommend that adjustment of the (digital) camera and the position or type of light source should be prepared with the specimen contained in a closed Petri dish as to restrict exposure to the fumes of chlorallactophenol to the shortest time possible. Mounting of cleared roots is carried out in a fume cabinet, using forceps and wearing gloves.

Chlorallactophenol is prepared by mixing 200 g chloral hydrate, 100 g lactic acid and 100 g solid phenol in a beaker placed in a water-bath at 60°C, in a fume cabinet. Transfer solution to a dark glass bottle that can be tightly capped to avoid oxidative browning.

ACKNOWLEDGEMENTS

The authors thank A Kondorosi (CNRS, Gif-sur-Ivette Cedex, France) for plasmid pPB12B and A 't Hoofd for the montage of photographic work. CL Díaz and HP Spaink were supported by the EU International Cooperation Project ICA4-CT-2001-10056.

REFERENCES

Bauer P, Ratet P, Crespi MD, Schultze M, and Kondorosi A. (1996) **Nod factors and cytokinins induce similar cortical cell divisions, amyloplanst** deposition and *MsEnod12A* expression pattern in alfalfa roots. *The Plant Journal* 10, 91-96.

Beeckman T, and Engler G. (1994) **An easy technique for the clearing of histochemically stained plant tissue**. *Plant Molecular Biology Reporter* 12, 37-42.

Jefferson RA, Kavanagh TA, and Bevan MW. (1987) **GUS-fusions: β-glucuronidase as a sensitive and versatile gene fusion marker in higher plants**. *EMBO Journal* 6, 3901-3907.

Leong SA, Willimans PD, and Ditta GS. (1985) **Analysis of the 5' regulatory region of the gene for δ-aminolevulinic acid synthetase of *Rhizobium meliloti*.** *Nucleic Acids Research* 13, 5965-5976.

Pichon M, Journet E-P, de Billy F, Dedieu A, Huguet T, Truchet G, and Barker DG. (1994a) ***ENOD*12 gene expression as a molecular marker for comparing *Rhizobium*-dependant and –independent nodulation in alfalfa**. *Molecular Plant-Microbe Interactions.* 7, 740-747.

Pichon M, Journe, E-P, Dedieu A, de Billy F, Truchet G, and Barker DG. (1994b) **Rhizobium meliloti elicits transient expression of the early noduline gene *ENOD*12 in the differentiating root epidermis of transgenic alfalfa**. *Plant Cell* 4, 1199-1211.

Scheres B, van de Wiel C, Zalenski A, Horvath B, Spaink H, Zwartkruis F, Wolters A-M, Gloudemans T, van Kammen A, and Bisseling T. (1990) **The *ENOD12* gene product is involved in the infection process during the pea-*Rhizobium* interaction**. *Cell,* 60, 281-294.

Trinh TH, Ratet P, Kondorosi E, Durand P, Kamaté K, Bauer P, and Kondorosi A. (1998) **Rapid and efficient transformation of diploid *Medicago truncatula* and *Medicago sativa* ssp. *falcata* lines improved in somatic embryogenesis**. *Plant Cell Reports* 17, 345-355.

Van Spronsen PC, Grønlund M, Pacios Bras C, Spaink HP, and Kijne JW. (2001) **Cell biological changes of outer cortical root cells in early determinate nodulation**. *Molecular Plant-Microbe Interactions* 14, 839-847.

Weijers D, Franke-van Dijk M, Vencken R-J, Quint A, Hooykaas,P, and Offringa R. (2001) **An Arabidopsis Minute-like phenotype caused by a semi-dominant mutation in a ribosomal protein *S5* gene**. *Development* 128, 4289-4299.

Chapter 3.2

EMBEDDING ROOT AND NODULE TISSUE IN PLASTIC (BMM)

Mette Grønlund, Adamantia Agalou, Maria C Rubio, Gerda EM Lamers, Andreas Roussis, and Herman P Spaink*

*Clusius Laboratory, Institute of Biology, Leiden University, Wassenaarseweg 64, 2333 AL Leiden, THE NETHERLANDS; *Corresponding author.*

Email: spaink@rulbim.leidenuniv.nl

Phone: +31 71 527 5063 Fax: +31 71 527 5088

Keywords: Plastic embedding; immunolocalisation; immunocytochemistry; *in situ* RNA hybridisation.

Immunolocalisation and in situ *hybridisation allow the detection of proteins and RNA respectively, in individual cells of different tissue types. Hybridisation on serial 5-8 μm sections is most frequently performed with tissue embedded in paraffin wax but it is often difficult to obtain high-resolution sections from soft tissue with this embedding process. To overcome this, alternative localisation protocols have been developed utilising plastic resins. We have used plastic embedded tissue from* Lotus japonicus *roots and young nodules successfully in immunolocalisation experiments and developed a protocol that can also be adapted for* in situ *RNA localisation studies. The different parameters tested are described, as well as the use of alkaline phosphatase- or fluorescently-conjugated secondary antibodies.*

INTRODUCTION

Functional analyses of genes require essential knowledge of their expression pattern at the cellular level and it is often valuable to compare the spatiotemporal distribution of both the transcriptional and translational products, since in some cases the RNA expression pattern of a given gene does not overlap with that of the encoded protein (Lucas, 1995).

In situ hybridisation and immunocytochemistry give the possibility to detect RNA and proteins respectively, in the individual cells of different tissues. Although, thin sections of tissue embedded in paraffin wax are routinely used for hybridisation studies, it has always been difficult to obtain high-resolution sections from soft tissues embedded in paraffin. As a result, alternative methodologies making use of plastic polymers suitable for embedding have been developed (Baskin et al., 1992; Kronenberger et al., 1993; Ruzin 1999; Scarpella et al., 2000).

A.J. Márquez (ed.), Lotus japonicus Handbook, 111-122.

In *Lotus japonicus,* the use of paraffin works well with nodule tissue, but root sections maintaining a reasonable preservation of the cellular structure are difficult to obtain. A wealth of information is rapidly accumulating with respect to the plant genes/proteins involved in the early recognition steps during the symbiosis between rhizobia and legumes (Schauser et al., 1999; Stracke et al., 2002, Radutoiu et al., 2003, Madsen et al., 2003). With this in mind, it would be valuable to adapt existing RNA and protein localisation protocols to allow for higher resolution root sections in order to follow the induction of such genes/proteins, for example, in root hair and epidermal cells where signal molecules are perceived.

We have successfully employed plastic embedding for immunolocalisation of *Lj*SBP1 on *L. japonicus* root tissue (Flemetakis et al., 2002), as well as for different superoxide dismutases (EC 1.15.1.1) "SODs" in young nodules. The particular protocol used in these experiments is described in detail in the following paragraphs and has been used for root and nodule tissue prepared for *in situ* RNA localisation studies. The use of plastic in such studies has been described for Arabidopsis (Kronenberger et al., 1993) and rice (Scarpella et al., 2000) but it has proven to be difficult to implement in *L. japonicus,* in terms of high signal intensity when compared to paraffin sections processed in parallel. Some modifications of both embedding and hybridisation procedures have been tried, but still further optimisation is needed for plastic embedded tissue to become the preferred starting material for *in situ* RNA hybridisation studies in our hands.

The detection method used in most of our immunolocalisation experiments and in all the *in situ* RNA hybridisation studies so far, is based on the 4-nitroblue tetrazolium chloride (NBT) and 5-bromo-4-chloro-3-indolyl-phophatase (BCIP) detection of alkaline phosphatase-conjugated anti-rabbit secondary and anti-DIG primary antibodies (Jackson et al., 1994). In some of the immunolocalisation experiments a fluorescently-labelled secondary antibody has been used instead, with subsequent monitoring of the signal by confocal microscopy, providing a useful alternative for a better resolution at the cellular level and it will also be tested for detection of transcripts. The use of different fluorescently-labelled (secondary) antibodies for detection of signals in the hybridisation procedures would furthermore enable simultaneous localisation of two proteins or transcripts in the same tissue.

MATERIALS AND PROCEDURE

Immunolocalisation

Fixation

Root and young nodule tissue was cut and placed in fixation buffer and then mild vacuum was applied for 30-60 minutes, until the tissue sank. The tissue was incubated overnight at 4°C, slowly rotating. Fixation buffers used were made from a fresh stock of paraformaldehyde diluted to a concentration of 4% in 1x PBS or 0.1M Na_2HPO_4 buffer with or without 0.25% glutaraldehyde, while for some experiments 2% paraformaldehyde and 2.5% glutaraldehyde in 1x PBS was used instead. All fixation buffers contained 10 mM DTT.

Dehydration

The tissue was dehydrated in ethanol series including 10 mM DTT: 70%, 80%, 90%, 96%, 2 times 100%, 30 minutes each step, slowly rotating at room temperature.

Infiltration and embedding

Tissue was embedded in the acrylic medium BMM, which, in contrast to most other plastics, can subsequently be removed from the obtained sections using acetone.

BMM

BMM is a mixture of 4 Butyl Methacrylate, 1 Methyl Methacrylate (Merck), 10 mM DTT and either as photo-catalyst for polymerisation: 0.5% (w/v) benzoin ethyl ether/benzoin methyl ether (both work equally well), combined with UV polymerisation at -20°C to 4°C, or a thermal-catalyst: 1% benzoyl peroxide, combined with heat (60°C) polymerisation (see Comments). After mixing, the BMM solution is bubbled with N_2 in order to displace dissolved oxygen, since oxygen will inhibit polymerisation.

Following dehydration, the tissue was gradually infiltrated with BMM in a series of: 3:1, 1:1, and finally 1:3 of 100% EtOH:BMM, left slowly rotating at 4°C for more than 1½ hours at each step, and finally at 4^0C overnight in 100% BMM. The following day the BMM solution was replaced with fresh and incubated another 2 hours at 4°C, before embedding.

Embedding was performed in BEEM capsules (Agar Scientific, UK), the important features of the capsules are that they are airtight and UV transparent.

UV-polymerisation

The UV lamp(s) (6-8 W, long-wave, Philips, The Netherlands) were placed 15 cm (for root tissue) or 20-25 cm (for nodule tissue) below the capsule holders (see Comments).

Sectioning

Sections of 5-8 μm were made with a Leica RM 2165 rotary microtome and floated on drops of sterile water on BIOBOND (EMS catalogue #71304) coated slides. Slides were allowed to dry on a hot plate for a few hours to allow the sections to adhere to the slides.

Immunolocalisation

Blocking, hybridisation, and washing was all done in the same *immuno-buffer* consisting of: 20 mM Tris-HCl pH=8.2, 0.9% NaCl, 0.01% $BSA\text{-}_c^{TM}$ (Aurion) (0.1% for hybridisation) and 0.02% fish skin gelatine (Sigma G7765). In some of the experiments the sections went through a denaturation step before the actual localisation protocol, a treatment that seemed not to be essential for the particular proteins tested, but might have an impact in general depending on the protein in question. Denaturation was performed in 0.4% SDS, 3 mM β-mercaptoethanol, 12 mM Tris-HCl pH=6.8.

The equilibration buffer for alkaline phosphatase/ NBT-BCIP detection was 0.2 M Tris-HCl pH 9.5, 10 mM $MgCl_2$ and it was also used to dilute NBT and BCIP before addition to the slides. 10 mg NBT was dissolved in 0.1 ml dimethylformamide and 0.1 ml buffer. 5 mg BCIP was dissolved in 0.1 ml dimethylformamide before mixing with 30 ml buffer. Alternative ready-made NBT-BCIP tablets (Roche) were dissolved in water and added to the slides.

Cover slips were mounted in *gelvatol* (20 g polyvinyl alcohol dissolved in 40 ml glycerine and 80 ml PBS (pH= 6 to 7), stirring overnight at 4°C, after centrifugation the supernatant was used as mounting medium). In the case of fluorescent signal, we used gelvatol containing the anti-fading agent DABCO (Sigma), 100 mg/ml.

Immunolocalisation protocol

1. The embedding medium (BMM) was removed from the sections by two rounds of 15 minute acetone treatment (see Comments)
2. Slides were washed in 1x PBS
3. Denaturation of proteins were performed in the buffer given above, incubation 20 minutes at room temperature (this step is optional).
4. Slides were blocked in immuno-buffer 3 times for 10 minutes at room temp.
5. Primary antibody was diluted in immuno-buffer and applied slides resting on a racked hybridisation container with water in the bottom, to keep the humidity high. Slides were incubated with primary antibody overnight at 4°C or at room temperature for a minimum of two hours.
6. Slides were washed 5 times for 10 minutes in immuno-buffer at room temperature
7. Secondary antibody, diluted in immuno-buffer was applied to the slides and incubated at room temperature for more than two hours in the dark (darkness is essential, when using fluorescently-labelled secondary antibodies).
8. Slides were washed 5 times for 10 minutes in immuno-buffer

Detection

For fluorescently-labelled secondary antibodies, slides were rinsed in H_2O, mounted in gelvatol containing DABCO, and examined with a confocal microscope.

For alkaline phosphatase/NBT-BCIP detection, slides were equilibrated 3 times for 5 minutes in equilibration buffer, before adding the NBT/BCIP substrate. The slides were incubated at room temperature, in the dark, until colour appeared and checked regularly under a light microscope. The colour reaction was stopped by rinsing several times in H_2O and coverslips were mounted with gelvatol before inspection.

In situ RNA hybridisation

Fixation

Root and young nodule tissue was cut and placed in fixation buffer and a mild vacuum was applied for 1-2 times for 30 minutes, until the tissue sank. The tissue was incubated overnight at 4°C, slowly rotating. Fixation buffers used were made from a fresh stock of paraformaldehyde diluted to a concentration of 4% in 1x PBS or 0.1M Na_2HPO_4 buffer with 0.25% glutaraldehyde.

Dehydration

The tissue was dehydrated in ethanol series containing DTT (see Comments): buffer, 10%, 30%, 50%, 70%, 90%, 3 times 100%, 30 minutes each step, slowly rotating at room temperature.

Infiltration and embedding

Following dehydration, infiltration and embedding was performed as described above (see Comments).

Sectioning

Sections of 5-8 µm were made with a Leica RM 2165 rotary microtome and floated on drops of RNase-free water on Super Frost R Plus slides (Menzel Glaser). Slides were allowed to dry on a hot plate for a few hours to allow the sections to adhere to the slides.

In situ RNA hybridisations

Hybridisations were carried out according to the protocol we use as standard on paraffin sections, essentially as described in Papadopoulou et al., 1996. In one comparative experiment, we tested also the protocol used for maize and rice (Langdale, 1994; Scarpella et al., 2000), but the different protocols did not show any significant differences in signal detection.

Glass ware was baked 4 hours at 180°C, hybridisation chambers and plastic slide holders were treated with 0.4 M NaOH and solutions were treated with 0.1% (v/v) DEPC prior to use, to be RNase-free.

Probe transcription and labelling

The RNA probes were DIG labelled, using a DIG RNA labelling mix from Roche, and a T7 transcription kit from Ambion. Approximately 1 µg of linearised cDNA template was mixed with 2 µl of DIG RNA labelling mix, 2 µl transcription buffer and 2 µl T7 RNA polymerase in a total reaction volume of 20 µl. Transcription was allowed to proceed for more than 60 minutes at 37°C, 1 µl DNaseI (Ambion) was added and incubated at 37°C for more than 30 minutes, to remove the cDNA template. The reaction was stopped and the product precipitated twice, according to the Ambion protocol, in order to remove un-incorporated nucleotides. The product yield was estimated on an agarose gel and approximately 0.5-1 µg DIG-RNA was used for hybridisation, after hydrolysis, per slide (see Comments). Hydrolysis of the transcripts was performed according to Papadopoulou et al. (1996), to yield probes of approximately 150 bp.

***In situ* RNA hybridisation protocol**

1. The embedding medium (BMM) was removed from the sections by two rounds of 15 minutes acetone treatment (see Comments). Sections were left to air dry.
2. Sections were hydrated in an ethanol series: 100%, 90%, 70%, 50%, 30%, 10%, 3 times H_2O, 1 minute each step
3. Proteinase K treatment was performed in 100 mM Tris-HCl pH=7.5, 50 mM EDTA, 1 mg/ml proteinase K, for 30 minutes at 37°C with gentle shaking (see Comments).
4. Slides were rinsed 3 times in H_2O.
5. To remove positive charges, slides were acetylated in 0.1 M triethanolamine, 0.25% (v/v) acetic anhydrite. Triethanolamine was dissolved in H_2O, acetic anhydrite added just before the slides and the solution was stirred until foam appeared, where after the slides were soaked in the solution and incubated 10 minutes at room temperature.
6. Slides were rinsed for 10 minutes in 2x SSC.
7. Slides were dehydrated in an ethanol series: 10%, 30%, 50%, 70%, 90%, 2 times 100%, 1 minute each step, and air dried in vacuum for at least one hour.
8. Meanwhile, the hydrolysed and precipitated probes were resuspended in 10 µl H_2O. 10 µl deionised formamide was then added, mixed with probe, denatured at 68°C for 10 minutes, cooled on ice, and mixed with 80 µl hybridisation mixture. The hybridisation solution was heated at 60°C for 3 minutes before it was added to the slide and a cover slip was applied. Slides were resting on a rack, with 2x SSC soaked paper below to keep the hybridisation chamber moist overnight.
9. Hybridisation was performed at 42°C overnight (12-16 hours).
10. Post-hybridisation, the cover slips were gently let to glide off in a washing buffer of 4x SSC containing 5 mM DTT, slides were washed 4 times for 10 minutes at room temperature.
11. RNase treatment was performed in 500 mM NaCl, 1 mM EDTA, 10 mM Tris-HCl pH=7.5, 50 µg/ml RNaseA, 30 minutes at 37°C, with subsequent washes in RNase buffer for 4 times for 15-20 minutes at 37°C.
12. Slides were washed in 2x SSC, 30 minutes at room temperature, prior to blocking, antibody incubation and detection.
13. Slides were incubated in B1 for 5 minutes at room temperature.
14. Blocking was performed first in B1 + blocking reagent and then in B2 + BSA, at least 30 minutes in each buffer at room temperature.
15. AP conjugated anti-DIG-antibody was diluted in B2+0.1% BSA, applied to slides and incubated for at least 2 hours at room temperature, in the dark.

16. Slides were washed in B2, 30 minutes and in B1 3 times for 20 minutes at room temperature before detection.

Hybridisation mixture

- 50% deionised formamide, 300 mM NaCl, 10 mM Tris-HCl pH=7.0, 1 mM EDTA, 0.02% Ficoll, 0.02% polyvinylpyrrolidone, 0.02% BSA, 10% Dextran sulphate, 70 mM DTT

Buffers

- B1: 100 mM Tris-HCl pH=7.5, 150 mM NaCl
- B2: 100 mM Tris-HCl pH=7.5, 150 mM NaCl, 0.3% Tween-20

Blocking buffers

- B1 + 0.5% (w/v) blocking reagent (Roche)
- B2 + 1% (w/v) BSA (0.1% BSA for antibody incubation)

Detection

Equilibration and detection was performed as described above for immunolocalisation. The colour reaction was stopped by rinsing the slides several times in H_2O. The slides were then dehydrated in an ethanol series, allowed to air dry and mounted with Eukit (Agar Scientific, UK) before inspection.

COMMENTS

For the *in situ* RNA hybridisation studies, we tested both UV polymerisation and heat polymerisation, in the embedding step. Using UV polymerisation, nodules do not polymerise properly in the central tissue. Better polymerisation of the nodules was obtained using heat polymerisation, but it turned out that with this procedure no signal could be obtained, whereas parallel experiments on UV polymerised nodules, gave signal, although weak. So it seems that heat polymerisation is incompatible with the protocols tested for *in situ* RNA hybridisation.

The optimal distance from the UV source, and thereby the speed of polymerisation should be tested for each specific tissue. We found that increasing the distance for nodule tissue/decreasing the speed of polymerisation, allowed the central tissue of the nodule to be polymerised in a bigger percentage of the samples.

For thicker sections, the slides may have to be incubated longer in acetone to remove the plastic completely. This does not seem to interfere with the following reactions.

For the dehydration, DTT was included in the EtOH steps, in some experiments 10 mM was added in each step, as used in immunolocalisation studies; in other experiments only 1 mM DTT was added in the steps until 100% EtOH, where 10 mM DTT was then added, as described in Kronenberger et al. (1993). No significant difference was observed.

The amount of probe found to be needed for obtaining only a weak signal for *in situ*

RNA hybridisation on BMM embedded tissue was approximately 10 times higher than for paraffin embedded tissue.

We have tried treating the sections longer with proteinase K (up to 1 hour), to see if that would allow better penetration of the probe, but we did not observe any difference in the final signal.

RESULTS

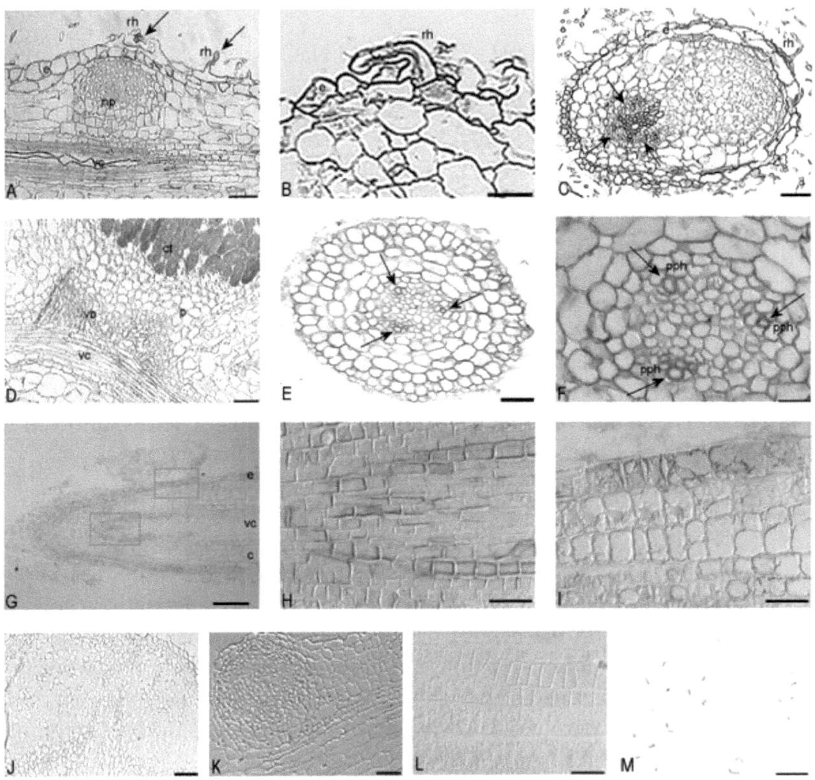

Figure 1. Localisation of LjSBP1 in root tips and roots. Immunolocalisation of the selenium binding protein (SBP) of Lotus japonicus *on roots, nodules and root tips. Detection was performed with alkaline phosphatase conjugated secondary antibodies. The embedding of the root tissue in BMM resulted in sections of high cellular resolution, enabling the detection of SBP in cellular structures that are otherwise difficult to preserve in detail, like root hairs (A), infection threads (B), and the phloem of the vascular cylinder (C). Furthermore, signal was detected in the vascular bundles and the infected cells of the central tissue in mature nodules (D). H and I are differential interference contrast microscopy pictures of the marked regions in picture G. Signal was observed in the cells of the epidermis and the outer cortex of the root tip (I) as well as on the vasculature (H). Transverse sections of that area allowed the identification of the protophloem cells of the stele as the cells with high expression levels of the protein (E and F). J-M sections incubated with pre-immune serum. A to D, G, J, K, bars=50μm. E, M, bars=25μm.*

F, H, I, L, bars=15μm. c, cortex; e, epidermis; np, nodule primordium; p, parenchyma; pph, protophloem; rh, root hair; vb, vascular bundle; vc, vascular cylinder.

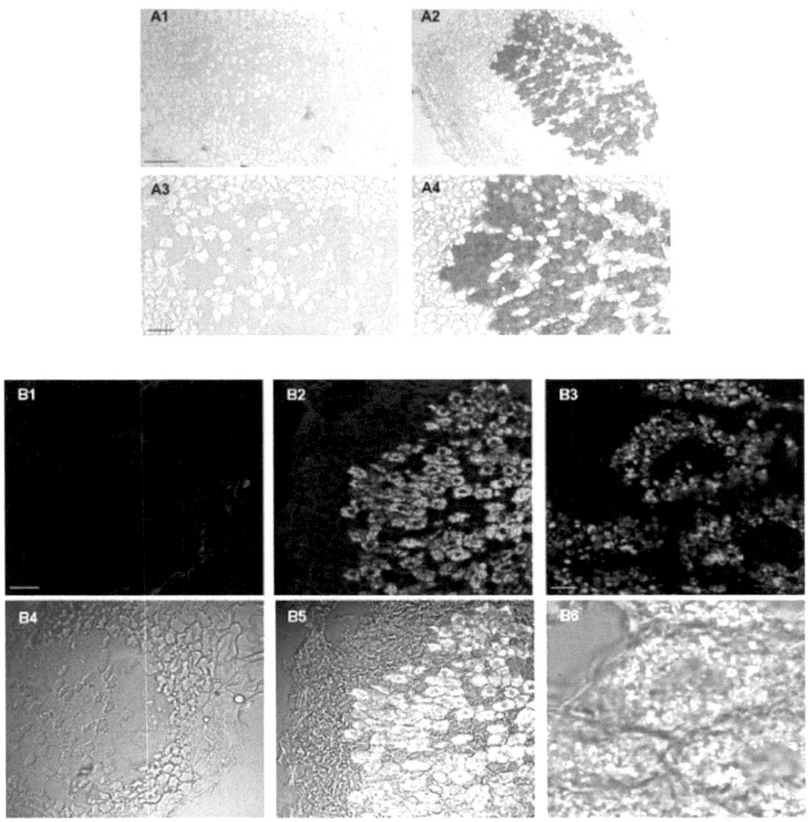

Figure 2. Comparison of results obtained with the same primary antibody (MnSOD) and either AP-detection (A) or fluorescent detection (B). Immunolocalisation of MnSOD in Lotus japonicus *nodules. Images represent longitudinal sections of 13 days post inoculation (dpi) nodules. Alkaline phosphatase-conjugated (A) and alexa488-conjugated (B) anti-rabbit immunoglobulin Gs were used as secondary antibodies. A3 and A4 are magnifications of A1 and A2, respectively. B1, B2, and B3, are the confocal microscope images showing the signal in the infected cells of the central tissue (background subtracted). B4, B5, and B6, are overlays of B1, B2, and B3, and the corresponding transmitted light images. B3 and B6 are a high magnification of the central tissue. No signal was observed in the controls (A1, A3, B1, B4). Bars = 50 μm (A1 and A2), 30 μm (B1, B2, B4, and B5), 20 μm (A3 and A4), and 3 μm (B3 and B6). See colour plate 3 (G) for a colour version of B1-B6.*

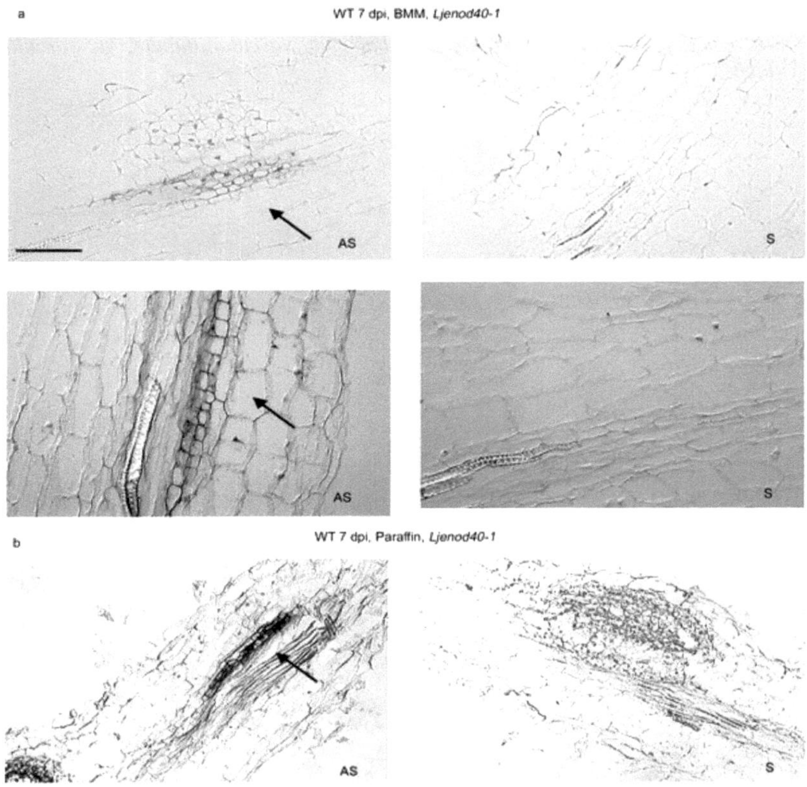

Figure 3. The weak signals obtained for RNA localisation of Ljenod40-1 transcripts are compared to signals obtained on paraffin sections. Ljenod40-1 localises in the pericycle cells at the base of dividing cortical cells. The signal obtained on BMM embedded tissue (a) is much lower, but the cellular resolution/tissue preservation better compared to the paraffin embedded tissue (b). AS, sections probed with antisense Ljenod40-1 transcript. S, sections probed with sense Ljenod40-1 transcript, as negative control. Arrows indicate signal. Detection was performed with alkaline phosphatase-conjugated anti-DIG primary antibody. Bar = 100 μm.

CONCLUDING REMARKS

As seen in Figures 1 and 2, immunolocalisation works well on plastic embedded *L. japonicus* root and nodule tissue, with only minor adjustments of the original protocol. Adjustments mainly concern the speed of polymerisation for different tissue types, as for example the need for slow polymerisation of nodules, to allow also the central tissue to polymerise. Furthermore, one should determine for each protein/antibody pair whether a denaturation step has to be included in the immunolocalisation protocol.

From Figure 3 it is evident that more parameters should still be tested in order to enhance the signals obtained on plastic embedded tissue used for *in situ* RNA:RNA hybridisation studies. However, even though we do not achieve signal strengths as high as those obtainable with paraffin sections, we believe it worth working to optimise the protocol, considering the much higher cellular resolution in sections of plastic embedded root tissue, in particular.

ACKNOWLEDGEMENTS

Adamantia Agalou was supported by a grant from the State Scholarships Foundation of Greece (IKY) and Mette Grønlund by the EU research training network HPRN-CT-2000-00086. Maria C Rubio was supported by a grant from the Spanish Ministry of Education and Culture (EX2002-0425) and by a Marie Curie Individual Fellowship (Contract HPMF-CT-2002-01943), and Andreas Roussis was supported by the EU research training network ERB-FMRX-CT98-0243.

REFERENCES

Baskin TI, Busby CH, Fowke LC, Sammut M, and Gubler F. (1992) **Improvements in immunostaining samples embedded in methacrylate: localisation of microtubules and other antigens throughout developing organs in plants of diverse taxa**. *Planta* 187, 405-413.

Flemetakis E, Agalou A, Kavroulakis N, Dimou M, Martsikovskaya A, Slater A, Spaink HP, Roussis, A. and Katinakis, P. (2002) ***Lotus japonicus* Gene Ljsbp Is Highly Conserved Among Plants and Animals and Encodes a Homologue to the Mammalian Selenium-Binding Proteins**. *Molecular Plant-Microbe Interactions* 15, 313-322.

Jackson D, Veit B, and Hake S. (1994) **Expression of maize KNOTTED1 related homeobox gene genes in shoot apical meristem predicts patterns of morphogenesis in the vegetative shoot**. *Development* 120, 405-413.

Kronenberger, J, Desprez, T, Hofte, H, Caboche, M, and Traas, J. (1993) **A Methacrylate embedding procedure developed for immunolocalization on plant tissues is also compatible with *in situ* hybrization**. *Cell Biology International* 17, 1013-1021.

Langdale, JA. (1994) ***In situ* hybridisation**. In: *The Maize Handbook* (Freeling M and Walbot V, Eds.). New York: Springer-Verlag, pp 165-180.

Lucas, WJ, Bouche-Pillon, S, Jackson, DP, Nguyen, L, Baker, L, Ding, B, and Hake, S. (1994) **Selective trafficking of KNOTTED1 homeodomain protein and its mRNA through plasmodesmata**. *Science* 270, 1980-1983.

Madsen, EB, Madsen, LH, Radutoiu, S, Olbryt, M, Rakwalska, M, Szczyglowski, K, Sato, S, Kaneko, T, Tabata, S, Sandal, N, and Stougaard, J. (2003). **A receptor-like kinase gene involved in legume perception of rhizobial signal molecules**. *Nature* 425, 637-640.

Papadopoulou K, Roussis, A, and Katinakis P. (1996) **Phaseolus ENOD40 is involved in symbiotic and non-symbiotic organogenetic processes: expression during nodule and lateral root development**. *Plant Molecular Biology* 30, 403-417.

Radutoiu, S, Madsen, LH, Madsen, EB, Felle, H, Umehara, Y, Grønlund, M, Sato, S, Nakamura, Y, Tabata, S, Sandal, N, and Stougaard, J. (2003). **Plant recognition of symbiotic bacteria requires two LysM receptor-like kinases**. *Nature* 425, 585-592.

Ruzin, SE. (1999) **Butyl/methyl methacrylate embedding**. In: *Plant Microtechnique and Microscopy*, New York Oxford: Oxford University Press, p.4.

Scarpella E, Rueb S, Boot KJM, Hoge JHC, and Meijer AH. (2000) **A role for the rice**

homeobox Oshox1 in provascular cell fate commitment. *Development* 127, 3655-3669.

Schauser, L, Roussis, A, Stiller, J, and Stougaard, J. (1999). **A plant regulator controlling development of symbiotic root nodules.** *Nature* 402, 191-195.

Stracke, S, Kistner, C, Yoshida, S, Mulder, L, Sato, S, Kaneko, T, Tabata, S, Sandal, N, Stougaard, J, Szczyglowski, K, and Parniske, M. (2002). **A plant receptor-like kinase required for both bacterial and fungal symbiosis.** *Nature* 417, 959-962.

Chapter 3.3

RNA ISOLATION USING PHASE EXTRACTION AND LiCl PRECIPITATION

Catherine Kistner and Manuel Matamoros*

*The Sainsbury Laboratory, Colney Lane, Norwich NR4 7UH, UK; *Corresponding author.*

Email: manumat@eead.csic.es

Phone: +34 976 716 062 Fax: +34 976 716 145

The extraction of good quality RNA from plant tissues is a critical step in gene expression studies. We have developed a protocol for RNA isolation from Lotus japonicus *roots and leaves that yields RNA that can be successfully used in downstream applications such as reverse transcription-polymerase chain reaction (RT-PCR). The use of an adequate extraction buffer is crucial to obtaining good quality RNA, especially when working with roots. The composition of the extraction buffer has been optimized to avoid contamination with phenolic compounds, which are very abundant in Lotus roots and can inhibit downstream enzymatic reaction such as reverse transcription. In our laboratory, this method yields highly purified RNA suitable for cDNA synthesis and is routinely used for gene expression studies in* L. japonicus *roots. For Northern blots, dirtier (i.e. phenolic contamination), higher yielding extraction methods might be suitable as well.*

PROCEDURE

1. Warm 10 mL of extraction buffer to 65°C in a water bath (do not forget to add β-mercaptoethanol before use).
2. Grind plant tissue in liquid nitrogen and transfer the frozen powder to an eppendorf tube (the volume of powder should not exceed 200 μL).
3. Add 600 μL of extraction buffer and 600 μL of phenol: chloroform: isoamylalcohol (25:24:1 pH 8.0). Mix by inverting the tube and shaking for 5-10 min at 55°C, then separate phases at room temperature by centrifugation 10 min (>10.000 rpm).
4. Collect upper phase and extract again with phenol: chloroform: isoamylalcohol.
5. Take upper phase and add LiCl to a final concentration of 2M. The RNA is precipitated overnight on ice and harvested by centrifugation at max speed 10 min at 4°C.

A.J. Márquez (ed.), Lotus japonicus Handbook, 123-124.

6. Discard the supernatant, wash the pellet in cold 2M LiCl, spin, and wash again twice in cold 80% ethanol.
7. Dry the pellet (but not over dry) and resuspend it in RNA-Secure-Resuspension Solution (Ambion).
8. Measure the RNA concentration. Ratio A_{260}/ A_{280} should be ~2. Optionally check RNA integrity on an agarose formaldehyde gel.
9. If necessary for downstream application perform DNase I treatment 30 min at 37°C (20 U of DNase I per 2 μg of RNA).
10. Inactivate DNaseI by adding 0.2 volumes of DNase Inactivation Reagent (Ambion) for 2 min at room temperature. Centrifuge at 13,200 rpm 1 min and use the supernatant for first strand cDNA synthesis.

EXTRACTION BUFFER

- 2% CTAB (hexadecyltrimethylammonium bromide)
- 2% PVP (polyvinylpyrrolidone 360.000)
- 100 mM Tris-HCl pH 8.0
- 25 mM EDTA
- 2 M NaCl

(Stir with moderate heating, autoclave)

- 2% β-mercaptoethanol (add just before use)

Chapter 3.4

RNA ISOLATION USING CSCL GRADIENTS

Lene Krusell, Thomas Ott, and Michael Udvardi*

*Max-Planck-Institute of Molecular Plant Physiology, Plant Nutrition Group, Am Mühlenberg 1, 14476 Golm, GERMANY; *Corresponding author.*

Email: udvardi@mpimp-golm.mpg.de

Phone: +49 331 567 8149 Fax: +49 331 567 8134

Through the years, different methods have been developed to extract high quality RNA from Lotus japonicus *that can be used for downstream applications such as northern blot analysis, RT-PCR, and cDNA probe synthesis for microarray analysis. Several companies have developed fast and simple RNA extraction kits (Qiagene, Nucleobond, etc.) suitable for extraction from small amounts of tissue (from 100-200 mg). However, RNA prepared with kits is commonly contaminated with genomic DNA, proteins, and polysaccharides. Here we describe a RNA extraction method that yields high quality RNA from different tissue types of* Lotus japonicus *(Sambrook* et al. *1989).*

PROCEDURE FOR LYSIS AND ULTRACENTRIFUGATION

The method involves two steps. In the first step, cellular lysis is promoted by the presence of the ionic detergent Sarcosyl in the extraction buffer. Sarcosyl together with the chelating agent EDTA and the strong reductant, β-mercaptoethanol inactivates ribonucleases (RNases) by disrupting protein tertiary folding. During the first step, RNA is liberated together with other cellular components such as genomic DNA and mitochondrial DNA.

The second step includes an isopycnic ultracentrifugation through a caesium chloride (CsCl) gradient that separates RNA from DNA and proteins based on differences in their buoyant densities.

Solutions for lysis and ultracentrifugation

For all solutions, use Diethyl pyrocarbonate (DEPC)-treated H_2O or, in the case of EDTA and NaCl, treat the entire solution with DEPC (Sambrook et al. 1989). Always use freshly made extraction buffer.

Extraction buffer

- 100 mM Tris-HCl, pH 7.5

A.J. Márquez (ed.), Lotus japonicus Handbook, 125-128.

- 50 mM EDTA, pH 8.0
- 500 mM NaCl
- 5% β-Mercaptoethanol

CsCl Cushion

- 96% CsCl
- 100 mM EDTA, pH 8.0

Sarcosyl (N-Lauryl-sarcosine) solution

- 14% Sarcosyl

NaCl solution

- 1M NaCl

Lysis and ultracentrifugation procedure

The following method is based on 0.5 g of starting material but it can be applied to larger tissue samples by increasing buffer volumes and the size of the ultracentrifuge tube.

1. Homogenize 0.5 g of tissue in liquid nitrogen and transfer the resulting powder to a 15 ml centrifuge tube.
2. Immediately add 3.33 ml of extraction buffer and 1 ml of 14% Sarcocyl, vortex briefly.
3. Add 0.65 g of CsCl powder, mix well, and incubate at 65°C for 15 min.
4. Centrifuge at 4°C for 20 min at 9500 rpm (Kontron A8.24 rotor).
5. Filter the supernatant through Miracloth (Calbiochem) and store on ice while preparing the CsCl cushion.
6. Add 1.6 ml of CsCl to a 5 ml ultracentrifuge tube.
7. Apply 3.4 ml of the filtered supernatant carefully to the top of the cushion
8. Centrifuge at 20°C for 20 hours at 40,000 rpm
9. The centrifugation results in two phases. Genomic DNA is located at the interface whereas RNA is located at the bottom of the tube below the cushion phase. Remove 1/3 of the upper phase and carefully add a similar volume of DEPC H_2O to the tube (referred to subsequently as a wash step).
10. Repeat this washing step 2 times removing more of the upper phase each time without disturbing the cushion. Make sure that the genomic DNA at the interface is completely removed to limit contamination of RNA.
11. Remove ½ of the cushion and wash with DEPC H_2O

12. Remove the second half of the cushion, re-suspend the RNA in 133 µl DEPC H_2O and transfer to a 1.5 ml eppendorf tube
13. Add 2.5 vol absolute ethanol (EtOH), mix, and incubate at -20°C for a minimum of 2 hours.
14. Centrifuge at 4°C for 30 min at 10.000 rpm in a microcentrifuge.
15. Wash the RNA pellet in 0.5 ml 70% EtOH. Repeat the centrifugation and re-suspend the pellet in 133 µl DEPC H_2O
16. Precipitate over-night by adding 1/10 vol of 1 M NaCl and 2.5 vol of absolute EtOH.
17. Centrifuge at 4°C for 30 min at 10.000 rpm, wash with 0.5 ml 70 % EtOH, air-dry at room temperature and re-suspend the pellet in 30-50 µl of DEPC H_2O
18. Store the RNA at -80°C

The concentration and quality of the RNA is determined by measuring the absorbance at both 260 nm and 280 nm. The purity of RNA is estimated by the ratio of absorbance at 260 to 280 nm with RNA absorbing at 260 nm and proteins absorbing at 280 nm. A ratio of 1.8-2.0 indicates high purity.

DENATURING GEL-ELECTROPHORESIS OF RNA

For downstream applications such as northern blot or simply for checking the quality of the RNA (besides the spectrophotometric measurement) analysis of the RNA on a denaturing gel is appropriate. Intact RNA is usually observed as two distinct bands on the gel representing the 28S and 18S ribosomal RNA. Smearing of bands indicates degradation of the RNA. Different electrophoresis methods have been developed, but in this section, we describe the use of a formaldehyde procedure.

Solutions for electrophoresis

RNA-gel

- 1.5% agarose
- 1x gel-running buffer
- 17% formaldehyde

10x gel-running buffer

- 0.2 M MOPS
- 50 mM Sodium Acetate
- 10 mM EDTA
- Adjust pH to 7.0 using NaOH. Do not autoclave the running buffer but use sterile DEPC treated H_2O for preparing the buffer.

RNA-loading buffer

- 20% Ficoll 400
- 0.1225% Bromophenol blue
- 0.1225% Xylene cyanol FF
- 4% Ethidium bromide

Dissolve in TE-buffer made with DEPC treated H_2O.

Electrophoresis procedure

The agarose is melted in H_2O, whereupon the running buffer and formaldehyde is added and the gel solidified. Due to the toxicity of the formaldehyde, all procedures after addition of formaldehyde are performed in a chemical hood.

The RNA samples are prepared by mixing the following:

- 10-20μg RNA (max volume 40 μl)
- 12 μl Formaldehyde
- 40 μl Formamide
- 8 μl 10x running buffer

The RNA samples are incubated at 56°C for 10 min and 10 μl of RNA loading buffer added before the samples are loaded onto the gel submerged in 1 x running buffer. The gel is run at 3-4 V/cm for 3-4 hours with circulation of the buffer.

The RNA is evaluated by UV illumination of the gel.

REFERENCES

Sambrook J, Fritsch EF, and Maniatis T. (1989). **Molecular cloning: A Laboratory Manual**, Ed.2. Cold Spring Harbour Laboratory, Cold Spring Harbour, NY.

Chapter 3.5

96-WELL DNA ISOLATION METHOD

Jillian Perry[1*] and Martin Parniske[2]

[1]*Sainsbury Laboratory, John Innes Centre, Norwich Research Park, Colney, Norwich, NR4 7UH, UNITED KINGDOM; *Corresponding author.* [2]*Institut für Genetik, LMU Münche,n Maria Ward-Str. 1a 80638 München, GERMANY.*

Email: jillian.perry@bbsrc.ac.uk

Phone: +44 1603 450 000   Fax: +44 1603 450 045

Keywords: High throughput, DNA isolation, High yield, DNA TILLING, Map based cloning, Leaf material

The following protocol provides a high throughput, low cost method of producing a superior DNA yield of high quality which is suitable for TILLING, map based cloning or any application which requires long term DNA storage. The protocol has been designed for DNA extraction from leaf material, preferably young leaf tissue should be used as this minimises samples being contaminated with polysaccharides and phenolics.

METHOD

1. Pre-heat extraction buffer to 65°C.
2. Label collection tubes (Qiagen Cat. No. 19560) and add a single tungsten carbide bead (Qiagen Cat no. 69997) to each tube.
3. Prepare ice bucket and tube of ethanol to wash forceps after each harvest.
4. Harvest material (3 growing tips) into a single collection tube.
5. Add 400µl of extraction buffer to each tube (use a multi-pipettor). Put on lids (Qiagen Cat. No. 19566).
6. Homogenise material on the mixer mill (Retsch MM300) for 2 min/30s.
7. Incubate at 65°C for 30min to 1 hour.
8. Centrifuge for 10 minutes at full speed (Sigma 4K15).
9. Label a new rack of collection tubes.
10. Remove 300µl of supernatant into a new collection tube (use a multi-pipettor set speed to slow) with extended length tips (Matrix Cat. No. 8252).

A.J. Márquez (ed.), Lotus japonicus Handbook, 129-131.

11. Carefully add 200µl of phenol:chloroform to each tube (THIS PROCEEDURE SHOULD BE CARRIED OUT IN THE FUME HOOD), use 200µl manual multi-pipettor with filter tips.
12. Put lids onto the tubes and invert several times, so the samples are well mixed. Centrifuge for 10-15 minutes.
13. Label a set of storage plates.
14. Using a manual multi-pipettor with filter tips very carefully remove 200µl of the upper layer to a new storage plate (AB Gene Cat. No. AB 0765) (THIS PROCEEDURE SHOULD BE CARRIED OUT IN THE FUME HOOD).
15. Using a multi-pipettor add 1/10th vol. of 3M sodium acetate (~20µl) and an equal volume of isopropanol (~220µl). Put on lids, mix well, and leave at -20°C for a maximum of 1 hour.
16. Leave plates on desk until the have reached room temperature, as spinning when still frozen at high speeds can cause the plate to crack. Centrifuge at 5600 rpm for 45 minutes.
17. Remove supernatant and add 100µl of TE containing RNAse A at a final concentration of 10µg/ml and incubate 30 min at 37°C
18. Repeat the precipitation step (step 15). After centrifugation at 5600 rpm for 45 minutes remove the supernatant and add 200µl 70% ethanol. Put on lids and leave for 15 minutes or overnight.
19. Remove ethanol and leave to air dry. Add 100µl of TE and store in fridge/freezer.

BUFFERS AND SOLUTIONS

Extraction Buffer

- 200 mM Tris-HCl pH 7.5
- 250 mM NaCl
- 25 mM EDTA
- 0.5% SDS

TE

- 10 mM Tris-HCl pH 8.0
- 0.1 mM EDTA pH 8.0

TE and RNAse A

- For 2 plates: 20 ml TE and 20µl DNAse-free RNase A (10 mg/ml)

References

Perry, JA, Wang, TL, Welham, TJ, Gardner, S, Pike, JM, Yoshida, S, and Parniske, M. (2003) **A TILLING reverse genetics tool and a web-accessible collection of mutants of the legume *Lotus japonicus***. *Plant Physiology* 131, 866-871.

SECTION 4

FUNCTIONAL GENOMICS

Chapter 4.1

GENOME STRUCTURE ANALYSIS

Satoshi Tabata

Kazusa DNA Research Institute, Department of Plant Gene Research, 2-6-7 Kazusa-kamatari, Kisarazu, Chiba 292-0818, JAPAN

Email: tabata@kazusa.or.jp

Phone: +81 438 52 3933 Fax: +81 438 52 3934

Keywords: *Agrobacterium rhizogenes*, *Rhizobium*, nodulation

A large-scale structural analysis of the genome of Lotus japonicus *is in progress. This project involves the interfacing of three approaches: EST (Expressed sequence tags) analysis, genome sequencing, and generation of DNA markers. Each approach is described here.*

EST ANALYSIS

EST analysis is performed to make a catalogue of expressed portions of the genome by developing anonymous partial cDNA sequences. So far, 110,000 Lotus ESTs have been deposited in the public EST database. These include ESTs from roots, nodule primordial, immature nodules, mature nodules, flower buds, pods, and whole plants (Asamizu et al. 2000, Endo et al. 2000, Asamizu et al. 2004). The sequence information and search results are available at the web site at http://www.kazusa.or.jp/en/plant/lotus/EST/. These ESTs as well as the corresponding cDNA clones are serving as valuable information and material resources for the functional analysis of individual genes and of the genome. The EST clones corresponding to ESTs can be requested at www.kazusa.or.jp/clonereq/.

GENOME SEQUENCING

Genome sequences, which complement the ESTs, provide a whole picture of the genetic information carried by living organisms. To understand the genetic systems carried by legume plants, large scale genome sequencing of *L. japonicus* accession Miyakojima MG-20 is in progress.

Genomic libraries of *L. japonicus* were constructed as a source for sequencing using a transformation competent artificial chromosome (TAC) and a standard BAC

A.J. Márquez (ed.), Lotus japonicus Handbook, 135-138.

pBeloBAC11 as vectors. Genome sequencing is performed from the multiple seed points of the genome, which are selected based on the EST information (Sato et al 2001). TAC clones are selected by screening the three-dimensional DNA pools of the TAC genomic libraries by PCR with oligonucleotide primers designed based on nucleotide sequences of the public ESTs and cDNAs from *L. japonicus*. The nucleotide sequence of each TAC insert is determined according to the shotgun method. Assignment of the protein-encoding regions and gene annotation were performed by combination of similarity searches and computer predictions. As of February 28, 2004, a total of 1,568 seed clones have been selected, and 320 clones covering 32,537,698 bp genomic regions have been sequenced, annotated, and released to public databases (Kato et al 2003).

The structural features of the 3,136 potential protein-encoding genes in *L. japonicus* were compared to those of *Arabidopsis thaliana* (Table 1) (Kato et al 2003). The genes of two plants share several common features: average length of the coding exons (296 bp vs 256 bp) and the average number of introns per gene (3.6 vs 4.0). However, the average length of genes including introns, (2,706 bp), was longer in Lotus than in Arabidopsis (1,918 bp), which is due to a longer average intron length in Lotus (378 bp vs 157 bp). The average gene density in Lotus was one gene in every 10.4 kb, half the density of Arabidopsis (4.5 kb).

	Lotus	**Arabidopsis**
Number of assigned genes	3,136[a]	6,451[b]
Gene length (bp) including introns	165-21,418 (2,706)	78-17,203 (1,918)
Product length (amino acids)	16-2,185 (450)	25-4,706 (427)
Genes with introns	2283 (73%)	4,906 (76%)
Number of introns/gene	0-37 (3.6)	0-48 (4.0)
Exon length (bp)	2-5,604 (296)	2-5,966 (256)
Intron length (bp)	30-8,862 (378)	23-2,989 (157)
GC content of exons	45%	44%
GC content of introns	33%	32%

Table 1. Structural Features of potential protein-encoding genes in L. japonicus *and* A. thaliana. *The 3,136 genes assigned in* L. japonicus[a] *and the 6,451 genes previously assigned in our* A. thaliana *genome sequencing project*[b] *were compared. Average values are shown in parentheses.*

LINKAGE MAPPING

The sequenced TAC clones are mapped onto a genetic linkage map by sequence repeat length polymorphism (SSLP) markers generated utilizing the sequence information of each clone. In the Lotus genome, Simple sequence repeats such as (AT)n, (GT)n, (AAT)n of equal to or longer than 15 bp occurred every 12 kb on average (Sato et al 2001). Genetic mapping of the generated DNA markers was performed using the 127 F2 mapping population derived from a cross between two Lotus accessions, Miyakojima MG-20 and Gifu B-129 (Figure 1). At the time of

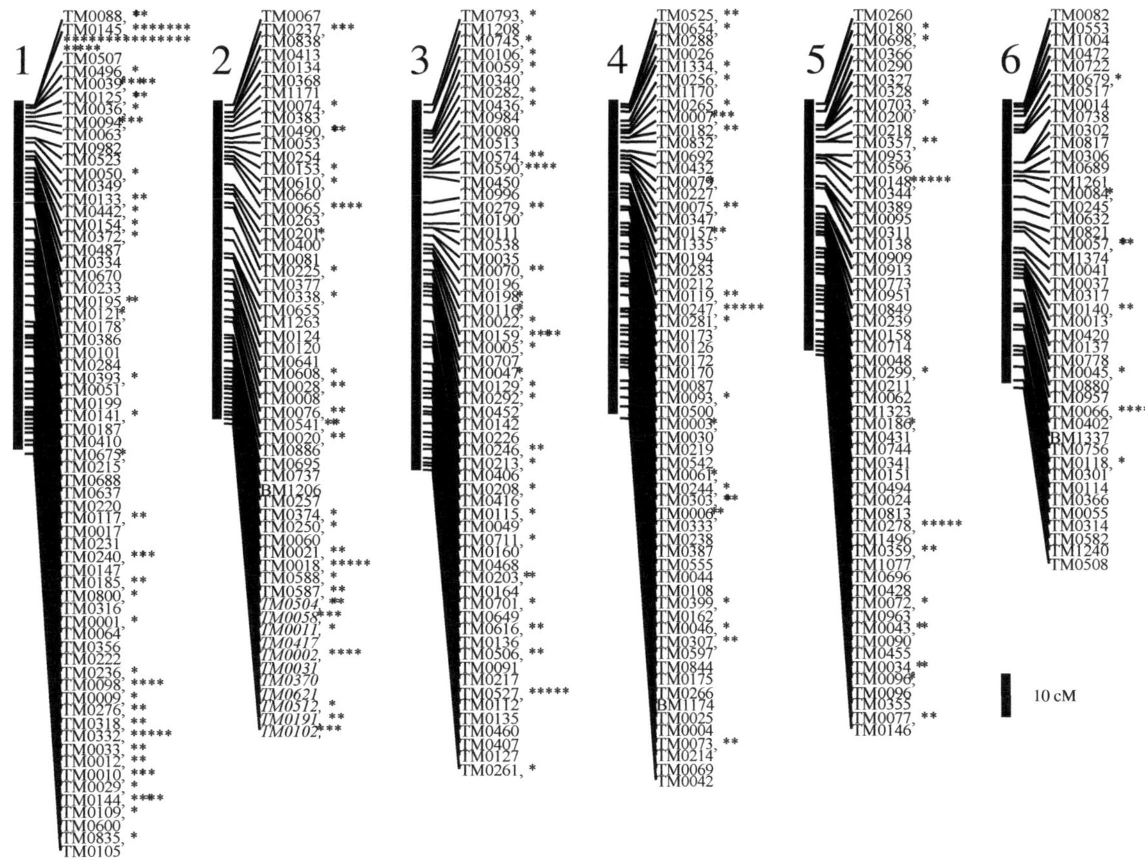

Figure 1. *Relative positions of the SSR and dCAPS markers on the genetic linkage map. Asterisks indicate redundant markers.*

writing, a total of 714 DNA markers including SSLP and some dCAPS (derived cleaved amplified polymorphic sequence) markers have been mapped onto the linkage map. These PCR-based markers, almost all of which are co-dominant markers, with the surrounding DNA sequences will accelerate the process map-based cloning in Lotus. The marker information is available through our web database at www.kazusa.or.jp/lotus/.

REFERENCES

Asamizu E, Nakamura Y, Sato S, and Tabata S. (2000) **Generation of 7137 non-redundant expressed sequence tags from a legume, *Lotus japonicus***. *DNA Research* 7, 127-130.

Asamizu E, Nakamura Y, Sato S, and Tabata S. (2004) **Characteristics of *Lotus japonicus* Gene Repertoire Deduced from Large-scale Expressed Sequence Tags (EST) Analysis**. *Plant Molecular Biology* 54, 405-414.

Endo M, Kokubun T, Takahata Y, Higashitani A, Tabata S, and Watanabe M. (2000) **Analysis of Expressed Sequence Tags of Flower Buds in *Lotus japonicus***. *DNA Research* 7, 213-216.

Sato S, Kaneko T, Nakamura Y, Asamizu E, Kato T, and Tabata S. (2001) **Structural analysis of a *Lotus japonicus* genome. I: Sequence features and mapping of fifty-six TAC clones which cover the 5.4 Mb regions of the genome**. *DNA Research* 8, 311-318.

Kato T, Sato S, Nakamura Y, Kaneko T, Asamizu E, and Tabata S. (2003) **Structural analysis of a *Lotus japonicus* genome. V: Sequence features and mapping of sixty-four TAC clones which cover the 6.4 Mb regions of the genome**. *DNA Research* 10, 277-285.

CHAPTER 4.2

TRANSCRIPTOME ANALYSIS USING cDNA ARRAYS

Thomas Ott[1], Lene Krusell[1], Susanne Freund, and Michael Udvardi*

*Max-Planck-Institute of Molecular Plant Physiology, Plant Nutrition Group, Am Mühlenberg 1, 14476 Golm, GERMANY; [1]These authors contributed equally to this publication; *Corresponding author*

Email: udvardi@mpimp-golm.mpg.de

Phone: +49 331 567 8149 Fax: +49 331 567 8134

Keywords: cDNA array, transcriptome, expression profiling

DNA array-based technologies make it possible to measure transcript abundance for thousands of genes in parallel, which can provide a global view of the transcriptional output, or transcriptome of an organism. Isolation and sequencing of thousands of Lotus japonicus *cDNA clones, to produce so-called Expressed Sequence Tags (ESTs), facilitated the production of cDNA arrays, which enable us, and others to study transcriptome dynamics in this species. We have produced two different cDNA arrays on nylon filters, which are available to the scientific community: The first array contains 11,600 cDNA amplicons each representing a different gene, kindly provided by Kazusa DNA Research Institute in Japan (11K LjKDRI array I); and a second array contains 9,600 cDNA amplicons, representing approximately 4,000 different nodule-expressed genes (4K nodule array). In this chapter, we describe the production and use of these filters for transcriptome analysis. Methods for analysing the data obtained from these arrays are described in the* LOTUS JAPONICUS *EXPRESSION DATABASE chapter.*

INTRODUCTION

Until a few years ago, gene expression analysis is confined to one or a few genes at a time, and it is virtually impossible to identify entire sets of genes involved in a common process that are co-ordinately regulated. Large-scale DNA sequencing projects, including genome and EST sequencing, together with development of technologies for high-throughput measurement of gene transcript levels have profoundly changed the way in which molecular biologists approach biological questions. Now it is possible to survey the expression level of thousands of genes in parallel, which not only enables identification of novel 'candidate' genes involved in

A.J. Márquez (ed.), Lotus japonicus Handbook, 139-148.

a given biological phenomenon, but also reveals sets of genes and the corresponding biochemical processes that act in concert to achieve a specific biological outcome.

Recently developed technologies for multi-parallel analysis of transcript levels include: massively parallel signature sequencing (MPSS); cDNA-AFLP; and DNA array-based methods (Aharoni and Vorst, 2001). DNA that is spotted onto arrays ranges in size from small oligonucleotides of 25-70 bp to longer PCR-amplicons of up to 1kb or more that are derived from cDNA or genomic DNA templates. Oligonucleotides can provide higher specificity in transcript detection, if they are designed to poorly conserved regions of homologues genes. However, the cost of thousands of oligonucleotides can be prohibitive, especially for individual research laboratories. PCR-amplicons derived from cloned cDNA are a less-expensive alternative to oligonucleotides. DNA is generally spotted robotically onto either glass slides or nylon membranes to produce high or low-density arrays, referred to as micro- or macro-arrays, respectively. Glass slide micro-arrays can accommodate tens to hundreds of thousands of oligos or PCR-amplicons, while nylon macro-arrays are generally limited to 10-20,000 DNA spots (Richmond *et al*., 1999).

Conceptually, DNA arrays are used in a reverse-northern manner to quantify transcript abundance, where DNA (not RNA) is immobilised onto a solid support and a probed with a complex pool of either radioactively or fluorescently labelled cDNA derived from RNA. Following probe hybridisation and washing, slides or filters are read by fluorescence or radioactivity detectors, which record a digital signal for each immobilised DNA spot that is later subjected to numerical and statistical analysis. In the following sections, we provide details of the production and use of nylon macro-arrays containing cDNA amplicons for Lotus transcriptome analysis. Bioinformatics software for management and analysis of transcriptome data are described in the *LOTUS JAPONICUS* EXPRESSION DATABASE chapter.

The non-redundant 11k LjKDRI-array contains PCR fragments from amplified EST clones that were obtained from different sources (mature nodules (Gifu), *Ljhar1-3* roots rich in nodule primordia, uninfected roots (MG-20), pods (MG-20)), whereas the templates for the redundant 10k LjNEST array are only nodule derived.

MATERIALS AND METHODS

Buffers and Solutions

10 x PCR buffer (final pH 8.4)

- 0.5 M KOAc
- 1% Tween 20
- 15 mM MgOAc
- 0.35 M TrisBase, pH 10.4
- 0.15 M TrisHCl, pH 4.8

SSarc buffer (total volume 1 litre)

- 240 ml Sarcosyl NL30
- 200 ml 20×SSC
- 8 ml 0.5M EDTA, pH 8.0
- 552 ml H_2O

Diluted SSarc buffer (total volume 1 litre)

- 100 ml SSarc
- 2 ml 0.5 M EDTA, pH 8.0
- 898 ml H_2O

20×SSC (adjust pH to 7.0 with 10 N NaOH)

- 3.0 M NaCl
- 0.3 M Na-citrate

Formamide hybridisation buffer

- 40 mM NaPP buffer (pH 6.8)
- 5 x Denhardt Solution
- 1% SDS
- 0.5% BSA
- 10μg/ml Herring Sperm (denatured)
- 50% Formamide
- 5 x SSC

Alkaline buffer for removing radioactivity

- 0.4 M NaOH
- 0.6 M NaCl

Neutralisation buffer

- 0.5 M Tris-HCl pH 7.5
- 1 x SSC
- 0.1 % SDS

Array construction

Robotic-spotting of PCR-amplified cDNA clones onto nylon membranes as detailed below generates lotus cDNA arrays.

PCR amplification of cDNA clones

Lotus cDNA clones contained in plasmid vectors are PCR-amplified either from *E. coli* colonies (11K LjKDRI clones) or from purified plasmids (4K nodule clones). For colony PCR, sets of 384 clones in *E. coli* are grown overnight on LB-plates with appropriate antibiotic selection, and then inoculated into PCR reactions in 96-well plates. For PCR-amplification from plasmids, purified DNA is diluted in H_2O to approx. 50 ng/µl, one microliter of which is used as template in PCR reactions. For pSPORT and pBlueScript vectors, the following *LacZ* gene primers, which flanked each cDNA insert, are used:

- LacZ1: GCTTCCGGCTCGTATGTTGTGTG
- LacZ2: AAAGGGGGATGTGCTGCAAGGCG

PCR is performed in 96 well plates in a reaction volume of 100 µl 1x PCR buffer (see above) containing 0.2 mM of each dNTP, 10 pmol Primer 1, 10 pmol Primer 2, 1.5 mM MgOAc, 1 unit Taq Polymerase, H_2O to 100 µl). The amplification program is:

- 95°C for 5 min
- 35 cycles of 95°C for 1 min
- 65°C for 1 min
- 72°C for 1.5 min
- One cycle at 72°C for 10 min

Four sets of 96 PCR reactions are combined into one 384-well plate following the pattern indicated in Figure1, using a HYDRA96 Robbins Scientific Robotic system.

Array spotting

The cDNA amplicons are spotted onto Hybond N^+ nylon membranes using a BioGrid robotic system with a 384er griding tool (BioRobotics, England). The spotting pins have a radius of 0.4 mm and each PCR reaction is transferred 10 times per spot, resulting in a total spotted volume of approximately 0.25 µl. Filters are saturated with 0.5 M NaOH denaturation buffer during spotting, then incubated for at least 5 minutes in a 1M Tris (pH 7,5)/ 1,5M NaCl neutralisation buffer. Finally, DNA is covalently linked to the membrane using a UV-cross linker (UV Stratalinker 1800, Stratagene) at 120.000 µJoules.

Forty-two 384-well plates (16,128 clones) can be spotted onto a 20x20 cm filter using the above robotic system. Each filter is divided into six tertiary grids, which in turn contain 384 secondary grids. Secondary grids consist of a 4x4 pattern representing seven clones spotted in duplicates, an empty spot for quantifying the local background (L), and a human desmin clone (D) used as a control for unspecific binding (Figure 1).

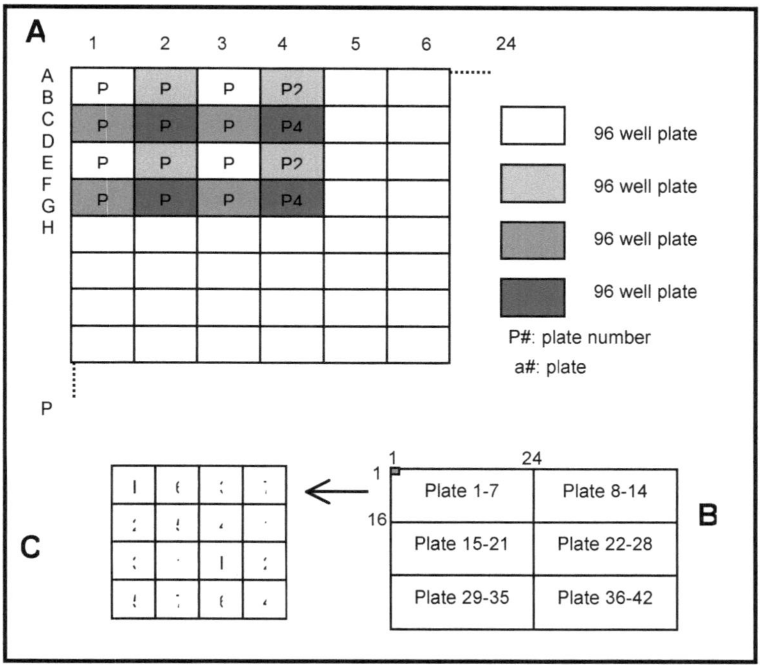

Figure 1. (A) Schematic distribution of four 96 well PCR plates after their content has been transferred to one 384 well plate. (B) Spotting scheme of the arrays used in our lab. Seven 384 well plates are spotted per subarray. (C) All samples are spotted in doubles within a 4x4 spot group.

Probe synthesis and array hybridisation

I. Reference hybridisation

Due to slight variations in PCR amplification and spotting efficiency it is necessary to estimate the amount of DNA applied in each spot. This is achieved by hybridising arrays with a ^{33}P-labelled 13-mer oligonucleotide (5'-TTCCCAGTCACGA-3') complementary to the vector part of each amplified cDNA. The signal obtained for each spot after washing of arrays provides a measure of the amount of DNA in each spot. Subsequently, each filter is used multiple times (up to 6 times) to measure transcript levels in different biological samples, in so-called complex hybridisations, as described in detail below.

Pre-hybridisation

Filters are incubated > 1 hour in 200 ml (1-2 filters) SSarc (see above) at 5°C.

Probe labelling and hybridisation (for 1-2 filters)

Reactions contain the following components:

- 60 pmol 13-mer oligonucleotide
- 3 μl 10 x T4 buffer (Roche)
- 50 μCi γ^{33}P-ATP
- 20 units T4 polynucleotide Kinase
- H_2O to 30 μl

Labelling reactions are carried out at 37°C for 30 min, and then mixed with 10 ml SSarc before being added to prehybridised filters in 200 ml SSarc, and incubated at 5°C overnight.

Filter Washing

Filters are washed at 5°C for 30 min in 200 ml (1-2 filters) SSarc buffer.

Data capture

Each filter is exposed to a BioImager screen over-night and scanned the following day in a BAS-1800 II BioImager (Fujifilm) with a resolution of 50μM×50μM and 16 bits per pixel. Evaluation of data is described below.

Removal of radioactivity

Bound radioactive probe is removed from filters by incubating them twice in an alkaline buffer at 42°C for 30 min each time. Filters are then neutralised for 30 min at room temperature in neutralisation buffer (see above). Exposing filters overnight to a BioImager screen confirms complete removal of radioactivity.

II. Complex hybridisation

Total RNA is used for the preparation of complex radioactive probes for array hybridisation. The radioactive signal associated with each spot on the array is used to calculate relative transcript abundance, after normalisation for DNA amount in each spot.

Pre-hybridisation

Filters are pre-hybridised in 25 ml of formamide-based hybridisation buffer (see above) for 4 hours at 42°C.

Labelling of cDNA

Radioactive cDNA probes are prepared from total RNA by reverse transcription, as follows. A solution of 10-25 μg of total RNA and 1.5 μg of oligo$(dT)_{12\text{-}18}$ primer in 12.5 μl DEPC-treated H_2O is heated at 70°C for 10 min to denature the RNA, then cooled on ice for 5 min. Subsequently, 16 μl of cold probe labelling solution (see below) is added and the sample incubated for a further 10 min at 4°C. The sample is then incubated at 42°C for 5 min before adding 1 μl of SuperScript III reverse transcriptase (200 units/μl; Invitrogen). The ensuing reaction is allowed to proceed for 1 hour at 42°C, before addition of 0.5 μl of 25 mM unlabelled dCTP, and further incubation at 42°C for 30 min. Reverse transcription is terminated by inactivation of the enzyme by heating at 70°C for 15 min.

Probe labelling solution (total volume 16 μl):

- 6 μl 5 x RT buffer (Invitrogen)
- 2.5 μl 0.1 M DTT (final conc. 8.3 mM)
- 0.5 μl 25 mM mix of dGTP, dATP, dTTP (final conc. 0.42 mM)
- 1 μl Rnasine inhibitor (40 units/μl)
- 6 μl α^{33}P-dCTP (10 μCi/μl)

RNA template is finally hydrolysed by adding of 10 μl of 1N NaOH and incubating the resulting solution at room temperature for 10 min, followed by 15 min at 50°C. The probe solution is then neutralised by adding 10 μl 1N HCl and 2 μl 1M NaPP (sodium phosphate) buffer (pH 7.2) and incubated at room temperature for 10 min.

Unincorporated radioactive nucleotides are removed by cleaning the probe over a silica membrane (nucleotide removal kit, QIAGEN) and eluted in 200μl elution buffer, following the protocol provided with the kit.

Labelling efficiency is determined by measuring the ^{33}P radioactivity of 1 μl probe in a LS6500 Scintillation counter (Beckman Coulter) scintillation counter.

Hybridisation

Between 5-15 million cpm of labelled cDNA is added to arrays in pre-hybridisation solution, which are then incubated at 42°C for approximately 24 hours. Addition of more than 15 million cpm occasionally led to problems in removal of radioactivity from the filters.

Washing

Non-specifically bound probe is removed by washing filters using the following procedure. Filters are sequentially washed for:

- 20 min in 100 ml 1 x SSC, 0.1% SDS, 4 mM NaPP pH 7.2 at 65°C
- 20 min in 100 ml 0.2 x SSC, 0.1% SDS, 4 mM NaPP pH 7.2 at 65°C
- 20 min in 100 ml 0.1 x SSC, 0.1% SDS, 4 mM NaPP pH 7.2 at 65°C

Data capture

Each filter is exposed to a BioImager screen over-night and scanned the following day in a BAS-1800 II phosphor imager (Fuji) with a resolution of 50μM×50μM per pixel (16 bits). Evaluation of data is described below.

Removal of radioactivity

Removing radioactivity form the filters is carried out as described above during the reference hybridisation.

Array image analysis and data evaluation

After capturing a digital image of the radioactivity associated with each spot on an array, images must be analysed with software that can identify each spot and measures its intensity.

Different programs are available for this purpose (Richmond and Somerville, 2000). We use AIS (InterFocus), a commercially available program.

As mentioned above, filters are scanned with a pixel size of 50×50μM and a digital resolution of 16 bits per pixel. The first step of image analysis is griding of the filter, a process in which each individual spot or empty position on the filter is overlayed with a virtual grid to locate the spot prior to measuring its the signal intensity. It is recommended that the user define different hierarchical levels in order to accommodate different distances between spots of different groups. In our case, we defined three different levels for each array. Level 1 represents the distance between individual spots within one spot group (4×4 spots), level 2 describes the distance between the various spot groups, and level 3 gives the distance between sub-arrays (Figure 2).

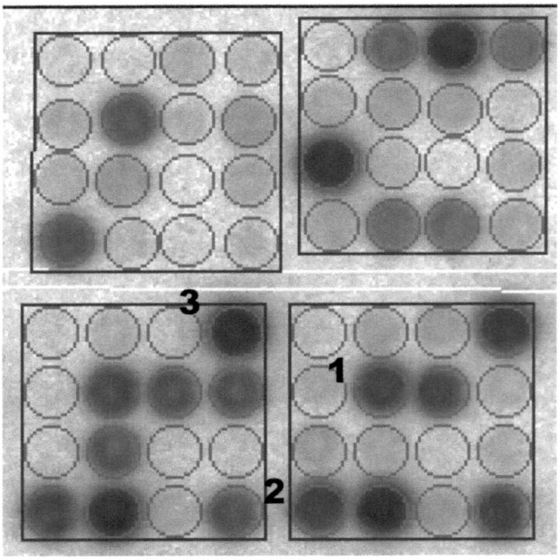

Figure 2. Picture of a complex hybridisation with the grid at the border of two subarrays. Arrows show the distance of the different levels. 1 = level 1 distance between spots within one spot group; 2 = level 2 distance between spots of two different spot groups; 3 = level 3 distance between spots of two subarrays. The distance 1 equals the search radius.

This approach eases filter analysis, gives a better orientation on the filter during griding, and, finally, saves a lot of time because all spots on the filter have to be manually checked to ensure correct griding.

In order to avoid areas of radioactivity that are not due to DNA-DNA interactions during the griding process, a search radius can be defined in which the program tries to set the next spot grid to a neighbouring spot. The search radius value therefore depends on the distance between spots within a spot group and should be measured by a straight line connecting the centres of the spots with each other (arrow 1, figure 2). In addition, a sensitivity value for the spot search can be chosen. The higher this value, the more sensitive the screen for spots during the automatic pre-gadding. However, it should be noted that local contaminations on the filter might be regarded as spots if the filters are grided with a sensitivity value that is too high, leading to local misgriding or even to displacement of the whole grid. Therefore, the sensitivity value should be adjusted before starting the griding.

Careful manual pre-adjustment of the entire grid at level 3 (sub-array level) prior to initiating the automatic overlay is strongly recommended as the quality of the overlay is dependent on this adjustment. After completing the alignment of the grid, an intensity value for each spot is calculated and can be exported as a text-file that is compatible with other programs or databases used for subsequent analysis and statistical evaluation of the data. One public source for this is the *Haruspex* database developed at the Max-Planck-Institute of Molecular Plant Physiology in Golm (Germany) (see Methods for analysing the data obtained from these arrays are described in the *LOTUS JAPONICUS* EXPRESSION DATABASE chapter 4.3); other available programs can be found in Richmond and Somerville (2000).

Technical replication and reliability of array data

Each hybridisation experiment must be repeated at least three times in order to perform the statistical T-test. One should bear in mind that there is a certain degree of technical variability between different filters. Therefore, in cases where more than one filter is used for a particular experiment, one should ensure that probes representing each of the biological treatments are hybridised to all of the filters, in order to reduce the number of false positive clones that are obtained after the statistical analysis.

Another way to check for reliability of the results is to compare the signal intensities of genes that are spotted more than once on the same filter. In addition, other techniques like real-time RT-PCR or RNA-blotting can be used to confirm the results obtained from the DNA array experiments.

REFERENCES AND RECOMMENDED READING

Aharoni A and Vorst O. (2001) **DNA microarrays for functional plant genomics**. *Plant Molecular Biology* 48, 99-118.

Richmond CS, Glasner JD, Mau R, Jin H, and Blattner FR. (1999) **Genome-wide expression profiling in *Escherichia coli* K-12**. *Nucleic Acids Research* 27, 3821-3835.

Richmond T and Somerville S. (2000) **Chasing the dream: plant EST microarrays**. *Current Opinion in Plant Biology,* 3, 108-116.

Colebatch G, Kloska S, Trevaskis B, Freund S, Altmann T, and Udvardi MK. (2002a) **Novel aspects of symbiotic nitrogen fixation uncovered by transcript profiling with cDNA arrays**. *Molecular Plant-Microbe Interactions* 15, 411-420.

Colebatch G, Trevaskis B, and Udvardi MK. (2002b) **Functional genomics: tools of the trade**. *New Phytologist* 153, 27-36.

Lipshutz RJ, Fodor SPA, Gingeras TR, and Lockhart DJ. (1999) **High-density synthetic oligonucleotide arrays**. *Nature Genetics* 21, 20-24.

Murphy D. (2002) **Gene expression studies using microarrays: principles, problems, and prospects**. *Advances in Physiology Education* 26, 256-270.

Yang YH, and Speed T. (2002) **Design issues for cDNA microarray experiments**. *Nature Reviews Genetics* 3, 579-588.

Journet EP, van Tuinen D, Gouzy J, Crespeau H, Carreau V, Farmer MJ, Niebel A, Schiex T, Jaillon O, Chatagnier O, Godiard L, Micheli F, Kahn D, Gianinazzi-Pearson V, and Gamas P. (2002) **Exploring root symbiotic programs in the model legume *Medicago truncatula* using EST analysis**. *Nucleic Acids Research* 30, 5579-5592.

Eisen MB, and Brown PO. (1999) **DNA arrays for analysis of gene expression**. *Methods Enzymology* 303, 179-205.

Eisen MB, Spellman PT, Brown PO, and Botstein D. (1998) **Cluster analysis and display of genome-wide expression patterns**. *Proceedings of the National Academy of Sciences U S A* 95, 14863-14868.

Kane MD, Jatkoe TA, Stumpf CR, Lu J, Thomas JD, and Madore SJ. (2000) **Assessment of the sensitivity and specificity of oligonucleotide (50mer) microarrays**. *Nucleic Acids Research* 28, 4552-4557.

Additional information is available at:

http://genome-www5.stanford.edu//restech.shtml

Chapter 4.3

LOTUS JAPONICUS EXPRESSION DATABASE

Sebastian Kloska[1], Peter Krüger[2], and Joachim Selbig[2,3*]

[1]*Scienion AG, Volmerstrasse 7a, D-12489 Berlin, GERMANY;* [2]*Max Planck Institute of Molecular Plant Physiology Am Mühlenberg 1, D-14476 Golm, GERMANY;* [3]*University of Potsdam, Institute of Biochemistry and Biology; *Corresponding author*

Email: selbig@mpimp-golm.mpg.de

Phone: +49 331 567 8208 Fax: +49 331 567 8236

Keywords: Lotus japonicus; Expression Profiling; Statistics; Database; Clustering

Haruspex *is a data management and analysis system developed at the Max Planck Institute of Plant Molecular Physiology that enables researchers to screen large sets of profiling data.* Haruspex *provides a common platform for the web-based storage, retrieval, and statistical analysis of expression profiles, while links to external databases provide fast access to comprehensive and up to date information.*

INTRODUCTION

Since the early days of high throughput biology (DeRisi et al., 1997) it has been apparent that data generated with these techniques can only lead to conclusions by employing computer aided methods (Eisen et al., 1998). This new field of bioinformatics deals with the development of exploratory tools, statistical methods (Slonim 2002), and data storage and retrieval systems.

When the Max-Planck Institute of Plant Molecular Physiology established a Microarray facility (Thimm et al., 2001), our group initiated the development of a workflow system for the management and analysis of gene expression profiles: *Haruspex* – the Arabidopsis Expression Database.

Haruspex was developed in cooperation with the Computer Science Department of the University of Potsdam as a web-based data storage and processing system. This program allows researchers to submit raw expression profiling data to be normalised, filtered, and annotated. The pre-processed values can be grouped together to form project-specific datasets that can be further analysed within *Haruspex* using statistical methods.

A.J. Márquez (ed.), Lotus japonicus Handbook, 149-154.

DATABASE

Haruspex has been implemented as a set of server side programs written in *C++* and *Perl*, a *mod_perl*-based web interface (Stein and MacEachern 2000) and a set of *Java* applets (Joy et al. 2000) serving as GUIs for more complex user interaction. This front end communicates with a relational database system (*Adabas D*). The system runs under Linux with an Apache web server (Ben and Laurie 1999).

After scanning and quantification of the raw data, the user can submit the results by logging into the system and transferring a data file to the server. Here one has to annotate the hybridization and choose a predefined scheme that describes the spotting pattern of the filter and represents the link between the spot coordinates and all available information about the underlying genes (Figure 1A). After submission, the newly generated dataset can be listed and filtered based on normalized values and annotation. In a second step, data from different hybridisations can be bundled to groups of experiments. These can be used to generate more stable values averaged over a set of replications or may serve as a starting point for statistical tests in order to search for genes with significantly altered expression level (Figure 1B). We have also implemented various clustering methods on the server side.

Haruspex provides links to publicly available resources on the net. If the set of target genes represented on the array are annotated with GenBank Accessions the user may directly navigate from the expression profiling result tables to the desired information. Anonymous sequences (e.g. ESTs) for which only raw sequences are available can be mapped via BLAST (Altschul et al. 1990) to the most similar sequence within the target organism. Direct linkage is provided for the GenBank database at the NCBI (Benson et al. 1998), and resources at MIPS (Schoof 2003) and (Huala et al. 2001).

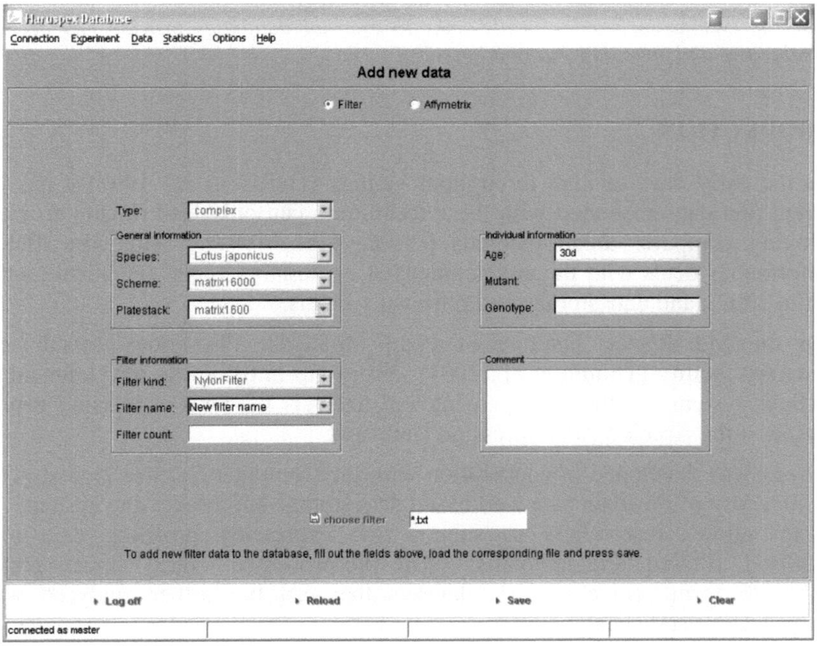

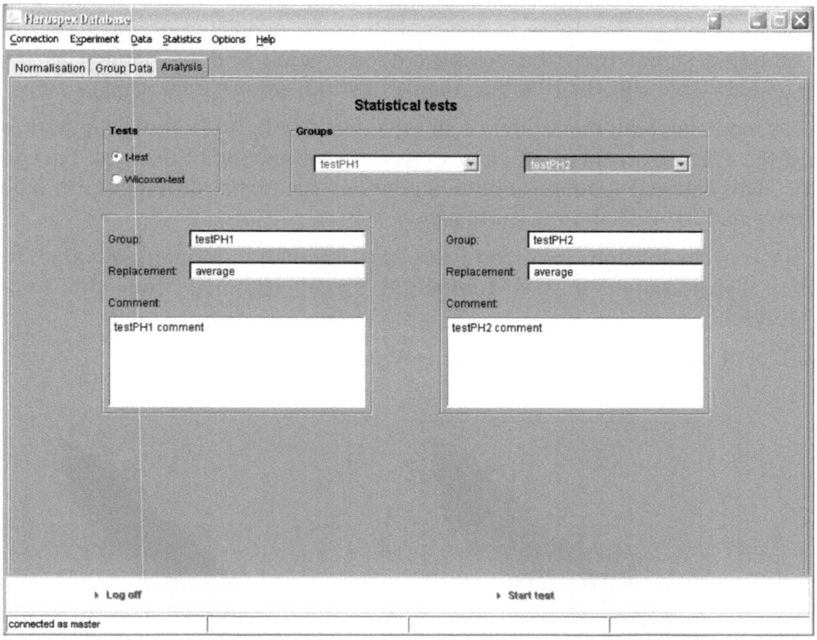

Figure 1. User Dialogs of the Haruspex Database front end. A. Previous page screen: Submission of new dataset representing a microarray experiment. The user is prompted for a name of the new hybridisation, a spotting scheme, and information about the experiment. After selecting a file on the local file system the data is transferred to the Haruspex server, normalised, and stored in the relational database management system (RDBMS). B. This page screen: Applying statistical tests to sets of hybridizations. After bundling experimental sets, one can apply statistical tests to the data by choosing two sets and specifying parameters for the tests.

NORMALISING AND FILTRATION

The first step towards comparable data sets is a data filtration and normalisation procedure. For hybridisations with radioactive probes on nylon membranes, which constitute the main source for expression profiling data in the institute, we have implemented this procedure entirely within the database system. Each filter spotted with PCR products is first hybridised with vector specific oligomers. This reference hybridisation provides us with a quantification of the amount of target DNA spotted on the filter, which has a strong influence on the final signal observed in the experiment. After submitting the reference hybridisation to the repository, the user may transfer hybridisation data from complex hybridisations performed with labelled RNA-derived probes on the same filter.

When both datasets are available, *Haruspex* performs a three-step filtration and normalisation process. First, the system searches for undetectable spots in both the complex and reference hybridisations. This is done by comparing the raw data value obtained for a given spot with the value of an empty spot placed within the centre of each sub grid. If the spot signal is not at least twice as high as this local background signal we call the given spot undetectable. Undetectable spots on the reference hybridisation tell us that there is not enough probe deposited on the filter and the corresponding signal from the complex hybridisation will be flagged as invalid, while undetectable signals from the complex hybridisations are treated as zero.

In the second step both the reference and the complex hybridisations are normalized against the average signal of all valid spots after subtraction of the local background signal. This internal normalisation compensates for global differences in labelling efficiency and hybridisation performance.

The final normalisation step correlates the internally normalised value of the complex hybridisation to the normalised signal obtained from the reference hybridisation.

The effect of the normalisation scheme is represented in Figure 2.

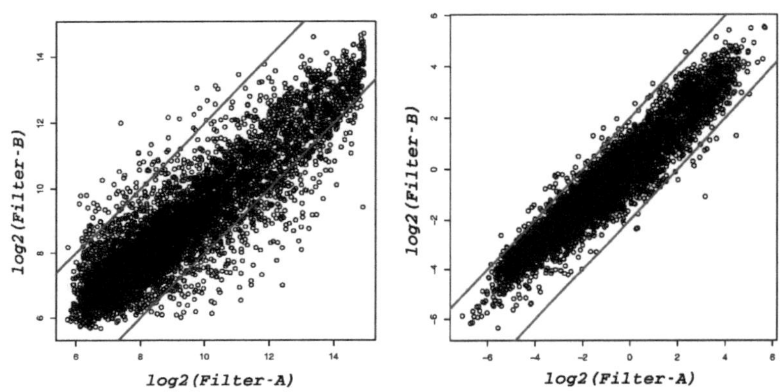

A **B**

Figure 2. Visualisation of the normalisation and filtration procedure. Logarithmic scatter plots of data from experimental replicates. (A) Represents raw data from a complex hybridisation. (B) Shows fully normalised values from the same hybridisation. Red lines denote the limits for 2-fold differences.

STATISTICAL TESTS AND DATA EXPLORATION

Data generated from a set of experimental replications can be bundled into a group. When combining such replicas the user might choose to discard certain values (e.g. such values flagged as being undetectable or zero). Two such groups may afterwards be compared employing a *t*-Test on a *per*-gene basis. The newly generated dataset can be browsed and further filtered for genes that show a low P-value and/or high expression ratio under the chosen conditions.

Another option provided by *Haruspex* is the generation of a hierarchical cluster dendrogram (Eisen et al. 1998). Here a hybridisation group should represent a series of experiments performed under different conditions (e.g. time series or dosage response). These can be pre-filtered by various limits like expression intensity or variance between individual hybridisations in order to restrict the analyses to those genes that show the highest dynamic. Afterwards, *Haruspex* allows the user to perform clustering on the data set specifying such parameters as distance measure or linkage type. The resulting data is reinserted into the repository and may be visually explored with a dendrogram navigation system written in Java.

CONCLUSION

Haruspex allows scientists to efficiently store, explore, and analyse their expression profiling data. It was initially designed to work with filter-based large volume experiments performed with the *A. thaliana* EST collection of the Michigan State University, but has since been adopted by other groups within the institute working with plant species like potato and *L. japonicus* (Colebatch et al. 2002).

Now that a full genome chip for *A. thaliana* is available from Affymetrix (Lipshutz et al. 1999), we are working towards inclusion of data generated by this platform. On the other hand the demand for custom designed filter based assays has not diminished, since there are still a large collection of organisms which although interesting for scientists do not generate demands large enough to make them interesting for commercial array providers.

Comparability of the data across different platforms and standardised data formats are still one of the major issues within the microarray domain. With a large set of public domain and commercial database systems (Gardiner-Garden and Littlejohn 2001) now available and the inherent complexity of high throughput experiments we still have a long way to go to make expression profiling data as easily accessible and comparable as sequence data at the public sequencing repositories.

REFERENCES

Altschul, SF, Gish, W, Miller, W, Myers, EW, and Lipman, DJ. (1990) **Basic local alignment search tool.** *Journal of Molecular Biology* 215, 403-410.

Ben L and Laurie P. (1999) **Apache: the Definitive**. O'Reilly Pub. Inc. (www.apache.org/).

Benson DA, Boguski MS, Lipman DJ, Ostell J, Ouellette BF. (1998) **GenBank.** Nucleic Acids Res. 26 (http://www.ncbi.nlm.nih.gov/).

Colebatch G, Kloska S, Trevaskis B, Freund S, Altmann T, Udvardi MK. (2002) **Novel aspects of symbiotic nitrogen fixation uncovered by transcript profiling with cDNA arrays.** *Molecular Plant-Microbe Interactions* 15, 411-420.

DeRisi JL, Iyer VR, and Brown PO. (1997) **Exploring the Metabolic and Genetic Control of Gene Expression on a Genomic Scale.** *Science* 278, 690-686.

Eisen, MB, Spellman, PT, Brown, PO, and Botstein, D. (2001) **Cluster analysis and display of genome-wide expression patterns.** *Proceedings of the National Academy of Sciences USA* 95, 14863-14868.

Gardiner-Garden, M and Littlejohn, TG. (2001) **A comparison of microarray databases.** *Briefings in Bioinformatics* 2, 143-58.

Huala E, Dickerman AW, Garcia-Hernandez M, Weems D, Reiser L, LaFond F, Hanley D, Kiphart D, Zhuang M, Huang W, Mueller LA, Bhattacharyya D, Bhaya D, Sobral BW, Beavis W, Meinke DW, Town CD, Somerville C, and Rhee SY. (2001) **The *Arabidopsis* Information Resource (TAIR): a comprehensive database and web-based information retrieval, analysis, and visualization system for a model plant.** *Nucleic Acids Research* 29, 102-105. (www.tigr.org).

Joy B, Steele G, Gosling J, and Bracha G. (2000) **Java™ Language Specification (2nd Edition).** Addison-Wesley Pub. Inc. (http://java.sun.com).

Lipshutz RJ, Fodor SP, Gingeras TR, and Lockhart DJ. (1999) **High density synthetic oligonucleotide arrays.** *Nature Genetics*: 21:20-24.

Schoof, H. (2003) **Towards interoperability in genome databases: the MAtDB (MIPS *Arabidopis thaliana* database) experience.** *Comparative & Functional Genomics*: 4, 255-258. (http://mips.gsf.de).

Slonim DK. (2002) **From patterns to pathways: gene expression data analysis comes of age.** *Nature Genetics* 32 Suppl, 502-508.

Stein L and MacEachern D. (2000) **Writing Apache Modules with Perl and C** (2nd Edition). O' Reilly Pub. Inc. (http://perl.apache.org).

Thimm 0, Essigmann B, Kloska S, Altmann T, and Buckhout TJ. (2001) **Response of *Arabidopsis* to Iron Deficiency Stress as Revealed by Microarray Analysis.** *Plant Physiology* 128, 1030-1043.

Chapter 4.4

PROCEDURES FOR MASS SPECTROMETRIC PROTEOME ANALYSIS

Gerhard Saalbach

Plant Research Department; Risø National Laboratory; POB 49, Frederiksborgvej 399, 4000 Roskilde; DENMARK

Email: G.Saalbach@risoe.dk

Phone +45 4677 4290 Fax +45 4677 4282

Keywords: proteome, proteomics, symbiosomes, membrane, mass spectrometry, LC-MS, sequence database

INTRODUCTION

Any kind of protein or protein mixture can be analysed by mass spectrometry (MS) to identify proteins. The proteins are cleaved with an endopeptidase, like trypsin, and the peptides are analysed by MS. The data are used for database screening. The classical approach uses 2D gel-electrophoresis to separate protein mixtures. Individual spots can be cut out and analysed. Other approaches omit the 2D gels and use chromatographic procedures to separate the peptides before the mass spectrometric analysis. Here, we describe the procedures for the analysis of spots from 2D gels (or bands from 1D gels) with an LCQ mass spectrometer (Finnigan, San Jose, USA) using nanospray sample application. Work for the analysis of protein mixtures and use of 2D-chromatography is in progress.

The procedures described here have successfully been used for the analysis of proteins of the peribacteroid membrane (PBM) isolated from root nodules of *Lotus japonicus*. The procedure for PBM preparation is described in a separate protocol in this handbook. Such preparations can be dissolved in SDS-buffer and run on a normal 1D SDS-gel. Bands from such gels can be cut out and analysed according to the following protocols.

A.J. Márquez (ed.), Lotus japonicus Handbook, 155-160.

MATERIALS

The described procedures are for use with an LCQ mass spectrometer (Finnigan, San Jose, USA) using nanospray sample application. Consumable as well as chemicals, buffers, solution etc. are described in the protocols in detail.

The peribacteroid symbiosome membrane was isolated from *Lotus japonicus* root nodules as described in the accompanying protocol (S Wienkoop, Chapter 4.5). Aliquots of approximately 50 µg protein were separated by 1D gel electrophoresis. Bands for MS analysis were cut out after standard Coomassie Blue staining.

Abbreviations used are as follows: DTT: dithiothreitol; ACN: acetonitril; FA: formic acid.

PROCEDURES

In-gel trypsin digestion and peptide purification

This protocol is for samples treated with dithiothreitol (DTT) and iodoacetamide. This is normally the case with 2D-gel spots. Bands from 1D-gels have to be treated with DTT and iodoacetamide separately. Gels should preferably be stained with Coomassie brilliant blue. Throughout the staining procedure the concentration of ethanol and acetic acid should be reduced to 30% and 5%, respectively. The whole procedure should be kept as short as possible. The following protocols are based on published procedures (Shevchenko et al. 1996, Gobom et al. 1999).

Washing the gel pieces

- Cut the spots from the gel using a sharp scalpel, transfer the gel pieces to 1.5 ml tubes, can be kept at -20 °C
- Wash the gel pieces with 200 µl water, vortex 10 min, remove liquid
- Wash the gel pieces with 200 µl 100 mM NH_4HCO_3, vortex 10 min, do not remove liquid
- Add directly 200 µl acetonitrile (ACN), vortex 10-30 min, remove liquid
- Dark blue spots: wash again 10-30 min with 50 % ACN/ 50 mM NH_4HCO_3
- Repeat step 5 until the gel pieces are de-stained to a large extend, then remove all liquid
- Add pure acetonitril, vortex 10 min
- Remove all liquid, dry completely in speed vac.

In-gel trypsin digestion

- Make 10 ml buffer: 25 mM $NH_4H_2CO_3$, 10% ACN, 1 mM $CaCl_2$ (the $CaCl_2$ must be added last to avoid precipitation)
- Dissolve 1 vial of 25 µg Trypsin-Sequencing grade (Roche, Mannheim, Germany) with 50 µl 1 mM HCl, keep on ice

- Take ca. 600 µl from the buffer and add 15 µl Trypsin, keep on ice, this gives 12.5 ng trypsin per µl and is sufficient for ca. 30-40 spots
- Add this mix to the dry gel pieces: ca. 10 µl to the small, 15 µl to the medium, 20 µl to the big spots
- Keep on ice for 45 min, but check the swelling from time to time and add more (5-10 µl) trypsin solution, if necessary
- After ca. 45 min, wash the gel pieces with the trypsin-free buffer: add 100 µl buffer to each piece and immediately remove all liquid again, then add 1 droplet of fresh buffer, just to cover the gel piece
- Close the lids of the tubes well, cover the whole rack of tubes in plastic wrap, incubate in an air incubator at 37°C overnight.

Extraction of the peptides

- To the trypsin treated gel pieces add 20 µl 5% formic acid (FA), vortex 5 min, then transfer the liquid into a fresh siliconised tube, keep the pipette tip also in this fresh tube
- To the gel piece add 20 µl of 1% FA, 5% ACN
- Vortex 15-30 min, remove liquid with the same pipette tip into the same fresh tube
- To the gel piece add 20 µl of 1% FA, 50% ACN
- Vortex 15-30 min, remove liquid with the same pipette tip into the same fresh tube
- To the gel piece add 20 µl of 1% FA, 90% ACN
- Vortex 15-30 min, remove liquid with the same pipette tip into the same fresh tube
- Discard the gel pieces.
- Dry down the collected extracts in a speed vac
- Re-dissolve in 15 µl 5% FA.

Purification of the peptides for nanospray

- Make a micro-column from a 10 µl Eppendorf GELoader Tip (0030 001.222), squeeze it at the very tip to close it (or narrow it), with another pipette tip fill it with 20 µl of 50% methanol/5% formic acid (FA)
- Add a small volume of POROS R2 (PerSeptive Biosystems, Framingham, USA) suspended in methanol
- Press liquid through the tip using a 1 ml syringe adapted to the tip with suitable tubing, watch the growing column in the tip, stop if it has reached a length of ca. 5-7 mm (NOT MORE!), then remove all the remaining supernatant (especially excess POROS) from the tip

- Wash this micro-column again with 20 µl of 50% methanol/5% FA
- Wash it with 2x 20 µl of 5% FA, finally leave a few µl of liquid in the column
- Add your sample into this purification tip; the whole amount can be applied, but this is mostly not necessary with Coomassie spots. It may be necessary with silver stained spots.
- After adding the sample press all liquid through with the syringe
- Wash 3x with 20 µl 5% FA, after last washing step press all liquid out
- Fix a nanospray capillary in the holder (see www.protana.com)
- Cut the purification tip 2 mm above the thin conical part (to make it fit into the centrifuge) and insert it into the nanospray capillary fixed in the holder
- Add 1-3 µl of 50% methanol/5% FA into the micropurification tip sitting in the nanospray capillary in the holder
- Spin the whole assembly in a suitable micro-centrifuge shortly at 5000 rpm
- Take out the micropurification tip and check whether all liquid has passed through, if not, repeat spin (depends on how tight you have squeezed the tip and on possible clogging)
- If the sample is eluted remove and discard the micro-purification tip.

Peptide analysis with LCQ-MS (ThermoFinnigan)

Mass spectrometry

- Take the capillary out of the holder, keep it always with the tip downwards, check that the eluate is sitting in the tip
- Insert the capillary into the holder which is already pre-mounted at the mass spec
- Connect high voltage cables and carefully adjust the tip to the heated capillary opening at the mass spec, apply some air pressure to the capillary with the syringe
- Break the tip by gently touching the heated capillary surface and by moving it up and down, breaking occurs mostly quickly and no big can be seen
- Check it by moving the needle back from the surface and applying some more pressure, then a small (!) droplet should be seen at the tip
- If this was successful, adjust the tip of the needle back to the inlet of the heated capillary
- Open a nanotune file where the capillary voltage is adjusted to 0.7 kV
- Start the LCQ (turn sheath gas off)
- Watch spectrum, stable peaks at intensities of 10^6 to 10^8 should appear

- Do automatic tuning on base peak, check the result, if negative, repeat tuning on a different peak or discard it and restart the original file
- If the spectrum is finally OK and stable, save it
- Individual peptides can now be analysed with zoom scanning and tandem MS manually, or automatically by running an Xcalibur sequence.

Database search (Mascot)

- Copy the raw files from the nanospray experiment to the Xcalibur/data directory on the C: drive
- Start Sequest and set up a directory,
- Create *.DTA files for all your single raw files (stupid work), or just the one (or few) from the automatic recording (better)
- These will be saved in the directory you have generated before (under C:/Sequest)
- To the same directory copy a *.bat file made according to Mascot:

  Here is a DOS batch file that does the job. It is very simple, with no error handling. The batch file and the *.DTA files to be merged must be in the same directory. The original files are not altered. The merged output is in a file called merge.txt. Just cut and paste the following lines into a text editor, (e.g. Notepad), and save as a file with the extension .bat (e.g. merge.bat):

```
ECHO OFF
ECHO. > blank.txt
ECHO. >> blank.txt
FOR %%i IN (*.dta) DO COPY /A %%i+blank.txt %%i.tmp
COPY /A *.dta.tmp merge.txt
DEL blank.txt
DEL *.dta.tmp
ECHO All done, output file is merge.txt
```

- Run this file; it generates a new file in this directory called merge.txt; this can be renamed according to the sample name; it contains all the data from the different DTA files in one file now (but a maximum of 300 DTA files can be sent to Mascot at one time.)
- Start the Mascot search at www.matrixscience.com/cgi/index.pl?page=../home.html, fill out the form, and select the NCBInr database (EST can only be screened **after** this NCBInr run, it takes much longer). The merge.txt file is the data file. Select the data format "Sequest(DTA)"!
- Start the search, and if it is running close this window (e.g. by going back to start the next search), the result will then be sent to your email address.

Here are examples of search results from two experiments:

- www.matrixscience.com/cgi/master_results.pl?file=../data/20000726/FTgolre.dat
- www.matrixscience.com/cgi/master_results.pl?file=../data/20000726/FTgofrc.dat

REFERENCES

Wienkoop S. **Isolating peribacteroid membranes for proteome analysis**, this handbook, chapter 4.5.

Shevchenko A, Wilm M, Vorm O, and Mann M. (1996) **Mass spectrometric sequencing of proteins silver-stained polyacrylamide gels.** *Analytical Chemistry* 68, 850-858.

Gobom J, Nordhoff E, Mirgorodskaya E, Ekman R, and Roepstorff P. (1999) **Sample purification and preparation technique based on nano-scale reversed-phase columns for the sensitive analysis of complex peptide mixtures by matrix-assisted laser desorption/ionization mass spectrometry.** *Journal of Mass Spectrometry* 34, 105-116.

Chapter 4.5

ISOLATION OF PERIBACTEROID MEMBRANES FOR PROTEOME ANALYSIS

Stefanie Wienkoop

Plant Research Department; Risø National Laboratory; POB 49, Frederiksborgvej 399; 4000 Roskilde; DENMARK

Email: Stefanie.Wienkoop@risoe.dk

Phone: +45 4677 4290 Fax: +45 4677 4282

Keywords: *Lotus japonicus*, symbiosomes, peribacteroid membrane, membrane isolation, two-phase system, proteome analysis

INTRODUCTION

Before doing proteome analysis on membranes they need to be purified from the tissue. Different ways of PBM preparation using Percoll density gradients (Rosendahl et al, 1992)) or an aqueous-polymer two-phase system has been described previously (Christiansen et al, 1995). The peribacteroid membrane (PBM) from *Lotus japonicus* root nodules was isolated and purified with an optimised procedure using the aqueous-polymer two-phase system as final wash step. The procedure is described in the following protocol. The isolated PBM can than be used for various protein extraction procedures, gel electrophoresis and mass spectrometric analysis.

MATERIALS

Buffers for PBM isolation

AP-buffer (adapted from Rosendahl et al, 1992)

25 mM morpholinoethanesulforic acid, 450 mM mannitol, 7 mM Na_2EDTA, 7 mM $CaCl_2$, 5 mM $MgCl_2$, 20 mM ascorbic acid, 10 mM dithiotreitol (DTT), 1 mM phenylmethane sulphonyl fluoride, "Complete" protease inhibitors (IP) (Roche), pH 7.0, adjusted with 1 M 1,3bis[tris/hydroxymethyl) methylamino]-propane).

A.J. Márquez (ed.), Lotus japonicus Handbook, 161-163.

H-buffer

0.33 M sucrose, 10 mM DTT, 0.1 mM Na_2EDTA, 5 mM Tris-HCl pH 7.8

Aqueouse-polymer two-phase system

An aqueous polymer, two-phase system (Larsson, 1985) was optimised by testing activities of enzymes such as the P-type ATPase (Christiansen et al, 1995), cytochrome c reductase (CCR) (Christiansen et al, 1995), and the succinate dependent cytochrome c oxidase (CCO) (Wienecke et al, 1982) in PBM fractions prepared with differed polymer concentrations. The optimal polymer concentration was found at 6.1% where the P-type ATPase activity was highest and no CCR and CCO activity was detected. Therefore, a 6.1% Dextrane-T500/PEG-3350 system was prepared containing 0.33 M sucrose, 5 mM potassium phosphate buffer pH 7.8, 1 mM DTT, and 0.1 mM Na_2EDTA and used for PBM preparation.

Plants and bacteria

- *Lotus japonicus*
- *Mesorhizobium loti* (strain R7A)

PROCEDURE

PBM isolation and purification with the aqueous two-phase system

Nodules were harvested from 7-week-old plants. All of the following procedures were carried out at 4°C:

1. Nodules were squeezed in a potter in AP-buffer. 100 ml buffer was used for 20 g nodules.
2. The homogenate was pressed through 4 layers of "KO-TON" filter (CHICOPEE, Holland) and centrifuged for 15 min at 200 x *g* to remove debris.
3. The supernatant was centrifuged for 15 min at 1500xg to collect the symbiosomes.
4. The pellet was resuspended in H-buffer in a glass homogeniser to disrupt the symbiosomes.
5. The homogenate was centrifuged at 7000x*g* for 10 min to pellet the bacteroids.
6. The supernatant was carefully taken up and centrifuged at 100,000x*g* for 1 h.
7. The pellet was re-suspended in H-buffer in a glass homogeniser.
8. The homogenate was layered on top of an aqueous-polymer two-phase system. A 16 g-system was used for 20 to 40 g of nodules. The partitioning was carried out according to Larson (1985).

9. Finally, the last upper aqueous phase was centrifuged at 100,000x*g* for 1 h and the pellet representing the PBM-fraction was stored at -80°C.

REFERENCES

Christiansen J.H, Rosendahl L, and Widell S. (1995) **Preparation and characterization of sealed inside-out peribacteroid membrane vesicles from *Pisum sativum* L. and *Glycine max* L. root nodules by aqueous polymer two-phase partitioning**. *Journal of Plant Physiology* 147, 175-181.

Larsson C. (1985) **Plasma membranes**. In: *Modern methods of plant analysis. New series vol. 1, Cell components* (Linskens HF and Jackson JF, Eds.), Springer, Berlin, 85-104.

Rosendahl L, Dilworth M.J, and Glenn AR. (1992) **Exchange of metabolites across the peribacteroid membrane in pea root-nodules**. *Journal of Plant Physiology* 139, 635-638.

Wienecke K, Glas R, and Robinson DG. (1982) **Organelles involved in the synthesis and transport of hydroxyproline-containing glycoproteins in carrot root discs**. *Planta* 155, 58-63.

Chapter 4.6

METABOLOME ANALYSIS USING GC-MS

Guilhem Desbrosses, Dirk Steinhauser, Joachim Kopka, and Michael Udvardi*

*Max-Planck-Institute of Molecular Plant Physiology, Plant Nutrition Group, Am Mühlenberg 1, 14476 Golm, GERMANY; *Corresponding author*

Email: *udvardi@mpimp-golm.mpg.de*
Phone: +49 331 567 8149 Fax: +49 331 567 8134

Keywords: Lotus, omics technologies, metabolism, metabolites, metabolome, GC-MS, expression profiling, systems biology, bioinformatics

Global changes in gene transcription (the transcriptome), and protein amount/activity (the proteome) ultimately effect metabolism and the metabolite content (the metabolome) of cells, tissues, and organs of plants. In the past, studies of plant metabolism focussed on one or a few genes, proteins, and/or metabolites. New tools have been developed to measure levels of thousands of gene transcripts and proteins in parallel (see Chapters 4.2-4.4), facilitating non-biased, systems-wide investigations. To complement these 'OMICS' technologies, we use gas chromatography coupled to mass-spectroscopy together with bioinformatics tools to monitor changes in the levels of hundreds of metabolites in different organs. The methods described in this chapter provide qualitative and quantitative information about the Lotus metabolome, which together with transcriptome and proteome analyses, enable new insights into the links between genotype and phenotype at the whole-system level.

INTRODUCTION

The metabolome of an organism is the complete set of metabolites produced by the organism during its life cycle under all possible conditions. Unlike the transcriptome and proteome, which can be defined quite precisely by reference to a whole genome sequence, the metabolome is difficult to come to grips with. The chemistry encompassed by the metabolome is orders of magnitude more complex than the relatively simple chemistry of nucleic acids and proteins. Thus, comprehensive analysis of an organism's metabolome is a daunting challenge, and one that has not been fully realised in any organism. Individual plant species contain tens of thousands of different metabolites, and the plant kingdom is believed to produce hundreds of thousands of distinct metabolites (Pichersky and Gang, 2000). No single analytical platform has yet been devised that can measure all

A.J. Márquez (ed.), Lotus japonicus Handbook, 165-174.

metabolites in an organism. The quest for the Holy Grail that is the metabolome has resulted in an almost bewildering array of extraction, separation, and detections systems, many of which have been optimised for the analysis of certain classes of compounds. These include separation systems like gas chromatography (GC) and liquid chromatography (LC), and numerous detection systems, including mass spectrometry (MS), nuclear magnetic resonance spectroscopy (NMR), UV, and visible light spectroscopy, and enzyme based assays (Tretheway et al., 1999; Fiehn et al., 2000b; Sumner et al., 2003; Weckwerth, 2003; Kopka et al., 2004). Coupling of powerful separation and detection systems, such as GC-MS, enables relatively non-biased analysis of thousands of known and unknown metabolites, which can provide novel insight into the biological system under study. This type of analysis has come to be known as metabolite profiling, as it falls short of complete analysis of the metabolome, called metabolomics. In this chapter, we describe methods that have been developed for routine metabolite profiling of Lotus and other plant species, using GC-MS.

Metabolite profiling with GC-MS involves six steps:

1. **Extraction** of metabolites from the biological sample, which should be as comprehensive as possible, while at the same time avoiding degradation or modification of metabolites.
2. **Derivatisation** of metabolites to make them volatile and amenable to gas chromatography. Trimethylsilylation is the most favoured of the many possible derivatisation treatments that are possible.
3. **Separation by GC,** which generally involves highly standardised conditions of gas-flow, temperature programming, and standardised column material to achieve reproducible results.
4. **Ionisation** of compounds as they are eluted from the GC, which is required for subsequent mass spectrometry. Electron impact (EI) ionisation is most widely used, as it is robust and produces reproducible fragmentation patterns and molecular ions, which simplifies the evaluation of resulting mass spectra.
5. **Detection** of molecular ions, which can be achieved with different mass-detection devices, including single quadrupole detectors (QUAD), ion trap technology (TRAP), and time-of-flight detectors (TOF). Depending on the detection system, 10-40 samples per day can be processed with these different systems, with GC-TOF-MS providing the highest throughput.
6. **Evaluation** of data begins by matching chromatographic retention times and mass-spectral fragmentation patterns to reference data in local and/or public databases. Software provided with GC-MS equipment facilitates this to a greater or lesser extent, depending on the platform used. The best software programmes support automated, comprehensive extraction of all mass spectra from a chromatogram, correction for co-eluting metabolites, calculation of GC retention time indices, and automated selection of suitable mass fragments for quantification.

These six steps are described in more detail by Kopka et al. (2004), and in the sections below.

To date, GC-MS profiling of metabolites in plants has largely been confined to hydrophilic compounds, recovered in the methanol-water phase after methanol-water/chloroform extraction of tissues. Although not all hydrophilic compounds can be volatilised by derivatisation, the following classes of compounds are detected routinely: amino-, organic-, and aromatic-acids, sugars up to trisaccharides, alcohols and polyols, sterols, diglycerides, and some mono-phosphorylated metabolites. GC-MS metabolite profiling of potato (Roessner et al., 2000; Roessner et al., 2001b; Roessner et al., 2001a), tomato (Roessner-Tunali et al., 2003), Arabidopsis (Fiehn et al., 2000b; Taylor et al., 2002), pumpkin (Fiehn, 2003), and alfalfa (Huhman and Sumner, 2002; Chen et al., 2003) have provided insight into the effects of genetic manipulation on plants and highlighted metabolic diversity amongst natural populations. We have combined GC-MS with transcriptome analysis to reveal aspects of metabolic differentiation that accompany global changes in gene expression during nodulation and symbiotic nitrogen fixation in Lotus (Colebatch et al., 2004).

MATERIALS AND METHODS

Plant metabolite extraction and derivatisation

Plant material that is harvested for GC-MS analysis is frozen immediately in liquid nitrogen and stored at -80°C until used. For each sample for metabolite analysis, between 25-50 mg (FW) frozen tissue is pulverized in a 2 ml Eppendorf tube containing a clean stainless steel metal ball (5 mm diameter) in a mixer-mill grinder (MM200, Restch, Haan, Germany) for 2 min at 30 cycles per sec. Grinding components of the mill are cooled with liquid nitrogen to keep the sample as cold as possible. Metabolites are extracted according to (Roessner *et al.*, 2000) with the following modifications: Ground samples are mixed with 360 µl methanol (-20°C) plus 30 µl ribitol in methanol (0.2 mg / ml) and 30 µl deuterated d_4-alanine in water (1mg/ml), which are included as internal standards. Samples are shaken for 15 min at 70°C before addition of 200 µl chloroform and further shaking at 37°C for 5 min. After addition of 400 µl water, samples are vortexed then centrifuged at 14.000 rpm for 5 min at RT. Two 80 µl aliquots of the aqueous phase are transferred to Eppendorf tubes and dried at RT by vacuum centrifugation (Centrivac, Hereaus, Hanau, Germany).

Metabolite samples are derivatized by methoxyamination, using a 20 mg/ ml solution of methoxyamine hydrochloride in pyridine, and subsequent trimethylsilylation, with N-methyl-N-(trimethylsilyl)-trifluoroacetamide, MSTFA, (Figure 1; Roessner et al., 2000; Fiehn et al., 2000b).

GC-MS metabolite profiles

In our institute, GC-MS spectra are typically obtained with a GC8000 gas chromatograph coupled to a Voyager quadrapol-type mass spectrometer, operated by MassLab software (ThermoQuest, Manchester, UK). Modifications to the initial GC-MS profiling method (Fiehn *et al.*, 2000b; Fiehn *et al.*, 2000a) include: injection of 1 µl sample in split-less mode; use of a 5°C min^{-1} temperature ramp with final temperature set to 320°C on a 30 m x 0.25 mm inner diameter Rtx-5Sil MS capillary column with an integrated guard column (Restek GmbH, Bad Homburg, Germany); and use of the C_{12}, C_{15}, C_{19}, C_{22}, C_{28}, C_{32}, C_{36} n-alkane mixtures for the determination of retention time indices (RI). Typical GC-MS profiles of different organs of Lotus are shown in Figure 2.

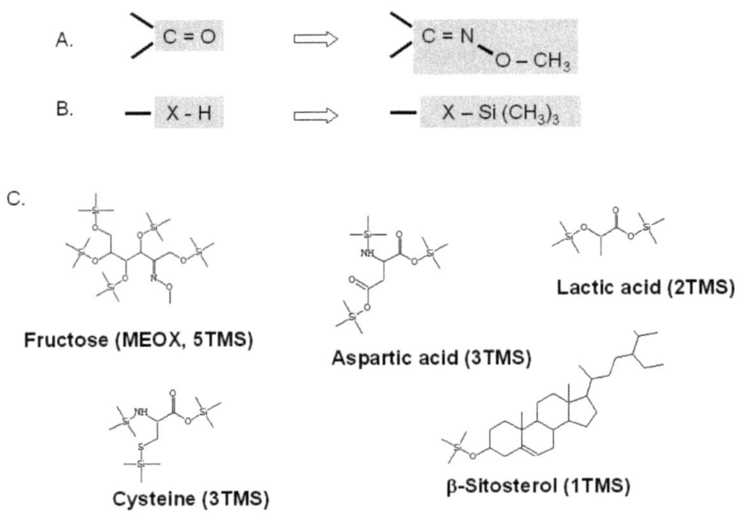

Figure 1. Chemical derivatisation and some typical metabolite derivatives. (A) Methoxyamination of carbonyl moieties. (B) Trimethylsilylation of acidic protons such as are present in hydroxyl-, sulfhydryl- carboxyl-, phosphate- or amino-groups. (C) Five typical metabolite derivates.

Mass-spectral tags for metabolites

Gas chromatography separates complex mixtures of metabolite derivatives into a series of compounds that enter the mass spectrometer and are subsequently ionized, fragmented, and detected. Each metabolite is, therefore, represented by one or more ionic fragments of precise mass, which together can serve as a 'tag' for that metabolite. We have termed these mass-spectral tags (MST), by analogy to expressed sequence tags (ESTs) of genes. Each MST has properties that facilitate unequivocal identification of the parent metabolite, following comparison to the pure reference compound (Wagner et al., 2003). The properties are (1) gas chromatographic retention, which is best characterised by a retention time index (RI), and (2) a specific composition of fragments, which are each characterized by a mass to charge ratios (m/z). A compendium library of MSTs has been collected from a set of Lotus organs using AMDIS software (Automated Mass Spectral Deconvolution and Identification System).

Fragments that belong to one MST have the same retention time index and occur in fixed relative abundance, independent of metabolite concentration. Therefore, any single fragment or set of fragments with identical retention time index can be used for the quantification of metabolites. As a rule, choice of fragments for quantitative purposes must be selective, i.e. only those fragments that are unique to a MST can be used. Fragments that are common to co-eluting MSTs or MSTs with similar retention time indices must be avoided for quantification purposes.

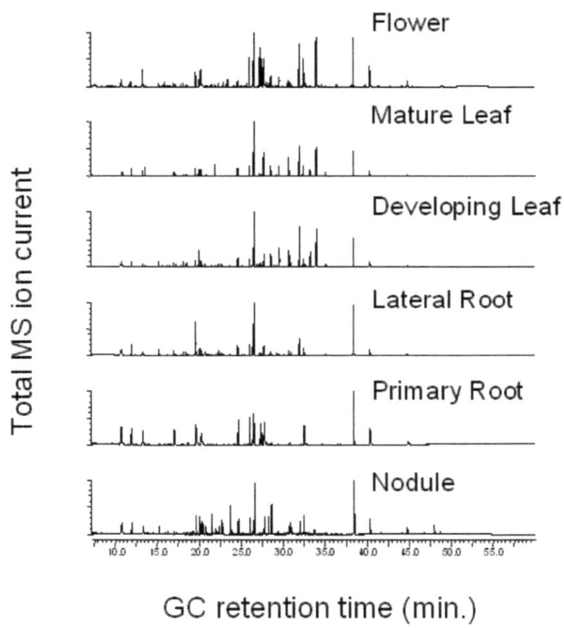

Figure 2. Typical GC-MS profiles of Lotus organs.

Fragments that are used for quantification are identified by mass to charge ratio (m/z), RI, and metabolite derivative. We use the following nomenclature for GC-MS fragments: mass to charge ratio followed by retention time index and name both separated by an underlined character, for example fragment 333_2014 Glucaric acid (6TMS) (Figure 3). MSTs that remain unidentified are classified tentatively by best matching mass spectra from a custom and a commercial NIST02 library (Institute of Standards and Technology, Gaithersburg, USA). A tentative match required a score >600 on a scale of 0-1000. We use the nomenclature, match value, and substance name of best fit, separated by a semicolon character and set into square brackets, for example 243_1930_[802; Methylcitric acid (4TMS)].

Compendium library of mass spectral metabolite tags

Mass spectral metabolite tags (MSTs) were obtained by automated de-convolution of GC-MS chromatograms using publicly available software (AMDIS, http://chemdata.nist.gov/mass-spc/amdis/; National Institute of Standards and

Technology, Gaithersburg, USA) (Stein, 1999). Mass spectra were collected above a threshold of 0.001% of total signal. Two independent GC-MS metabolite profiles of each sample type were processed. RIs of all MSTs were determined by AMDIS software. RI-annotated MSTs were uploaded into a non-supervised custom NIST02 mass spectral library (NIST02 mass spectral search program, http://chemdata.nist.gov/mass-spc/Srch_v1.7/index.html; National Institute of Standards and Technology, Gaithersburg, USA) (Ausloos *et al.*, 1999). Mass spectra of low relative amount were rejected. The resulting MST library of *Lotus japonicus* root and shoot organs is available as Supplementary Material, NIST02 custom library file Q_LJA_NS (Figure 4; see also: http://csbdb.mpimp-golm.mpg.de/gmd.html).

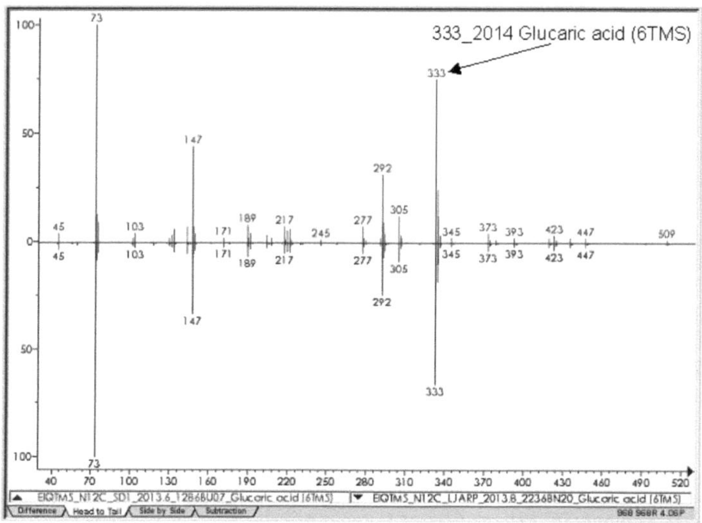

Figure 3. Head to Tail matching of glucaric acid mass spectra from the pure reference compound (above) and from a preparation of Lotus primary roots (below). NIST02 software and customized user libraries of mass spectra were used.

Identification of MSTs

MSTs are identified by manual comparison with MSTs derived from commercially available pure standard compounds (Figure 2). Standard compounds are processed through standard addition experiments in order to obtain respective MEOX- and MSTFA-derivative mass spectra and corresponding retention time indices. Required criteria for MST identification are chromatographic co-elution within an RI window of ± 2.5, and high mass spectral similarity, with matching value >700, on a scale of 0-1000. MST identification is supported by comparison with mass spectral entries from the commercial NIST02 library (Institute of Standards and Technology, Gaithersburg, USA). The current supervised and non-redundant MST Lotus library is available as Supplementary Material, NIST02 custom library file Q_LJA_ID.

Identified metabolites and additional information can be found at the Golm Metabolome Database (GMD; http://csbdb.mpimp-golm.mpg.de/gmd.html).

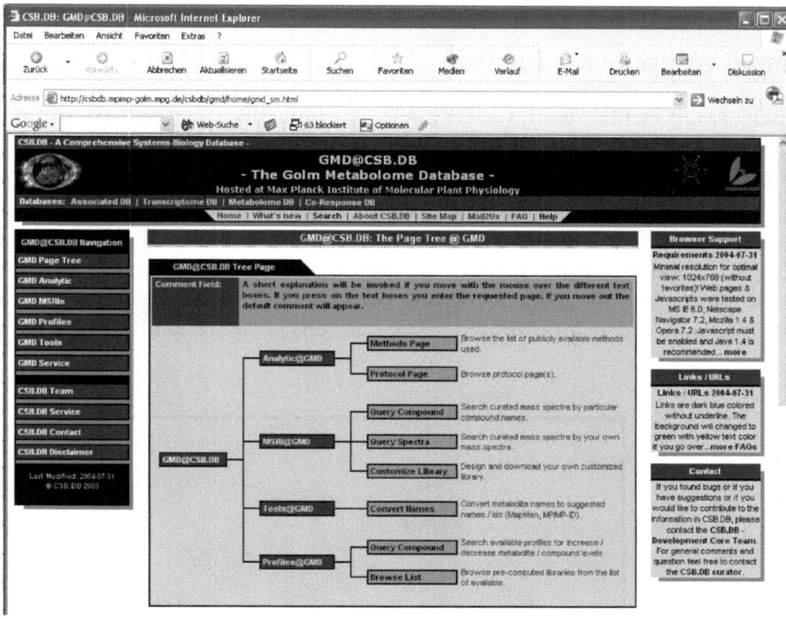

Figure 4. Resources for GC-MS Mass Spectral Metabolite Identification – The Golm Metabolome Database (GMD) http://csbdb.mpimp-golm.mpg.de/gmd.html

Lactic acid and benzoic acid are typical laboratory contaminants. Oligomethyl-cyclosiloxanes are also common contaminants, and accumulate with time in the MSTFA reagent used for derivatisation.

Generation of a metabolite response matrix

For each metabolite that is identified in the non-supervised MST library, a specific mass fragment and corresponding retention-time window is selected. The find algorithm of the MassLab software (ThermoQuest, Manchester, UK) is then used to automatically retrieve peak areas from GC-MS metabolite profiles. Correct peak integration is monitored manually. Peak areas with low intensity are rejected.

Numerical analysis

Peak areas, X, are equivalent to what we call fragment responses (X_i of fragment i). Fragment responses are normalized to the fresh weight of the sample and response of the internal standard, ribitol, ($N_i = X_i * X_{ribitol}^{-1} * \text{fresh weight}^{-1}$). This procedure corrects pipette errors and slight differences in sample amount. In a further step, the relative response of a fragment, R_i, is divided by the average normalised responses of all samples in one experiment or, alternatively of all non-treated control samples of an experiment ($R_i = N_i * avgN^{-1}$) to facilitate comparisons between samples. All response ratios are combined into a single matrix that described the complete set of metabolite response ratios of all samples in an experiment or group of experiments.

Principal components analysis

Principal components analysis (PCA) is typically performed after $\log_{10}$ transformation of data in a metabolite response matrix. Missing values are defined as 0 after $\log_{10}$ transformation. This substitution procedure assumes that missing values are unchanged compared to the co-analysed standard samples. The S-Plus 2000 software package standard edition release 3 (Insightful, Berlin Germany) is used for PCA and visualization. PCA analysis can reveal differences between the metabolic phenotype of different organs of wild type Lotus plants (Figure 5), or between the same organs of different plant genotypes (Taylor et al., 2002).

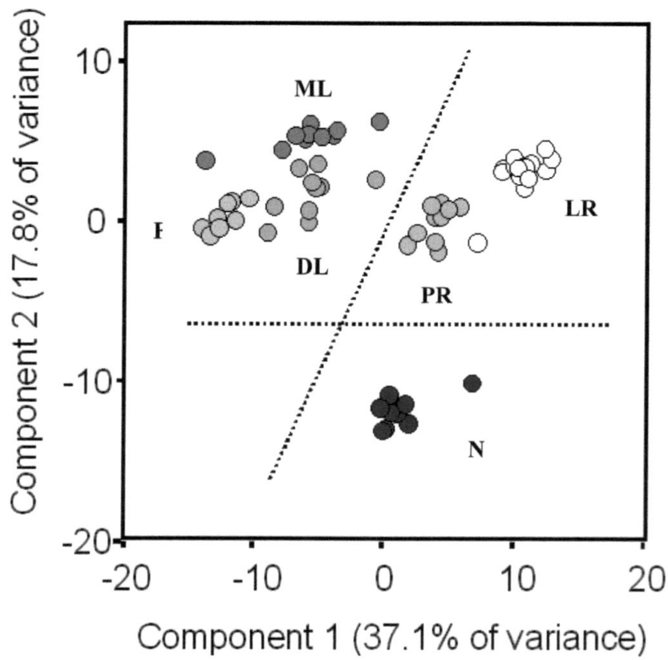

Figure 5. PCA analysis of GC-MS data from different Lotus organs. Filled circles represent individual samples from different organs: F, flowers; DL, developing leaves; ML mature leaves; LR, lateral roots; PR, primary roots; N, nodules. Clustering of samples from the same organ and separation of samples from different organs are obvious.

Hierarchical cluster analysis(HCA)

Hierarchical cluster analysis (HCA) is used to classify samples and MSTs, which represented identified metabolites and a selection of non-identified metabolites, according to their relative abundance in different samples or sample types. For this purpose, average normalized responses ($avgN_i$) are calculated for each MST and organ. Missing data are substituted by the normalized response at detection limit. HCA is performed after range normalization using Euclidian distance and average linkage. All procedures including analysis of variance and visualisation are performed with EXCEL software and the S-Plus 2000 software package standard edition release 3 (Insightful, Berlin Germany).

REFERENCES

Castrillo JI, Hayes A, Mohammed S, Gaskell S.J, and Oliver SG. (2003). **An optimized protocol for metabolome analysis in yeast using direct infusion electrospray mass spectrometry**. *Phytochemistry* 62, 929-937.

Chen F, Duran AL, Blount J.W, Sumner LW, Dixon RA, and Huhman D.V. (2003). **Profiling phenolic metabolites in transgenic alfalfa modified in lignin biosynthesis**. *Phytochemistry* 64, 1013-1021.

Colebatch G, Desbrosses G, Ott T, Krusell L, Kloska S, Kopka J, and Udvardi MK. (2004) **Global changes in transcription orchestrate metabolic differentiation during symbiotic nitrogen fixation in *Lotus japonicus***. *The Plant Journal* 39, 487-512.

Fiehn O. (2003). **Metabolic networks of *Cucurbita maxima* phloem**. *Phytochemistry* 62, 875-886.

Fiehn O, Kopka J, Trethewey RN, and Willmitzer L. (2000a). **Identification of uncommon plant metabolites based on calculation of elemental compositions using gas chromatography and quadrupole mass spectrometry**. *Analytical Chemistry* 72, 3573-3580.

Fiehn O, Kopka J, Dormann P, Altmann T, Trethewey RN, and Willmitzer L. (2000b). **Metabolite profiling for plant functional genomics**. Nat Biotechnol 18, 1157-1161.

Griffin JL, Cemal CK, and Pook MA. (2004). **Defining a metabolic phenotype in the brain of a transgenic mouse model of spinocerebellar ataxia 3**. *Physiological Genomics* 16, 334-340.

Huhman DV, and Sumner LW. (2002). **Metabolic profiling of saponins in *Medicago sativa* and *Medicago truncatula* using HPLC-coupled to an electrospray ion-trap mass spectrometer**. *Phytochemistry* 59, 347-360.

Kopka J, Fernie A, Weckwerth W, Gibon Y, and Stitt M. (2004). **Metabolite profiling in plant biology: platforms and destinations**. Genome Biology 5, 109.

Pichersky E. and Gang D. (2000). **Genetics and biochemistry of secondary metabolites: An evolutionary perspective**. *Trends in Plant Sciences* 5, 439-445.

Roessner U, Wagner C, Kopka J, Trethewey RN, and Willmitzer L. (2000). **Technical advance: simultaneous analysis of metabolites in potato tuber by gas chromatography-mass spectrometry**. *Plant Journal* 23, 131-142.

Roessner U, Willmitzer L, and Fernie AR. (2001a). **High-resolution metabolic phenotyping of genetically and environmentally diverse potato tuber systems. Identification of phenocopies**. *Plant Physiology* 127, 749-764.

Roessner U, Luedemann A, Brust D, Fiehn O, Linke T, Willmitzer L, and Fernie A. (2001b). **Metabolic profiling allows comprehensive phenotyping of genetically or environmentally modified plant systems**. *Plant Cell* 13, 11-29.

Roessner-Tunali U, Hegemann B, Lytovchenko A Carrari F, Bruedigam C, Granot D, and Fernie AR. (2003). **Metabolic profiling of transgenic tomato plants overexpressing hexokinase reveals that the influence of hexose phosphorylation diminishes during fruit development**. *Plant Physiology* 133, 84-99.

Sumner LW, Mendes P, and Dixon RA. (2003). **Plant metabolomics: large-scale phytochemistry in the functional genomics era**. *Phytochemistry* 626, 817-36.

Taylor J, King RD, Altmann T, and Fiehn O. (2002). **Application of metabolomics to plant genotype discrimination using statistics and machine learning**. *Bioinformatics* 18, S241-248.

Trethewey RN, Krotzky AJ, and Willmitzer, *L. (1999)* **Metabolic profiling : a Rosetta stone for genomics?** *Current Opinion in Plant Biology* 2, 83-85.

Wagner C, Sefkow M, and Kopka J. (2003). **Construction and application of a mass spectral and retention time index database generated from plant GC/EI-TOF-MS metabolite profiles**. *Phytochemistry* 62, 887-900.

Weckwerth W. (2003). **Metabolomics in Systems Biology**. *Annual Review of Plant Biology* 54, 669-689.

SECTION 5

MUTANTS, GENE-TAGGING, AND GENETIC ANALYSIS

Chapter 5.1

MUTAGENESIS

K Judith Webb[1*], Mark Robbins[1], Trevor L Wang[2], Martin Parniske[3], and Antonio Márquez[4]

*[1]Institute of Grassland and Environmental Research, Aberystwyth, UK; [2]John Innes Centre, Norwich, UK; [3]Sainsbury Laboratory, Norwich, UK; [4]Universidad de Sevilla, SPAIN, *Corresponding author.*

Email: judith.webb@bbsrc.ac.uk

Phone: +44 197 082 3124 Fax: +44 197 082 3243

Keywords: Ethyl methanesulphonate (EMS); gamma (γ-) irradiation; *Lotus japonicus*; mutagenesis; mutants; symbionts; secondary metabolism; tannins; starch

Mutants are central to elucidating gene function in many plant species. A range of symbiotic and other mutants have been created in legumes using both gamma (γ-) irradiation and chemical mutagenesis. Advantages and disadvantages of these two methods are presented, along with an outline of procedures for and indicators of successful mutagenesis in L. japonicus. *Screening procedures for symbiotic mutants, tannins and starch mutants are also described.*

INTRODUCTION

Both γ-irradiation and chemical mutagenesis cause mutations in plants, however, their mode of action and effectiveness differs. Gamma (γ-) irradiation causes lesions in DNA strands whilst the effects of chemical mutagens such as ethyl methanesulphonate (EMS) are subtler. EMS mainly causes single base pair changes (G/C to A/T) as opposed to large deletions or rearrangements (Sharma and Sharma, 1986), although it can cause small deletions (Schreck, 2002) and deletions and rearrangements (Barratt et al., 2001) depending on the position of the mutation.

Symbiotic and other mutants have been induced in legumes using a range of mutation approaches, including chemical mutagenesis using EMS (Blixt, 1972; Carroll et al., 1986; Duc and Messager, 1989; Kneen and Larue, 1988; Morandi et al., 2000; Nakata and McConn, 2000; Sagan et al., 1993; Wang et al., 1990) or N-nitroso-N-methyl urea (NMU; (Sharma and Sharma, 1986; Wang et al., 1990); γ-rays (Sharma and Sharma, 1986; Wang et al., 1990); and retrotransposition through

A.J. Márquez (ed.), Lotus japonicus Handbook, 177-186.

tissue culture (Krusell et al., 2002; Schauser et al., 1998; Schauser et al., 1999; Stracke et al., 2002).

In *Lotus japonicus,* successful chemical mutation programmes have been undertaken in different laboratories and mutants originating from them have been described in detail (Bonfante et al., 2000; Imaizumi-Anraku, 2003; Imaizumi-Anraku et al., 1997; Krusell et al., 2002; Orea et al., 2002; Senoo et al., 2000; Szczyglowski et al., 1998). These programmes have yielded a range of mutants in symbiotic interactions and secondary product biology. One of the most elegant uses of EMS mutants relates genotypic changes back to a phenotype using TILLING (Perry, 2003).

Here we present a broad outline of procedures for *L. japonicus* that can be adapted to suit particular purposes and local conditions. We strongly recommend that a pilot study be done first to determine the germination rate and plant fertility in treated seeds. It is well known that organism and tissues are extremely diverse in their response to mutagens. For example, in lentil not only were cultivars were found to differ in their sensitivity to mutagen, but roots were more sensitive than shoots (Sharma and Sharma, 1986). Variegated cotyledons and leaves (Wang et al., 1990) are used as indicators of successful mutagenesis. In an EMS mutagenesis programme with pea, treatment with 44 mM (0.55% v/v) EMS for 4 h resulted in 78% seed germination, 8.4% chlorophyll deficient seedlings and 16.4% plants reaching maturity and producing seed. We describe our experiences with γ-irradiation and EMS in *L. japonicus,* presenting results of pilot studies and a generalised procedure that can be adapted as required.

MATERIALS AND PROCEDURES

Control seed

Ensure a supply of good quality, highly inbred (F_9) seed of appropriate ecotypes or transformants of *Lotus japonicus* (Regel) Larsen. Test each new batch of seeds (100mg) to ensure high percentage (>90%) germination.

γ-irradiation

Seed can be irradiated using Cobalt 60 or Caesium 137.

Seed Batche	Krad	% Germ. (M_1)	% Survival (M_1)	% Albino (M_2)
Control	0	70	99.4	0.0
Batch I	5	71	96.0	0.6
Batch II	10	69	96.0	0.4
Batch III	20	78	89.0	0.4
Batch IV	30	72	80.0	0.4

Table 1: Effect of different γ irradiation doses on percentage germination and survival of different batches of L. japonicus *seed. 5000 seeds in each group.*

For the treatments described here, dry seed was irradiated with Cobalt 60 at the *Hospital Universitario Duque del Infantado* (Sevilla) by Prof Juan Carlos Mateo. Seed should be sown as soon as possible and progeny harvested.

Ethyl methanesulphonate (EMS)

Before working with EMS, check your local/national COSSH and disposal regulations for mutagenic materials. EMS is volatile and highly mutagenic. Chill EMS stock flask before opening to reduce fumes to a minimum. Carry out all handling and preparation of EMS materials in fume hood wearing protective clothing including rubber gloves. All solutions and equipment contaminated with EMS should be neutralised in 1M KOH or NaOH solution and allowed to react overnight before discarding. Do not release any EMS solution or any of the first washing steps into the environment. Whenever possible use plastic-ware and incinerate after contact with EMS. Purchase fresh stock of ethyl methanesulphonate (Sigma) since its activity is age dependent. Do a pilot study of EMS treatment on a small amount of seed (100 mg) to ensure appropriate germination of treated seed (50-80%). Wash EMS treated seed thoroughly by placing seeds in a closed nylon or muslin bag and transferring them to large volume of fresh, stirred water x 10 over a period of at least 3 h.

Seed Batch	**EMS Treatment**		**Scarified**	**% Germ. (M_1)**	**% Albino (M_2)**
	Percentage	**Hours**			
Control	0	0	+	75	0
EMS I	1.0	16	-	50	0
EMS II	0.3	16	+	76	3.8
EMS III	0.5	16	+	74	2.7
EMS IV	1.0	16	+	30	3.0
EMS V	1.0	8	+	63	2.6
EMS VI	1.0	4	+	80	1.7

Table 2. Effect of EMS concentration and length of treatment on indicators of mutagenesis in L. japonicus. *Groups of 1000 seeds were treated with EMS at the indicated concentrations and length of times. Results of mutagenesis on germination in M_1 generation and production of albinos in M_2 generation are shown. (Deduced germination in M_1 generation of* L. japonicus *with extended exposure to the same concentration (1% v/v) EMS.)*

Pilot treatment

1. Weigh seeds (100 seeds = approximately 110 mg).
2. Calculate volume EMS solution required according to weight of seed. For example, for 1g (dry weight) of seed, use 100 μl of EMS in 10 ml water for 1% v/v solution. ***Note***: When scaling up, increase the volume of solution used to keep the ratio between seed dry weight : μl EMS constant.

3. Prepare and scarify seeds. Test seed germination using method a) first, since some batches of seed do not need any treatment with acid.
   a. Rub 0.5 g batches seed between 2 pieces of 320 grit emery paper for 30 sec, **OR** Add concentrated H_2SO_4 (98% v/v) to seeds in a glass beaker and mix for 5-45 min depending on seed batch. Remove as much acid as possible, add a large quantity of water, and quickly remove this first wash since diluted acid can penetrate the scarified seed coat and kill the embryo. Add the removed acid slowly to a large volume tap water to reduce generation of heat (in a fume hood). Wash immediately and extensively with tap water.
4. Imbibe seeds in water for 3 h at 20°C and wash in a minimum of 50 ml water x 10 times and remove excess water of final wash. (This extensive washing step is to remove all organic matter and debris released from the seed coat. The surrounding water will be yellow after seed is fully imbibed. If not removed, the organic matter will react with EMS reducing the active concentration of EMS.)
5. Add 10 ml water followed by calculated volume EMS directly to immersed seed as it is unstable in water.
6. Incubate with gentle shaking 3 - 16 h in EMS at 20°C fume hood.
7. Remove as much EMS solution as possible. Inactivate the first two washes as described in general comments section 1) above.
8. Wash extensively (general comments section 4) and sow seeds in 9 cm Petri dishes with 10 x 9 cm Whatman No. 1 filter paper. Incubate at 20°C, 18 h photoperiod at 200 $\mu M\ m^{-2}\ sec^{-1}$. Check which treatment gives a percentage germination that falls within the desired limits and, if time permits, transfer to pots and grow plants to maturity to check their survival and fertility.

Full treatment

1. Repeat procedure for a larger number of seeds, remembering to keep the ratio between seed dry-weight : µl EMS constant.
2. Treat a small population (1g) control non-EMS treated seeds.

Scoring of phenotype and harvesting of M_2 seed

1. Sow scarified seeds in either trays or pots with a) or b) as a growing medium as required
   a. Inert clay material in trays or washed vermiculite (medium grade), watered weekly with nutrients with low levels of nitrogen (5mM).
   b. Levington F1 (Scotts, Suffolk IP8 4BZ; a peat-based compost with low nutrient content and fine structure). This has been used very successfully for the starch screen and for defective nodulation, especially if sown in small plugs that generate pot-bound plants at the time of screening (6-8 weeks after sowing).

2. Score seedlings (M_1 generation) for obvious phenotypes, especially appearance of variegated leaves after 50 days.
3. Harvest M_2 seed from individual plants, sow, and score for appearance of albino plants.

Note: In one EMS treatment (EMS 3; Table 3), 6g seed was mutagenised with 1% (v/v) EMS for 16 h; 1000 seeds were treated similarly but not subjected to EMS. The seeds were sown on Whatman No. 1 filter paper and germinated seedlings manually transferred to vermiculite watered with 5 mM nitrogen. After 50 days, seedlings were scored for obvious phenotypes, especially appearance of variegated leaves. Seed germination was reduced to 54%, 2.8% M_1 seedlings had variegated leaves (3rd to 6th leaves), 15.5% of the original EMS treated M_1 seed population reached maturity and produced M_2 seed (Fig. 1). Fig. 2 shows the cumulative percentage of variant phenotypes in M_1 and M_2 generations: variegated leaves (2.8%), symbiosis (0.45%), condensed tannin mutants (0.32%), and other morphological mutants (0.23%). Of these M_2 seedlings, 1.8% were albino, which is similar to observations for pea (Wang et al., 1990). In other batches, treating *L. japonicus* seeds with 1% EMS for 16h produced M_2 albino plants ranging from 0.3% to 3.6% of the population (Table 3).

Mutagenesis Procedure	% Albino (M_2 generation)
EMS 1	1.4
EMS 2	3.6
EMS 3	1.8
EMS 4	0.8
EMS 5	3.0
EMS 6	1.1
EMS 7	0.3

Table 3. Proportion of albino plants found in the M_2 of seven different EMS mutagenesis processes carried out using 1% EMS for 16 h. Batches of 1000 seeds were treated.

Screening procedures

Nodulation

1. Inoculate seedlings with appropriate *Mesorhizobium loti* strain grown on Yeast Mannitol agar slopes (Vincent, 1970) at 28°C for 24 h and mixed with water or nutrient solution. To ensure maximum nodulation, inoculate twice at 2 and 4 weeks after sowing.
2. Examine roots for presence or absence of nodules (nodulation mutants) or white nodules (nitrogen fixation mutants), 4-6 weeks after nodulation.
3. Re-pot potential mutants in low nutrient compost. By not providing extra nutrients, but re-potting putative mutants twice, there is a second opportunity to screen the putative nodulation mutants. Repeat as necessary.

Note: Of the six stable symbiotic mutants identified in EMS 3 (Table 3; Figures 1 and 2), one of the nodulation mutants *Ljsym*4-2 was also defective in its ability to sustain arbuscular mycorrhiza infection (Bonfante et al., 2000).

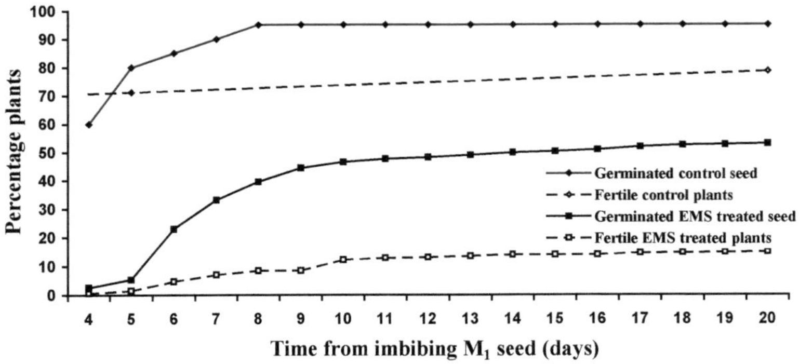

Figure 1. Percentage seedlings germinating and percentage fertile (M_1) plants generated from control and EMS treated seed (M_1) of L. japonicus, *plotted against time of germination of control and EMS treated M_1 seed.*

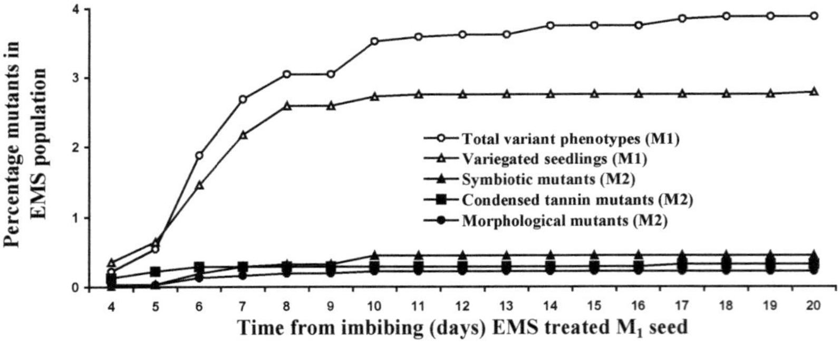

Figure 2. Cumulative percentage of mutant plants identified in EMS-treated M_1 and M_2 populations of L. japonicus *plotted against time of germination of EMS treated M_1 seed.*

Arbuscular mycorrhiza

1. Maintain stock cultures of arbuscular mycorrhiza fungus *Glomus intraradices* (BEG ID 195) on chive plants and use as nurse cultures (Wegel et al., 1998).
2. Sow seeds around a single chive plant with established mycorrhiza.
3. Harvest at least 1 cm sections of root and clear in 10% w/v KOH at 90°C for 2 h.

4. Rinse roots with water and soak in 1% v/v HCl 30 min.
5. Transfer roots to acidified trypan blue solution (0.05% w/v trypan blue, 50% v/v glycerol, 0.5% v/v HCl) at 90°C for 2 h.
6. De-stain by repeated soaking in acidified glycerol (50% v/v glycerol, 0.5% v/v HCl).
7. Mount roots on glass slides and examine at x200 magnification for presence and appearance of surface hyphae, penetration points and internal hyphae, vesicles and arbuscules.

Note: This screen is both time and labour consuming, requiring careful sample tracking. Make sure that EMS mutagenised population has a sufficiently high hit rate before embarking on such a project.

Condensed tannins in petals and leaves

1. Decolourise petals of leaves overnight in methanol, then subject to butanol-HCl hydrolysis.
2. Identify and quantify condensed tannins by measuring levels of anthocyanin in hydrolysates (Stafford, 1980).
3. To confirm condensed tannin phenotype in leaves, stain specifically for condensed tannin-containing cells using dimethylamino-cinnamaldehyde (Li, 1996) and observe under microscope.

Note: Leaves of *L. japonicus* contain no tannin cells in mesophyll tissues so the screen is for the presence of condensed tannins. Gain-of-function condensed tannin mutants are reported when leaf tissues consistently give rise to anthocyanins after hydrolysis. Some lines give rise to colour before boiling in butanol-HCl but these can be scored as negative since this phenotype is consistent with the release of anthocyanin and 3,4 diol conjugates rather than the hydrolysis of condensed tannin polymers (Skadhauge, 1996).

Starch (Harrison, 1998)

1. Remove chlorophyll pigments in hot aqueous ethanol.
2. Stain leaves with Lugol's iodine solution (Sigma), which stains most starches blue/black.

Note: If collected at the end of the day or late in the day, leaves will be full of starch and will consequently stain blue/black, mutants will appear lighter or yellow. If collected at the beginning of the day after the dark period, normal leaves will be depleted of starch and hence will appear yellow; mutants will appear blue/black. Thus plants that are defective in their ability to synthesise starch or to break it down at night can be readily detected based on the contrasting colours of mutant and non-mutant leaves. *L. japonicus* is particularly convenient for use in this screen as the leaflets fit 96-well plates and a high throughput can be achieved relatively easily using minimum materials (Wang et al., 1990).

Photorespiration (Orea et al. 2002 , Márquez et al. 2005)

1. Grow M_2 plants in a high-CO_2 atmosphere (0.7% v/v), in which photorespiration is practically suppressed.
2. After 3-4 weeks, transfer the plants from high CO_2 conditions to ordinary air (active photorespiration). In 2 or -3 days, some of the plants will start to develop symptoms of chlorosis, after 1 week the leaves will be dry and fall.
3. Place these plants back under high CO_2 conditions; those recovering a normal phenotype can be considered as putative photorespiratory mutants.

Note: Under the conditions used for mutagenesis (1% v/v EMS for 16 h) the percentage of germination in M_1 plants dropped to 45-65% while final viability was in the range 6-20 % of the sown seed. The screening of 27,100 M_2 plants arising from 660 independent M_1 plants yielded 24 putative photorespiratory mutants from which 3 have been identified as being affected in glutamine synthetase.

Chlorate resistance (Orea, 1999, Márquez et al. 2005)

1. Grow M_2 plants for about 1 month in Hornum nutrient medium (Handberg and Stougaard, 1992).
2. Deplete the plants of nitrogen for 4 days by transfer to N-free Hornum nutrient medium supplemented with 3 mM KCl.
3. Feed the plants for 1-2 weeks with 1 mM $KClO_3$, in Hornum nutrient medium containing 1 mM KNO_3 as the N-source. Most plants will die, but plants that remain green can be rescued in ordinary Hornum nutrient medium lacking chlorate.

Note: Chlorate resistance has been a common way to isolate mutants deficient in nitrate assimilation, either on nitrate reductase activity or nitrate uptake. The likely explanation for that is that nitrate reductase, in the presence of chlorate, is also able to catalyse the reduction of chlorate to chlorite, which is toxic to the cell. An ample variety of selection conditions were tried in our case for *L. japonicus,* by varying nitrate and chlorate concentrations, in the presence ar absence of organic C sources, both in Petri dishes or seed trays filled with vermiculite. In all cases, the conditions described above were found as possibly the most satisfactory. Around 250 plants were initially selected from examination of 100,000 M_2 / M_3 plants arising from 6,000 M_1 plants produced by mutagenesis with 1% v/v EMS treatment for 16 h. Most of the selected plants died while they were recovering from chlorate. None of the selected plants showed deficiency on nitrate reductase activity. Only two mutant plants showed genetic inheritance of the chlorate resistance mutations in selfed progeny.

REFERENCES

Barratt DHP, Barber L, Kruger NJ, Smith AM, Wang TL, and Martin C. (2001) **Multiple, distinct isoforms of sucrose synthase in pea.** *Plant Physiology* 127, 655-664.

Blixt S. (1972) **Mutation genetics in pea.** *Agri Hortique Genetica,* 1-293.

Bonfante P, Genre A, Faccio A, Martini I, Schauser L, Stougaard J, Webb J, and Parniske M. (2000) **The *Lotus japonicus* LjSym4 gene is required for the successful symbiotic infection of root epidermal cells.** *Molecular Plant-Microbe Interactions* 13, 1109-1120.

Carroll BJ, McNeil DL, and Gresshoff PM. (1986) **Mutagenesis of soybean (*Glycine max* (L) MERR.) and the isolation of non-nodulating mutants.** *Plant Science* 47, 109-114.

Duc G and Messager A. (1989) **Mutagenesis of pea (*Pisum sativum* L.) and the isolation of mutants for nodulation and nitrogen fixation.** *Plant Science* 60, 207-213.

Harrison C, Hedley CL, and Wang TL. (1998) **Evidence that the rug3 locus of pea (*Pisum sativum L.*) encodes plastidial phosphoglucomutase confirms that the imported substrate for starch synthesis in pea amyloplasts is glucose6phosphate.** *The Plant Journal* 13, 753-7562.

Imaizumi-Anraku H. (2003) **Legume nodulation and mycorrhiza formation-common genes are involved in both symbiotic interactions.** *Nippon Nogeikagaku Kaishi-Journal of the Japan Society for Bioscience Biotechnology and Agrochemistry* 77, 121-123.

Imaizumi-Anraku H, Kawaguchi M, Koiwa H, Akao S, and Syono, K. (1997) **Two ineffective-nodulating mutants of *Lotus japonicus* - Different phenotypes caused by the blockage of endocytotic bacterial release and nodule maturation.** *Plant and Cell Physiology* 38, 871-881.

Kneen BE and Larue, TA. (1988) **Induced Symbiosis Mutants of Pea (*Pisum-Sativum*) and Sweetclover (*Melilotus-Alba-Annua*).** *Plant Science* 58, 177-182.

Krusell L, Madsen LH, Sato S, Aubert G, Genua A, Szczyglowski K, Duc G, Kaneko T, Tabata S, de Bruijn F, Pajuelo E, Sandal N, and Stougaard J. (2002) **Shoot control of root development and nodulation is mediated by a receptor-like kinase.** *Nature* 420, 422-426.

Li Y-G, Tanner GJ, and Larkin PJ. (1996) **The DMACA-HCl protocol and the threshold proanthocyanidin content for bloat-safety in forage legumes.** *Journal of Science, Food and Agriculture* 70, 89-101.

Márquez AJ, Betti M, García-Calderón M, Pal'ove-Balang P, Díaz P, Monza J (2005) **Nitrate assimilation in *Lotus japonicus*.** *Journal of Experimental Botany* (in press).

Morandi D, Sagan M, Prado-Vivant E, and Duc G. (2000) **Influence of genes determining supernodulation on root colonization by the mycorrhizal fungus *Glomus mosseae* in *Pisum sativum* and *Medicago truncatula* mutants.** *Mycorrhiza* 10, 37-42.

Nakata PA, and McConn MM. (2000) **Isolation of *Medicago truncatula* mutants defective in calcium oxalate crystal formation.** *Plant Physiology* 124, 1097-1104.

Orea A (1999) **Nitrate assimilation in the model legume *Lotus japonicus*.** *Ph. D. Thesis, University of Seville (Spain).*

Orea A, Pajuelo P, Pajuelo E, Quidiello C, Romero JM, and Márquez AJ. (2002) **Isolation of photorespiratory mutants from Lotus japonicus deficient in glutamine synthetase.** *Physiologia Plantarum* 115, 352-361.

Perry JA, Wang TL, Welham TJ, Gardner S, Pike JM, Yoshida S and Parniske M. (2003) **A TILLING Reverse Genetics Tool and a Web-Accessible Collection of Mutants of the Legume *Lotus japonicus*.** *Plant Physiology* 131, 866-871.

Sagan M, Ney B, and Duc G. (1993) **Plant Symbiotic Mutants as a Tool to Analyze Nitrogen Nutrition and Yield Relationship in Field-Grown Peas (*Pisum-Sativum L*).** *Plant and Soil* 153, 33-45.

Schauser L, Handberg K, Sandal N, Stiller J, Thykjaer T, Pajuelo E, Nielsen A, and Stougaard J. (1998) **Symbiotic mutants deficient in nodule establishment identified after T-DNA transformation of *Lotus japonicus*.** *Molecular and General Genetics* 259, 414-423.

Schauser L, Roussis A, Stiller J, and Stougaard J. (1999) **A plant regulator controlling development of symbiotic root nodules.** *Nature* 402, 191-195.

Schreck S. (2002) **The role of plastidial phosphoglucomutase in carbon partitioning.** *PhD thesis. University of East Anglia UK.*

Senoo K, Solaiman MZ, Kawaguchi M, Imaizumi-Anraku H, Akao S, Tanaka A, and Obata H. (2000) **Isolation of two different phenotypes of mycorrhizal mutants in the model legume plant *Lotus japonicus* after EMS-treatment.** *Plant and Cell Physiology* 41, 726-732.

Sharma SK and Sharma B. (1986) **Mutagen sensitivity and mutability in lentil.** *Theoretical and Applied Genetics* 71, 820-825.

Skadhauge B. (1996) **Genetics and Biochemistry of proanthocyanidin biosynthesis and their biological significance in crop plants.** *PhD thesis. The Royal Vetinary and Agriculture University, Copenhagen, Denmark.*

Stafford HA, and Cheng T-Y. (1980) **The procyanidins of douglas fir seedlings, callus and cell suspension cultures derived from cotyledons.** *Phytochemistry* 19, 131-135.

Stracke S, Kistner C, Yoshida S, Mulder L, Sato S, Kaneko T, Tabata S, Sandal N, Stougaard J, Szczyglowski K, and Parniske M. (2002) **A plant receptor-like kinase required for both bacterial and fungal symbiosis.** *Nature* 417, 959-962.

Szczyglowski K, Shaw RS, Wopereis J, Copeland S, Hamburger D, Kasiborski B, Dazzo FB, and de Bruijn FJ. (1998) **Nodule organogenesis and symbiotic mutants of the model legume *Lotus japonicus*.** *Molecular Plant-Microbe Interactions* 11, 684-697.

Vincent J. (1970) **A Manual for the Practical Study of Root-nodule Bacteria.** *Blackwell Scientific Publications, Oxford and Edinburgh,* 164.

Wang TL, Hadavizideh A, Harwood A, Welham TJ, Harwood WA, Faulks R, and Hedley CL. (1990) **An Analysis of Seed Development.** In: *Pisum sativum* XIII The Chemical Induction of Storage Product Mutants. *Plant Breeding* 105, 311-320.

Wegel E, Schauser L, Sandal N, Stougaard J, and Parniske M. (1998) **Mycorrhiza mutants of Lotus japonicus define genetically independent steps during symbiotic infection.** *Molecular Plant-Microbe Interactions* 11, 933-936.

Chapter 5.2

A MUTANT CATALOGUE OF *LOTUS JAPONICUS*

Eloisa Pajuelo

Department of Microbiology and Parasitology, Faculty of Pharmacy, University of Sevilla. Prof. García González s/n, 41012-Sevilla, SPAIN

Email: epajuelo@us.es

Phone: +34 954 559 895 Fax: +34 954 628 162

Below are tables describing symbiotic, morphological, metabolic, and photorespiratory mutant lines (alleles). A more exhaustive catalogue of published and unpublished mutants and their gene mapping positions is now being prepared by Stougaard and Sandal and will soon be published elsewhere.

SYMBIOTIC MUTANTS

Mutant line	Phenotype	Description	References
LjNfr1= LjSym1-1 Ljsym1-2	Nod-, Ami+	No first responses after application of purified Nod factors (No depolarization nor extra-cellular alkalinisation). *Nfr1* encodes a receptor kinase involved in the recognition of Nod factors. The protein has a LysM domain in the extracellular moiety, involved in the recognition of the backbone of Nod-factor, a transmembrane domain, and a Ser/Thr kinase domain in the intracellular part. The simplest model is that NFR1 act together with NFR5 in the recognition of the Nod-factor.	Radutoiu et al. (2003), Nature 425, 585-592. Madsen et al. (2003), Nature 425, 637-640.
LjNfr5= LjSym5	Nod-, Ami+	No first responses after application of purified Nod factors (no depolarization or extracellular alkalinization). *Nfr5* encodes a receptor kinases involved in Nod factor recognition. The *Lj*NFR5 PK domain is atypical since it lacks the activation domain.	Radutoiu et al. (2003), Nature 425, 585-592. Madsen et al. (2003), Nature 425, 637-640.

A.J. Márquez (ed.), Lotus japonicus Handbook, 187-195.

282-287 (LjSym2-1), 282-288 (LjSym 2-2) Allelic to cac41.5 (allelic to LjEMS61 and 282-287) LjEMS61	Coi-, Nod-	Blocked infection by *Glomus intraradices* within rhizodermal cells, hyphae swelling, and deformation. Exaggerated swelling and deformation of root hairs after *Rhizobium* infection, not root hair curling, no infection threads formation. Gene cloned: SYMRK, encodes a symbiosis receptor kinase with three domains: a signal peptide, an extracellular domain comprising leucine-rich repeats, a transmembrane and an intracellular protein kinase domain. *Lotus SYMRK* is required for both rizobial and mycorrhizal symbiosis.	Schauser et al., (1998), Mol. Gen. Genet. 259, 414-423. Wegel et al. (1998), MPMI 11 , 933-936. Stracke et al. (2002), Nature 417, 959-962.
Ljsym21-1= LJEMS34, Ljsym21-2= LJEMS61 (allelic to Ljsym2)	Nod-	Fungal infections arrested in epidermal cells, hyphae swelling, and deformation. Exaggerated swelling and deformation of root hairs after *Rhizobium* infection, not root hair curling, no infection threads formation. Gene cloned: *SYMRK*, symbiosis receptor kinase.	Szczyglowski et al. (1998), MPMI 11, 684-697 Stracke et al. (2002), Nature 417, 959-962
Sym 16 = har = 1-3, (allelic to LjEMS102 = har 1-1, and Ljsym78)	Nod+++, fix+, Nts	Hyper-nodulating, nodulation resistant to nitrate, fails in autoregulation of nodulation, short and much branched root system. Gene cloned: The *har1* gene is involved in long-distance signalling of nodulation auto-regulation (shoot controlled auto-regulation of nodule number) and lateral root development. *Har1* encodes a receptor kinase with a LRR repeat, a transmembrane domain and an intracellular serine/threonine kinase domain with homology to Arabidopsis CLAVATA1	Schauser et al. (1998), Mol. Gen. Genet. 259, 414-423. Wopereis et al. (2000) The Plant Journal 23, 97-114. Krussell et al (2002), Nature 420, 422-426. Nishimura et al. (2002) Nature 420, 426-429.
LjEMS102 =Ljsym34-1 (har1-1)	Nod+++, brl, Nts	Hyper-nodulating, nodulation resistant to nitrate, fails in autoregulation of nodulation, short and much branched root system. Gene cloned: *HAR1*, receptor-like kinase involved in shoot autoregulation of nodule number.	Szczyglowski et al. (1998) MPMI 11, 684-697. Wopereis et al. (2000) The Plant Journal 23, 97-114. Krussell et al. (2002), Nature 420, 422-426. Nishimura et al. (2002) Nature 420, 426-429. Jiang and Gresshoff, (2002) Functional

			Plant Biology 29, 1371-1376.
Ljsym78 -1 Ljsym 78-2 (alellic to Ljsym16 and har1)	Nod+++, Arb++	RS: hyper-nodulating, branched root system AMS: Stimulated growth and activity of arbuscules, but decreased growth of hyphae and vesicles	Kawaguchi et al. (2002), MPMI 15, 17-26. Solaiman et al. (2000) Journal of Plant Research, 113, 443-448.
Nin	Nod-	Arrested at the stage of bacterial recognition, nodule inception, no formation of infection threads or nodule primordia, but normal root hair curling. T-DNA tagged. Gene cloned: *Nin*, plant growth regulator of nodule initiation with a DNA binding domain. NIN protein has regional similarity to transcription factors, and the predicted DNA-binding/dimerisation domain identifies and typifies a consensus motif conserved in plant proteins with a function in nitrogen-controlled development	Schauser et al. (1999), Nature 402, 191-195.
Ljsym77= astray	Nod++	Double number of nodules on a wider area of root, early initiation of nodule development. Affected in light and gravity responses, agravitropic lateral roots. Gene cloned: *LjBzf* homologous to *A. thaliana hy5* encodes a basic leucine zipper (bZIP)-type zinc finger protein with different roles in shoot (photomorphogenic development) and roots (lateral root formation and number of nodules).	Nishimura et al. (2002), Plant Cell Physiol. 43, 853-859. Nishimura et al. (2002) PNAS 99, 15206-15210.
Ljcbp1 / T-90	Nod-	Transgenic line isolated by promoter tagging, no nodulation, affected in calcium binding protein homologue	Webb et al. (2000), MPMI 13, 606-616.
Sym8	Fix-	Small white bumps, infection threads arrested at the 2nd or 3rd cell layer into the root, ending in a balloon-like structure. T-DNA tagged.	Schauser et al. (1998) Mol. Gen. Genet. 259, 414-423.
Sym13	Fix- Ami+	Affected late in nodule development. Small pink non-fixing nodules, growth of the nodules resumes after 6 weeks. T-DNA tagged.	Schauser et al. (1998), Mol. Gen. Genet. 259, 414-423.
Ljsym 3-1 LjSym 3-2	Coi-, Nod-	Blocked infection by *Glomus intraradices* within rhizodermal cells, hyphae deformation. No nodule formation.	Schauser et al. (1998), Mol. Gen. Genet. 259, 414-423. Wegel et al. (1998).

			MPMI 11, 933-936
LjSym 4-1, Ljsym4-2	Coi-, Nod-	Blocked infection by *Glomus intraradices* within epidermal cells, hyphae deformation, disorganisation, and disassembly of microtubules and microfilaments during fungal penetration attempts. No root hair curling or infection thread formation.	Schauser et al. (1998) Mol. Gen. Genet. 259, 414-423. Wegel et al. (1998) MPMI 11, 933-936. Bonfante et al. (2000) MPMI 13, 1109-1120. Genre and Bofante, (2002). Protoplasma, 219, 43-50.
Sym 6-1, Sym 6-2	Fix-	Small white bumps	Schauser et al. (1998) Mol. Gen. Genet. 259, 414-423.
Sym7	Fix-, Ami+	Small white bumps	Schauser et al. (1998) Mol. Gen. Genet. 259, 414-423.
Sym9	Fix-	Small white bumps	Schauser et al. (1998) Mol. Gen. Genet. 259, 414-423.
Sym10	Fix-	Small white bumps	Schauser et al. (1998) Mol. Gen. Genet. 259, 414-423.
Sym 11	Fix-	Young pinkish nodules, no nitrogen fixation	Schauser et al. (1998) Mol. Gen. Genet. 259, 414-423.
Sym12	Fix-	Pink non fixing nodules, sot sever symptoms of nitrogen starvation: light green leaves but no red purple stems	Schauser et al. (1998) Mol. Gen. Genet. 259, 414-423.
Sym14	Fix-	Pink non-fixing nodules. After 6 weeks, the main root degenerates, and adventitious roots form at the hypocotyl.	Schauser et al. (1998) Mol. Gen. Genet. 259, 414-423.
Sym 15	Nod↓ Fix+	Reduced number of nodules and development is slower than in wild type nodules.	Schauser et al. (1998) Mol. Gen. Genet. 259, 414-423.
Ljsym22-1= LJEMS46, (allelic to Ljsym4-1 and Ljsym4-2), LjEMS40	Nod-	No macroscopically visible nodules on normal looking roots. No calcium speaking upon inoculation with *Rhizobium.* The double mutant LjEMS40 exhibits a bushy root phenotype and it is non-nodulating.	Szczyglowski et al. (1998) MPMI 11, 684 697. Harris et al. (2003), MPMI 16, 335-341.

(digenic)			
Ljsym23-1= LJEMS70, Ljsym23-2= LjEMS167	Nod-	No macroscopically visible nodules on normal looking roots	Szczyglowski et al., (1998) MPMI 11, 684 697.
Ljsym24= LJEMS76	Nod-	No macroscopically visible nodules on normal looking roots	Szczyglowski et al. (1998) MPMI 11, 684 697.
Ljsym25= LJEMS223	Nod-	No macroscopically visible nodules on normal looking roots	Szczyglowski et al. (1998) MPMI 11, 684 697.
LjEMS237	Nod-	No macroscopically visible nodules on normal looking roots	Szczyglowski et al. (1998) MPMI 11, 684 697.
Ljsym26= LJEMS247	Nod-	No macroscopically visible nodules on normal looking roots	Szczyglowski et al. (1998) MPMI 11, 684 697.
Ljsym27= LjEMS81	Nod-*	Occasional presence of few, very small, white round nodule-like structures, leaky phenotype	Szczyglowski et al. (1998) MPMI 11, 684 697.
Ljsym28 = LjEMS236	Nod-*	Occasional presence of few, very small, white round nodule-like structures, leaky phenotype	Szczyglowski et al. (1998) MPMI 11, 684 697.
LjEMS45 = LjEMS88 = LjEMS217	Nod+, Fix-	Extensive nodulation in the upper and lower part of the root, but no nitrogenase activity	Szczyglowski et al. (1998) MPMI 11, 684 697.
LjEMS126 = Ljsym30	Nod+, Fix-	Infrequent, small, white nodule-like structures without nitrogenase activity Calcium speaking but no root hair deformation upon inoculation with *Rhizobium*	Szczyglowski et al. (1998) MPMI 11, 684 697. Harris et al. (2003) MPMI 16, 335-341.
LjEMS75= Ljsym31	Nod+, Fix +/-	White nodules and some pink wild type nodules, no nitrogenase activity	Szczyglowski et al. (1998) MPMI 11, 684 -697.
LjEMS208 = Ljsym32	Nod+, Fix +/-	Small pink nodules with low nitrogenase activity, stunted and chlorotic plants	Szczyglowski et al. (1998) MPMI 11, 684 -697.

LjEMS79	Nod+, Fix +/-	Leaky phenotype: some plants developed no nodules and some plants developed white, nodule like structures or few pink nodules	Szczyglowski et al. (1998) MPMI 11, 684 -697.
Ljsym70	Nod-, Ami+	Non-nodulating, but small bumps form of roots after prolonged incubation with *M. loti*. Advanced stage in the map-based cloning approach, HEGS/AFLP saturation mapping	Kawaguchi et al. (2002) MPMI 15, 17-26. Murakmi et al. (2002) Plant and cell Physiology, special issue 43, S124.
Ljsym71-1, Ljsym71-2	Nod-, Coi-	AMS: overproduction of deformed appressoria, arrested hyphae penetration in the exodermis	Senoo et al. (2000) Plant cell Physiol. 41, 726-732. Kawaguchi et al. (2002) MPMI 15, 17-26.
Ljsym72	Nod-, Coi-	AMS: poor development of mycelium, branched appressoria, nor hyphal penetration in the epidermis, neither formation of arbuscules or vesicles.	Senoo et al. (2000) Plant Cell Physiol. 41, 726-732. Kawaguchi et al. (2002) MPMI 15, 17-26.
Ljsym73	Nod-	Nodulation at very low frequency	Kawaguchi et al. (2002) MPMI 15, 17-26
Alb1 = *Alb 2*= Ljsym74	Hist-	Empty nodules lacking vascular system, bacteria remain in enlarged infection threads and fail to differentiate into bacteroids	Imaizumi-Anraku et al. (1997) Plant and Cell Physiology 38, 871-881. Kawaguchi et al. (2002) MPMI 15, 17-26. Imaizumi et al (2000) Mol. Gen. Genet. 264, 402-410.
sen1 = Ljsym75	Fix-	Stationary endosymbiont nodule, ineffective nodules, twice the nodule number, rhizobial cells fail to develop into bacteroids, many small vacuoles, early nodule senescence. Retarded plant growth.	Kawaguchi et al. (2002) MPMI 15, 17-26. Suganuma et al. (2003) Mol. Genet. Genomics 269, 312-320.
fen1 = Ljsym76	Fix-	Bacteria differentiate into bacteroids, but nodule size fails to increase due to cell	Imaizumi-Anraku et al. (1997) Plant and

		expansion, low nitrogenase activity	Cell Physiology 38, 871-881. Kawaguchi et al. (2002) MPMI 15, 17-26.
crinkle = Ljsym79	Hist-	Affected in infection thread growth, and in pollen tube growth. Many white small bumps and only few pink nodules, crinkly trichomes, short seed pods with some aborted embryos, swollen root hairs	Kawaguchi et al. (2002) MPMI 15, 17-26. Tansengco et al. (2003) Plant Physiol. 131, 1054-1063. Tansengco et al. (2004) Plant and cell Physiology, 45, 511-520.
Sym81	Fix-	Many small vacuoles in infected cells	Kawaguchi et al. (2002) MPMI 15, 17-26.
Sym82	Nod-, Ami-	Required for symbiosis of rhizobia and arbuscular mycorrhiza	Hayashi et al. (2004) Plant and Cell Physiology, Special Issue 45, S138.
Bel		Beading nodules	Hayashi et al., unpublished results, cited in VandenBoschs and Stacey (2003) Plant Physiol. 131, 840-865
Trinity	Nod-, Ami-	*Trinity* controls nodulation and micorrhization concomitantly	Imaizumi-Anraku et al. (2004) Plant and Cell Physiology Special Issue. 45, S137.
56M			Kumagai et al. (2004) Plant and Cell Physiology, Special Issue 45, S187.

MORPHOLOGICAL MUTANTS

Mutant line	Phenotype	Description	References
slp	nod↓	Slippery root, lacks root hairs	Kawaguchi et al. (2002) MPMI 15, 17-

			26
rdo1	nod↓	Affected in radial organisation, thick and short root, counter clockwise-directed leaflets	Kawaguchi et al. (2002) MPMI 15, 17-26.
Pfo	Flower mutant	Flowers are replaced by a large number of sepal-like organs, without sepals or stamens, and carpels are almost absent. Mutation causes sterility. Ac-tagged mutant, *pfo* has been isolated. Itencodes a F-box protein showing similarity to *Fimbriata* (*Antirrhinum*) and UNUSUAL FLOWER ORGANS (*Arabidopsis*) and to the protein Stamina pistilloida from *Pissum*. Pfo regulates floral organ identity.	Zhang et al. (2003) The Plant Journal, 33, 607 – 619.
sleepless		Affected in nyctinastic movement, unable to close leaflets toward the adaxial side at night. Pulvini at the leaflet base are replaced by a petiole-like structure.	Kawaguchi (2003) J. Plant Res. 116, 151-154.
Stipule-less mutants			Ishikawa et al. (2004) Plant and Cell Physiology, Special Issue 45, S163.
Stem	539 mutants	TILLING	Perry et al. (2003) Plant Physiology 131, 866-871.
Root	29 mutants	TILLING	Perry et al. (2003) Plant Physiology 131, 866-871.
Architecture	189 mutants	TILLING	Perry et al. (2003) Plant Physiology 131, 866-871.
Flower	440 mutants	TILLING	Perry et al. (2003) Plant Physiology 131, 866-871.
Fruit	775 mutants	TILLING	Perry et al. (2003) Plant Physiology 131, 866-871.
Leaf	1268 mutants	TILLING	Perry et al. (2003) Plant Physiology 131, 866-871.
Leaflets	1008 mutants	TILLING	Perry et al. (2003) Plant Physiology 131, 866-871.

Nodules	627 mutants	TILLING	Perry et al. (2003) Plant Physiology 131, 866-871.

METABOLIC MUTANTS

Mutant line	Phenotype	Description	References
Tan1	Accumula-tion of leaf proantho-cyanidin.	The gene *Tan1* is highly homologous to basic-helix- loop-helix (*myc*-like) genes; Expression of *TAN*1 correlates well with proanthocyanidin content in leaves. The gene is interrupted in the mutant.	Gruber et al. 2003, meeting abstract
Viridicaulis		Anthocyanine accumulation	T Aoki, unpublished results, cited in van den Boschs and Stacey (2003) Plant Physiol. 131, 840-865.

PHOTORESPIRATORY MUTANTS

Mutant line	Phenotype	Description	References
LjPR1, LjPR2	Photorespi-ratory mutants	Poor growth in air. Symptoms of chlorosis and some necrotic spots in leaves. Growth is restored in high CO_2. Affected in chloroplastic glutamine synthetase GS2	Orea et al. (2002), Physiol. Plant. 115, 352-361.

Chapter 5.3

TILLING

Jillian Perry[1], Tracey Welham[2], Soizic Cheminant[1], Martin Parniske[1], and Trevor Wang[2*]

*[1]The Sainsbury Laboratory, [2]John Innes Centre; Colney Lane, Norwich, NR4 7UH, UK; *Corresponding author.*

Email: trevor.wang@bbsrc.ac.uk

Tel: +44 1603 450 283 Fax: +44 1603 450 014

Keywords: TILLING, CEL I, reverse genetics, mutants, functional genomics, legumes, mutations

TILLING (Targeted Induced Local Lesions in Genomes; McCallum et al. *2000) is a reverse genetic tool for the identification of point mutations in genes of interest within EMS mutagenised populations (Till* et al. *2003, Wienholds* et al. *2003, Smits* et al. *2004), facilitating the efficient screening of a considerable range of mutant alleles for their relation to gene function. TILLING employs the mismatch specific endonuclease CELI which enzymatically identifies any single nucleotide polymorphism between PCR products by recognising sites of mismatch and cleaving the DNA. Several TILLING projects have been instigated in a variety of organisms such as* Arabidopsis thaliana, Medicago truncatula, *pea, soybean, maize, nematode, rat and zebrafish. In the following chapter, an outline of the procedure and methodology is described, and the initial results from the* Lotus japonicus *TILLING project are presented.*

INTRODUCTION

Over the last decade there has been a push to sequence the genomes of more and more organisms (the EU Arabidopsis Genome project 1998, the *Caenorhabditis* Sequencing Consortium 1998; International Human Genome Sequencing Consortium 2001; http://www.kazusa.or.jp/en/database.html; http://medicago.org/ genome/) resulting in the initiation of several large scale projects which are generating a plethora of genomic data, but little information as to the function of individual genes. As detailed in chapter 5.1, "mutants are central in elucidating gene function in many plant species" and TILLING provides a way of identifying these individuals. TILLING utilises the mismatch-specific CELI endonuclease (Yang *et al.* 2000) that cleaves at sites of mismatch and provides a simple high throughput method of screening a large EMS mutagenised population of plants. The technique is carried out in a 96-well format and allows up to eight-fold pooling of DNA samples (Colbert *et al.* 2001). This permits

A.J. Márquez (ed.), Lotus japonicus Handbook, 197-210.

the screening of several thousand plants, for mutations in a specific gene. Our aim was to set up a mutation machine that utilises the TILLING procedure for *Lotus japonicus*, an important model legume in the study of root symbiosis with rhizobia and arbuscular mycorrhizal fungi (see chapter 2.3).

TILLING

Populations available to TILL

We have generated two populations directly accessible for TILLING. A general purpose TILLING population, biased against the occurrence of severe developmental phenotypes, consisting of approximately 5000 plants each representing an independent family. We also carried out a forward screen on the M2 siblings to identify plants that were defective in root nodule symbiosis and assembled these into a smaller subpopulation. This population consisted of 616 plants and is enriched for alleles functionally impaired in symbiosis. Furthermore, to facilitate similar TILLING approaches for other traits, we collected seed from a variety of developmental mutants, and trait-specific populations *e.g.* altered leaf or flower morphology. Phenotypic descriptions and photographs of the various mutants have been entered into a web-accessible database http://www.lotusjaponicus.org/finder.htm.

RESULTS

Since June 2003, the *Lotus japonicus* TILLING facility (Perry *et al.* 2003) has been available to the research community to identify plants carrying point mutations for any gene of interest. In a pilot experiment, we analysed the frequency of point mutations in the symbiosis defective population (384 individuals), by TILLING the kinase domain region of the SYMRK (symbiosis receptor kinase) gene, which is required for root symbioses (Stracke *et al.* 2002). Using this population, 17 mutations were identified that relate to six independent alleles, thus demonstrating proof of concept (Perry *et al.* 2003).

Subsequently, we have steadily increased our TILLING population. Initially we had a population size of 2304 individuals but we now have the capacity to screen a population of up to 5000 individual M2 plants. Moreover, the size of our symbiosis-defective population has been increased to 616 plants.

Presently, the DNA samples are pooled four times (although other groups have successfully pooled eight times; Till *et al.* 2003) which we have found is the optimum pool size when using the ABI 377 sequencer in combination with non-commercial CELI, permitting the screening of 384 plants per gel run. Utilising primers that amplify a 1-1.5 kb region results in the interrogation of 384-576 kb of DNA per gel run. Thus, we can analyse 2304 plants from our main TILLING population by screening six 96-well plates.

In our TILLING runs, for each 1-1.5kb fragment, 2 to 15 mutations per 2304 individuals (2.3 Mb of sequence) have been identified, which gives an average of six mutations. This suggests that the main TILLING population has on average 1302 mutations per genome (haploid *Lotus* genome ~500 Mb; Hayashi *et al.* 2001). From the first 17 ca. 1kb fragments, 97 mutations have been identified of which 1% are truncations, representing changes that cause a premature stop codon or affect intron/exon splice sites. Sixty-four percent result in missense changes, causing amino acid change within the coding region, 35% cause silent changes, which are either amino acid changes which occur within non-coding regions, or those base changes which do not cause a change in the amino acid. Additionally, 29 1-1.5kb fragments from 15 genes have been screened using the symbiosis defective population (616 plants), identifying 64 mutations. Fourteen percent of these mutations are truncations, 62.5% missense mutations and 23.5% silent mutations.

PROCEDURE

Seed was chemically mutagenised using ethylmethanesulphonate (EMS) (Figure 1A), resulting in numerous random point mutations (see Chapter 5.1). The ensuing M1 plants were self-fertilised and M2 seed collected individually. Approximately 20-30 M2 seed from each family were sown (Figure 1B). Each plant was scored phenotypically for a variety of morphological characters such as mutant nodules, leaves and flowers, or mutants affecting metabolism (starch breakdown/biosynthesis). DNA was extracted from a single viable member of each mutagenised M2 line (Figure 1C) to generate the 5000 TILLING population, along with plants identified as symbiosis defective to generate the forward screen pool. M2 DNA samples were normalised and selectively pooled four times, and then amplified with fluorescently labelled gene specific primers (Figure 1D). The DNA was heated and cooled to form heteroduplexes (Figure 1E), enabling the identification of mismatches induced by the occurrence of point mutations. Cleavage of heteroduplexes formed between wild type and mutant PCR products was achieved using CELI, a plant endonuclease that preferentially cleaves at mismatches (Figure 1F). These cleavage products were analysed on sequencing gels (Figure 1G), the differential end labelling of amplification products permitting the characterisation of the pool (Figure 2). The method was repeated with the individual pool members in the presence of wild type DNA to identify the individual carrying the mutation.

Protocol

EMS mutagenisation

Lotus japonicus Gifu seed was EMS mutagenised as detailed in Chapter 5.1.

DNA Isolation

DNA was isolated in a 96-well format as in chapter 3.3. There are numerous methods to isolate genomic DNA; several kits are available that can be processed by

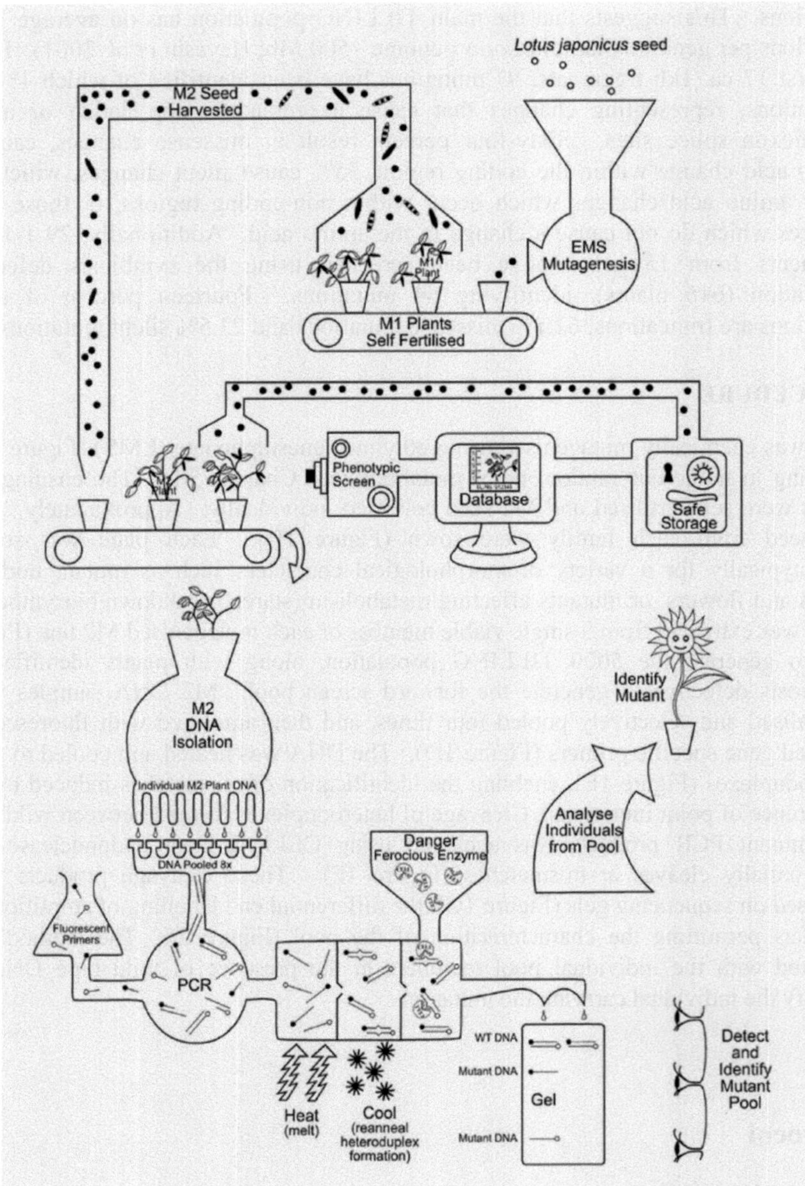

Figure 1. The 'mutation machine' detailing the TILLING process. (A) Seed is chemically treated with EMS to generate random point mutations throughout the genome. (B) The mutagenised seed is sown and the resultant individual M2 seed harvested. (C) Around 20-30 M2 seed is sown per family. (D) Each plant is phenotypically screened and the results archived. M3 seed from plants displaying a

nodulation defective phenotype are harvested. (E) One healthy wild type looking plant from each family is chosen to represent that family, its DNA is harvested, and seed collected. The DNA is quantified and normalised and then can be pooled up to 8x. (F) PCR is carried out using fluorescently labelled gene specific primers. The amplified products are heated and cooled to form heteroduplexes. (G) The heteroduplexes are incubated with CEL I a plant endonuclease, which cleaves the amplified products at the site of mismatch. (H) After incubation, the fragments are run out on a polyacrylamide denaturing gel were the fluorescently labelled cleavage products can be observed, the pool containing the mutation is then identified and the process repeated with the individuals of that pool in the presence of wild type DNA. (I) When these CEL I products are run out onto a polyacrylamide gel the individual can be identified which carries the mutation and is confirmed by sequencing.

robots to provide a quick and routine procedure. However we used a basic protocol, incorporating a phenol:chloroform step, as this method provided a superior yield of high quality DNA (Chapter 3.5).

DNA quantification

Quantification of the DNA is a vital step, as an imbalance in the quantity of DNA for any individual in a pool will lead to the loss of information for that individual or the others in the pool. Several methods are available to determine DNA quantification, such as running DNA on agarose gels next to quantity markers (Till *et al.* 2003) or spectrophotomeric methods. We quantified our DNA by spotting a diluted quantity of our RNase treated DNA onto ethidium bromide agarose gels and estimating DNA quantity against DNA standards (Sambrook *et. al.* 2001). We found this a quick and reliable method of quantifying DNA, which was then normalised and pooled appropriately.

1. Prepare 100ml of 0.8% agarose/TAE in the presence of 10µl of 10mg/ml ethidium bromide
2. Pour mixture into a gel tray and allow to set.
3. Produce a grid impression onto the gel by pressing a 96-well plate onto the agarose, this acts as a guide when pipetting the samples.
4. Using a multi-channel pipette load 0.5µl of 1/10 diluted RNAsed DNA onto the grid impression in the presence of 0.5µl of the following quantity standards (100, 80, 60, 50, 40, 20, 10, 5, 2 ng/µl).
5. Leave for 30 minutes and then view under the transilluminator (see figure 2).

Normalisation and Pooling of DNA

Add the appropriate quantity of TE (10 mM Tris-HCl pH 8.0, 0.1 mM EDTA pH 8.0) to each well to normalise the plate to 5ng/µl of DNA. Re-quantify the plate to check dilutions.

1. Once the plates are normalised, pool the DNA four times. Using a 12 channel pipette, pipette 20µl from rows A-E into row A of a fresh plate and repeat for rows F-H into row B until you have pooled all four of your plates.
2. Pipette 1µl of DNA from your pool plate into a number of new 96-well plates and store in freezer these are then ready to PCR.

The following methods are adapted from protocols generated in the Henikoff laboratory (Till *et al.* 2003):

Primer design

1. Generate primers using the CODDLE (Codons Optimized to Discover Deleterious Lesions) input utility www.proweb.org/input/ CODDLE generates a gene model and a protein conservation model, to identify regions which are most likely to cause deleterious mutations.
2. Order from the manufacturer of preference. It is advised that unlabelled primers be ordered and tested before the more expensive labelled primers are ordered.

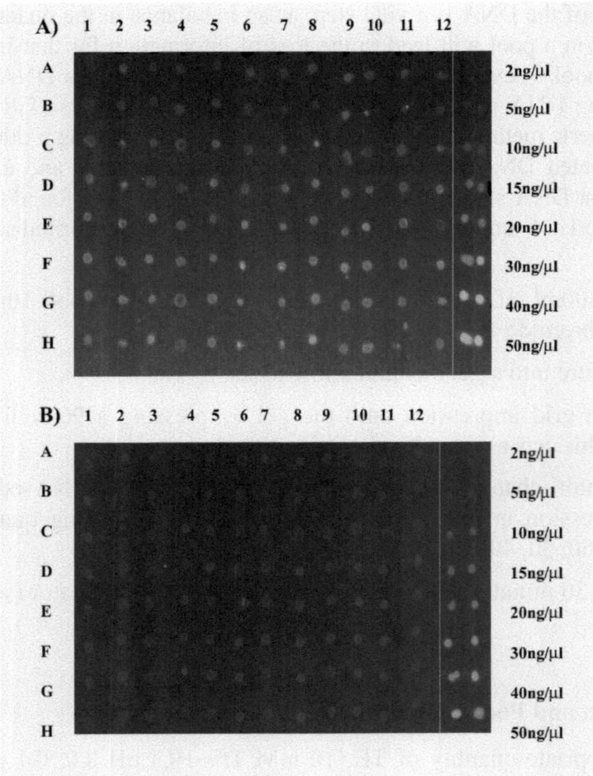

Figure 2. (A) Gel photo showing 0.5µl of 1/10 diluted DNA in 96-well format next to quantity markers. (B) Shows a gel photo of the normalised DNA to 5ng/µl.

3. Once amplification has been successful, primers can be ordered with the appropriate fluorescent label, the forward primer should be labelled with 6-FAM (6-carboxyfluorescein) and the reverse primer with TET (tetrachlorofluorescein) for use on the ABI 377 sequencer.

PCR reaction

1. Thaw the frozen stock solutions, quickly mix and spin. Thaw the relevant pool plate and label the plate appropriately, place in 96-well cool block (pre-chilled at -20°C).
2. Prepare Mastermix as follows (for 1 x 96-well plate):
    - 5mM dNTPs 44μl
    - 10x PCR buffer 110μl
    - Forward primer (2μM) 110μl
    - Reverse primer (2μM) 110μl
    - Water 605μl
    - Taq (5000U/ml) 11μl
3. Pipette 9μl of mastermix into each well using a multichannel pipette.
4. Seal using adhesive films, spin quickly, and place onto the PCR machine.
5. Run the following program:
    - 95°C for 2 min
    - 94°C for 20s
    - T_m +3°C to T_m -4°C for 30s, -1°C per cycle
    - gradient to 72°C at 0.5°C /s
    - 72°C for 1 min
    - 'GOTO' 2 for 7 cycles
    - 94°C for 20s
    - T_m -5°C for 30s
    - gradient to 72°C at 0.5°C /s
    - 72°C for 1 min
    - 'GOTO' 7 for 44 cycles
    - 72°C for 5 min
    - 99°C for 10 min
    - 70°C for 20s, -0.3°C per cycle
    - 'GOTO' 14 for 69 cycles

- 15°C for ever
- end

6. Run 1 μl of PCR product on an agarose gel along with quantity standards to confirm the success and quantity of the PCR product.

CEL I isolation

CEL I was isolated using a modified protocol from Oleykowski *et al.* 1998.

1. Juice extraction from 6 kg celery stalks, buffered in Tris/HCl pH 7.7.
2. Concentrate with 20 to 70 % saturated ammonium sulphate precipitation.
3. Affinity chromatography - batch elution from ConA Sepharose column.
4. Dialysis against 50 mM Tris/HCl, pH 7.8, 0.2 M KCl, 10 μM PMSF, 1 mM $ZnCl_2$ (final buffer used as in published method).

CEL I digestion

1. Thaw 10x CEL I buffer and a frozen aliquot of CEL I enzyme. Add the following reagents to a tube on ice (for 1x 96-well plate):
   - 2.4 ml water
   - 420μl 10x CEL I (see below)
   - 15μl CEL I
2. Place the PCR plate into a 96-well cool block (pre chilled at -20 °C), pour the CEL I mixture into a reservoir and add 20μl to each well using a multi channel pipette.
3. Quickly spin the plate and cover with adhesive PCR film. Incubate at 45°C for 30 – 45 min depending on amount of PCR product (around 200ng, incubate for 30 min, around 400ng incubate for 45 min). Meanwhile prepare loading dye/standards and the spin plates (see below).
4. After incubation place the plate in the 96-well cool plate, stop the reaction by adding 5μl of 0.15M EDTA (pH 8). Quickly spin the plate.
5. Transfer the 35μl samples onto the spin plate being careful not to touch the surface of the Sephadex columns.

10x CEL I Buffer

- 5ml 1M HEPES [4-(2-hydroxyethyl)-1-piperazineethanesulfonic acid (pH 7.5)]
- 5ml 1M Mg SO_4
- 100μl 10% Triton X-100
- 5μl 20 mg ml^{-1} Bovine Serum Albumin

- 2.5 ml 2M KCl
- 37.5 ml water

Loading dye standards

To identify lanes in which samples are loaded, size markers (Genescan 2500, size range 37–2500 and Genescan 500, size range 35–500) are added to the loading dye; this allows easy scoring of the wells that contain a mutation and also helps to determine the position of the mutation.

Markers	2500	500	0
Dye	8µl	24µl	31µl
Standard	3µl	9µl	-
Formamide	17µl	51µl	68µl
Water	-	-	12µl

1. In a 96-well plate, in an 8-well strip labelled from A – H, dispense 28µl into the following wells
    - A: Genescan 2500
    - C, E, and G: Genescan 500
    - B, D, F, and H: no size standard, only loading dye (blue dextran, 50mg/ml in 25 mM EDTA)
2. Dispense 2 µl from the above wells A-H into rows A-H of a new plate for each of the 12 columns, using the multichannel pipette.

Spin Plates

Prepare spin plates in advance and store in the fridge for no more than a week. Keep the plates in a plastic bag in the presence of a moist towel.

1. Load dry Sephadex (G50 fine) into all 96-wells of the column loader.
2. Remove excess powder off the top of the loader with the plastic scraper.
3. Place the multiscreen plate (Millipore multiscreen filter plates) upside down on top of the column loader.
4. Invert both the multiscreen plate and column loader.
5. Tap on top of the column loader to release the powder.
6. Using a multichannel pipette add 300µl of sterile water to each well. Store in fridge.
7. Before adding the sample, place a centrifuge alignment frame onto a waste plate and attach the spin plate on top.
8. Place this assembly in the centrifuge and spin for 4 minutes at 460g.
9. Remove the waste plate.

10. Take a labelled plate containing 2μl loading dye, attach the centrifuge alignment frame and firmly place the spin plate on top making sure that position A1 matches position A1 on the loading dye catch plate. Secure this with masking tape.
11. Load 35μl of the sample onto the column.
12. Put in the centrifuge and spin for 4 minutes at 460 g.
13. Dry samples down to 1μl by heating at 90 °C in the fume hood (approx. 30 mins.).

Preparing Gels for the ABI 377 sequencer

1. Prepare gel mix (prepares 2 x 12cm gel):
    - 9g urea
    - 4.4ml 40% acrylamide (29:1)
    - 12.5 ml water
    - spatula of resin beads
2. Stir for 30 mins until urea dissolved. In the meantime, set up the plates.
3. Using a 0.45 μm filter, set up filter pump. Prime the filter with 2.5 ml 10x TBE. Attach to the water pump and allow to flow through.
4. Filter the acrylamide mix. Make up to a final volume of 25 ml with water.
5. Add the following reagents depending on the gel mix amount.
    - 125 μl 10 % APS (0.1g in 1 ml)
    - 17 μl of TEMED
6. Gently mix.
7. Suck up the mixture with a syringe place the end of the syringe into one side of the comb area and slowly push down on plunger. Move the syringe backwards and forwards along the comb area at the same time tapping the glass plates to prevent bubbles forming. Once the gel mix reaches the bottom of the plates insert the well forming comb. Allow the gels to set for 2 hours before using.
8. After the gel has set, remove the clips and well forming comb and wash the gel thoroughly, clean out the comb region with deionised water from any acrylamide and soak up any dry any liquid from the comb area. Wipe the plate to remove any water and arylamide, use a clean paper towel each time to avoid smearing the acrylamide pay particular attention to the laser region. Put the gel into the frame and lock onto the sequencer securely.
9. Run a plate check to make sure the laser region is clear from debris. Add the upper TBE chamber. Prepare 1300ml of 1xTBE and fill up the bottom reservoir. Pre-run the machine to warm the gel to 51°C.
10. Denature CEL I samples at 95 °C for 2 minutes.
11. Load 0.3 μl of sample onto the membrane combs.

12. Fill the comb area with 1% Ficoll, insert the membrane comb and add the rest of 1xTBE to the upper reservoir, run the sequencer for 3 minutes. Remove the comb and rinse out the comb area with TBE from the upper reservoir, replace the reservoir lid and resume the run.

Scoring of mutations

1. Open the gel image. To get separate blue/ green images click on 'Gel' and go to 'adjust gel image'. Adjust accordingly. Save the individual blue or green images and the blue or green image in the presence of the size markers.
2. Save the images as PICT files and open these file using Adobe Photoshop. Choose one gel file, which shows either the individual blue or green image, and one that shows the reciprocal green or blue individual image in the presence of the size standard.
3. Click on an image and go to Layer, click on duplicate layer and choose the reciprocal image from the pull down menu. Click on the image called background copy and flip between the images by moving the arrow on the opacity slide in the layers box. By using the wand to from the tools box, a band in one layer can be identified by clicking on the image and visualised on the other layer by changing the opacity level. This helps to identify cleavage products in both images, helping to determine if reciprocal products can be observed and also helps to determine the size of each of the identified cleavage products (Figure 3).
4. Once pools have been identified to contain a mutation the individuals of the pool can go through the TILLING process again in the presence of wild-type (WT) DNA. This allows the identification of the individual from that pool which carries the mutation.
5. After identification of the individual, sequence the mutation.

Identification of homozygous mutants in the M3 generation

1. From the sequencing data, it can usually be determined whether the mutation in the TILLING plant of interest is homozygous or heterozygous.
2. *If the mutation is heterozygous,* 20 M3 seed from this individual are sown. This should statistically result in approximately 6 homozygous individuals if normal segregation is occurring
3. For genotyping the individuals from a segregating line, DNA from each of the 20 M3 individuals is isolated. Several SNP detection methods can be used to determine zygosity of the individuals. We found restriction enzyme analysis the quickest and most robust technique, but it is not applicable to all sites. Alternative methods include mini-sequencing and mass spectroscopy based genotyping.
4. Using single nucleotide polymorphism (SNP) detection methods homozygous lines can be identified. We routinely keep all individuals that are homozygous for the mutation, in addition 2 heterozygous individuals and 2 individuals that are homozygous WT for the mutation. This spectrum should be sufficient to

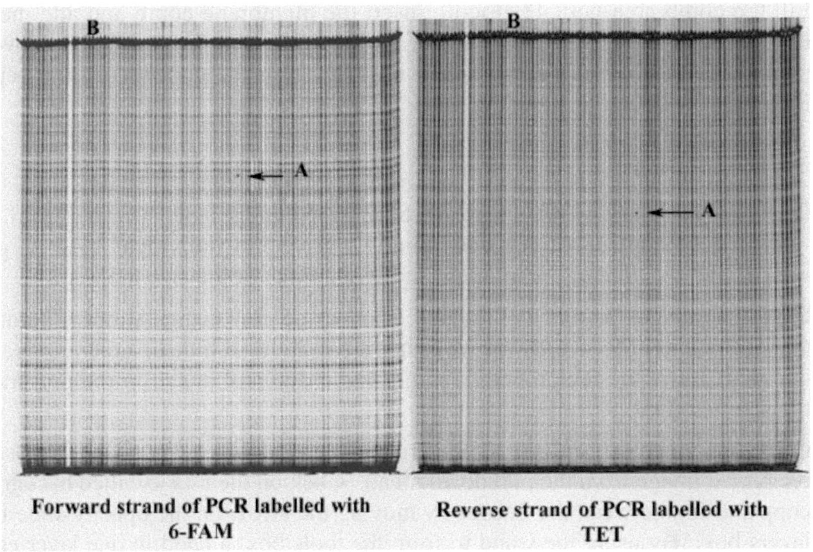

Figure 3. Separation of enzymatically cleaved and uncleaved PCR products. Products were amplified with fluorescently labelled primers and run on a polyacrylamide gel using an ABI377 sequencer. The arrow at A shows cleavage products identifying the positions of a mutation in both the forward and reverse strand. B represents uncleaved amplification products.

facilitate comparisons in the M4 generation of WT and mutant phenotype and to test for co-segregation of genotype and phenotype. The additional harvested seed also allows the maintenance of the line for future screens.

5. If feasible, a non-destructive phenotypic screen can be carried out on the M3 plants. It is advisable that genotyping of the plants is carried out first and that the plants maintained for M4 seed production. We therefore avoid all growth conditions that limit plant growth such as low nutrient supply used to promote symbiotic phenotypes.
6. If the M2 plant is homozygous for the mutation, only four seeds are sown to provide seed-producing plants so that sufficient individuals for detailed phenotyping can be obtained. When quantitative phenotypes are studied, it may be advisable to obtain sibling seed homozygous wild type for the mutation under study, to control for the influence of background mutations. If no siblings are immediately available, a backcross can be initiated that will deliver a population segregating for the mutation of interest.

Controlling the effect of background mutations

EMS causes multiple mutations per genome, and we have estimated our TILLING population has around 1300 on average. Therefore it is important to design phenotyping experiments in such a way that the influence of background mutations can be controlled. This issue needs different measures depending on the type of data that is desired from the mutants.

Strong phenotype, qualitative phenotyping only

When the phenotype conferred by the mutation is strong, for example when a mutation leads to a non-nodulating or non-starch accumulating plant and this level of qualitative data is sufficient, the following genetic tests can be performed to ensure the mutation under study is causing the phenotype.

- Analyse co-segregation between genotype and phenotype in a population segregating the mutation under study. For this, 93 plants plus three controls (homozygous wild-type; homozygous mutant; heterozygous individual) should be grown and the phenotype of each plant scored. DNA is prepared from the plants in a 96-well format and the genotype of each plant determined. If complete co-segregation occurs, the mutation causing the phenotype is within less than 1cM of the mutation genotyped.
- Obtain multiple mutant alleles in the candidate gene. If more than one independent allele results in a similar phenotype, the statistical chance is greatly increased that the phenotype observed is indeed caused by the mutation in the candidate gene.
- Test complementation between two independent mutant alleles of the same gene. Homozygous mutant individuals representing two different alleles are crossed and the resulting F1 is genotyped. If no complementation occurs, the statistical chance that both plants both plants carry a second site mutation responsible for the phenotype is dramatically reduced. Crossing success should be monitored by genotyping for heterozygosity of both mutations in the F1.

Weak phenotype, or quantitative data required

The large number of background mutations often causes a change in general plant fitness affecting physiological parameters and growth behaviour. If the quantitative effect of a particular mutation on, for example, plant growth is to be determined, larger populations and statistical methods need to be applied to control the influence of background mutations.

- Compare phenotypes of siblings with similar mutational loads. Quantitative phenotyping can be done on a population of siblings segregating the mutation of interest. Correlation between mutation and phenotype can be established. In addition, several homozygous WT and mutant lines can be grown and the phenotype compared between these lines. More than one line from each is necessary for better representation of the background mutational load in a particular family.
- Backcrossing. The classical, but for most projects prohibitively lengthy procedure, is to do a series of backcrossing steps to statistically 'purify' the line from background mutations. One backcrossing step reduces the mutation load by

50%, so to reach an asymptotic level of reduction one would need a minimum of 6 backcrosses, which exceeds the duration of an average research grant.

REFERENCES

The *C. elegans* Sequencing Consortium (1998) **Genome Sequence of the Nematode *C. elegans*: A Platform for Investigating Biology** *Science* 282, 2012-2018.

Colbert T, Till BJ, Tompa R, Reynolds S, Steine MN, Yeung AT, McCallum CM, Comai L, Henikoff S. (2001) **High-throughput screening for induced point mutations**. *Plant Physiology* 126, 480-484.

The EU Arabidopsis Genome project (1998) **Analysis of 1.9 Mb of contiguous sequence from chromosome 4 of *Arabidopsis thaliana*** *Nature* 391, 485-488.

Hayashi M, Miyahara A, Sato S, Kato T, Yoshikawa M, Taketa M, Hayashi M,. Pedrosa A, Onda R, Imaizumi-Anraku H, Bachmair A, Sandal N, Stougaard J, Murooka Y, Tabata S, Kawasaki S, Kawaguchi M, and Harada K. (2001) **Construction of a genetic linkage map of the model legume *Lotus japonicus* using an interspecific F2 population**. *DNA Research* 8, 301-310.

International Human Genome Sequencing Consortium (2001) **Initial Sequencing and analysis of the human genome**. *Nature* 309, 860-921.

Mc Callum CM, Comai L, Greene EA and Henikoff S. (2000) **Targeted screening for induced mutations** *Nature Biotechnology* 18, 455-457.

Perry JA, Wang TL, Welham TJ, Gardner S, Pike JM, Yoshida S and Parniske M. (2003) **A TILLING reverse genetics tool and a web accessible collection of mutants of the legume *Lotus japonicus***. *Plant Physiology* 131, 866-871.

Oleykowski CA, Bronson Mullins CR, Godwin AK and Yeung AT (1998) **Mutation detection using a novel plant endonuclease** *Nucleic Acids Research* 26, 204597-4602.

Sambrook J, Fritsch E, and Maniatis T. (2001) **Molecular Cloning: a Laboratory Manual.** 3rd edition Cold Spring Harbor Press, Cold Spring Harbor, New York A8.19.

Smits BM, Mudde J, Cuppen E, Plasterk RH. (2004)**Target-selected mutagenesis of the rat**. *Genomics* 83, 332-334.

Stracke S, Kistner C, Yoshida S, Mulder L, Sato S, Kaneko T, Tabata S, Sandal N, Stougaard J, Szczyglowski K, Parniske M. (2002) **A receptor-like kinase required for both bacterial and fungal symbiosis**. *Nature* 417, 959-962.

Till BJ, Reynolds SH, Greene EA, Codomo CA, Enns LC, Johnson JE, Burtner C, Odden AR, Young K, Taylor NE, Henikoff JG, Comai L, Henikoff S. (2003) **Large-scale discovery of induced point mutations with high-throughput TILLING**. *Genome Research* 13, 524-530.

Till BJ, Colbert T, Tompa R, Enns L, Codomo C, Johnson J, Reynolds SH, Henikoff JG, Greene EA, Steine M, Comai N, and Henikoff S. (2003) **High-throughput TILLING for functional genomics**. In: *Plant Functional Genomics: Methods and Protocols* (Grotewald E, Ed.) , Humana Press, 236; 205-220.

Wienholds E, Van Eeden F, Kosters M, Mudde J, Plasterk RH, and Cuppen E (2003) **Efficient target-selected mutagenesis in zebrafish**. *Genome Research* 13, 2700-2007.

Yang B, Wen X, Kodali NS, Oleykowski CA, Miller CG, Kulinski J, Besak D, Yeung JA, Kowalski D, and Yeung AT (2000) **Purification, Cloning and Characterization of the CEL I Nuclease**. *Biochemistry* 39, 3533-3541.

Chapter 5.4

DS GENE-TAGGING

Leila Tirichine, José A. Herrera-Cervera, and Jens Stougaard*

*Laboratory of Gene Expression, Department of Molecular Biology, University of Aarhus, Gustav Wieds Vej 10, 8000 Aarhus DENMARK; *Corresponding author*

Email: stougaard@mb.au.dk

Phone: +45 8942 5011 Fax: +45 8620 1222

Keywords: transposon-tagging, *Ds* trap, *Ds* activation.

A transposon tagging strategy is described, based on modified Ds elements, and aimed at tagging genes located in regions of the genome where map based cloning is virtually impossible; for example, near centromeres or in sub-telomeric regions. The launching vectors constructed have the Ds elements and the transposase gene within the same T-DNA. Two categories of Ds elements are used: Ds trap and Ds activation.

INTRODUCTION

Transposon insertional mutagenesis is a powerful tool for addressing gene function through analysis of the mutant phenotypes and for isolation of the mutated gene. In Lotus, this method was used successfully after transfer of the maize transposable element *Ac*. A plant developmental regulator (*Nin*) required in the early stages of nodule organogenesis (Shauser *et al*., 1999) and a proliferating flower organ (*Pfo*) gene involved in flower development (Zhang *et al.* 2003) were identified from *Ac* tagged mutants. Here we describe another transposon tagging strategy designed to enhance the frequency of transposition and effective reinsertion in *Lotus japonicus*.

TAGGING STRATEGY

The *Ac/Ds* element family is a two-component system with autonomous (*Ac*) and non-autonomous (*Ds*) components. The *Ac* element encodes a transposase that binds to the terminal inverted repeat ends of both *Ac* and *Ds*. The *Ac* transposase protein recognises the ends of *Ds* elements and catalyse their transposition to new chromosomal locations. In our launching vectors, the *Ac* transposase gene is cloned

A.J. Márquez (ed.), Lotus japonicus Handbook, 211-215.

within the same T-DNA as the *Ds* element. The *Ds* element is carrying a positive selectable marker to select for *Ds* insertions and *Ac* transposase gene is cloned next to a counterselectable marker to select transposase free lines. *Ds* constructs are introduced into *Lotus* genome by T-DNA mediated transformation (see protocol). The primary transformants are self-pollinated (Figure 1). In order to obtain the highest efficiency of independent reinsertions the *Ds* element should be mobilised during or after meiosis and insert into different regions of the genome.

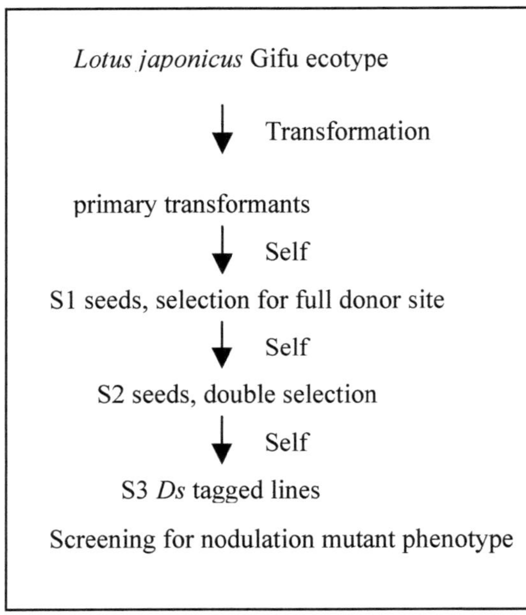

Figure 1. Propagating scheme for independent transpositions of the Ds element. S2s are subjected to the double selection. S3 seeds are screened for nodulation mutant phenotypes.

Ds trap construct

Ds trap constructs carry a promoter-less GUS reporter gene near their termini. Reporter gene expression occurs only when *Ds* element inserts within a transcribed chromosomal gene creating a transcriptional fusion. The *Ds* elements carry the ß-glucuronidase (GUS) gene as a reporter and the Neomycin phosphotransferase (NPTII) gene (conferring resistance to Kanamycin) or a phosphinothricin resistance gene (BAR) gene conference resistance to glufosinate (Figure 2).

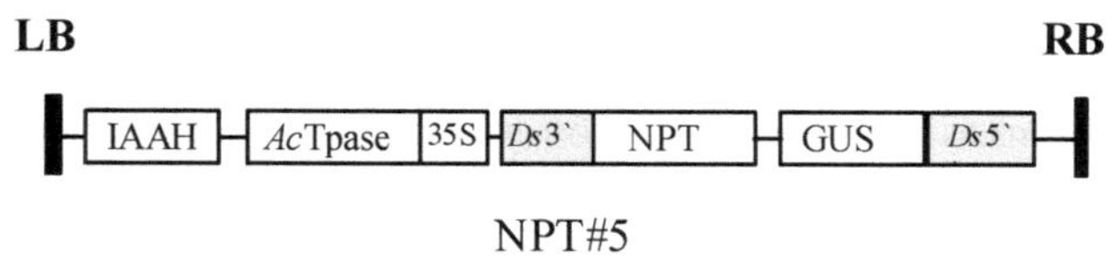

Figure 2. Example of Ds launching construct. IAAH, indole acetamide hydrolase gene (confers sensitivity to naphtalene acetamide NAM); NPTII, neomycin phosphotransferase gene (confers resistance to kanamycin); GUS, ß-glucuronidase reporter gene; Actpase, transposase gene, 35S promoter, LB and RB left and right borders respectively.

Ds activation construct

These constructs are designed to enable screening for gain of function by deregulation of an endogenous gene. Typically, an activation construct carries a constitutively expressed 35S promoter or a strong enhancer at one of the termini (Suzuki *et al.*, 2001). The *Ds* insertion near a chromosomal gene, resulting in expression from the 35S promoter or enhanced expression, could reveal a gain of function of a silent, subtle or downstream gene.

Selection for transposition

The tagging system we are using relies on a single T-DNA with a positive selection for the disarmed *Ds* element (NPTII) and a negative selection for the *Ac* transposase gene (IAAH) that confers sensitivity to naphthalene acetamide (NAM). The principle is that seedlings surviving this positive-negative selection should carry *Ds* reinsertions genetically unlinked to the T-DNA. The seeds from the primary transformants are selected on media containing Geneticin (an analogue of Kanamycin) to select for lines containing a single copy of the full construct. These lines segregate resistance 3:1 and are allowed to self-pollinate which results in the S2 generation. The S2 seeds are subjected to the double selection to recover only the lines that are free of transposase and contain a single reinserted *Ds* element. Each selected *Ds* insertion is thus stable but can later be remobilized by the appropriate cross to a line that express the *Ac* transposase. The double resistant S2 seedlings Nam^R Kan^R are allowed to self-pollinate. S3 seeds are harvested and screened for mutant phenotypes (Figure 1).

Molecular analysis of transposants

Each *Ds* line that has a phenotype of interest can be analysed and the gene carrying the *Ds* element can be cloned and sequenced. Flanking genomic DNA or cDNA corresponding to the gene nearest the insertion site can be rapidly obtained by PCR amplification using for example IPCR (Earp *et al.* 1990). The resulting product can be sequenced or used as hybridisation probe for further gene analysis.

Genetic analysis of transposants

In many cases, insertion of *Ds* transposons will be associated with a mutant phenotype. However, spontaneous mutations could be found in the *Ds* tagged lines. It is then important to determine whether the *Ds* insertion is responsible for the observed phenotype. One possibility is to perform a cosegregation analysis using DNA flanking the *Ds* element as outlined in the chapter on mapping. Another possibility is a reversion analysis.

Reversion

Insertion of *Ds* transposon results in the duplication of 8 bp of target sequence immediately flanking the insertion site. When *Ds* excises, the target duplication is partially removed resulting in small insertions and deletions at the original locus. If the *Ds* inserts into the coding region of the gene, only the events restoring the reading frame and resulting in a functional protein will revert the mutant phenotype back to the wild type. Reversion is accomplished by crossing mutant plants to transgenic plants carrying the transposase gene. The F1 plants are allowed to self-pollinate. In the F2 progeny, the transposon causing the mutation will excise resulting in somatic revertant sectors of tissue that have lost the *Ds* element. This is good evidence that the mutation is caused by *Ds* insertion.

Reinsertion

It happens that insertion of a transposon doesn't disrupt a gene. This is the case for example when the insertion is within intron that could be spliced from the RNA transcript without a consequential phenotypic effect. It can be then useful to obtain a secondary insertion into the nearest exon by inducing a short-range transposition. This could be achieved since a high proportion of *Ds* transposition is to closely linked sites. The *Ds* element is remobilized by crossing to a transposase line and then selecting against the transposase-linked marker in the next generation to stabilise the *Ds* element.

Positive-negative selection of *L japonincus* seedlings

1. Use surface sterilisation protocol to sterilise *L japonicus* seeds.
2. Prepare agar plates (half-strength B5, 0.8% noble agar, pH 5.5) containing 25 μg/ml Geneticin.
3. Prepare Top agar (Half strength B5, 0.4% noble agar, pH 5.5), containing 25 μg/ml Geneticin.
4. Mix seeds with the top agar and poor the entire content on the surface of the agar plates)
5. Incubate plates under the condition used for seeds germination: continuous light at 26°C.

For the BAR selection, use 20 μg/ml of Glufosinate ammonium. For NAM

(Naphtalene acetamide) selection, use 3 μM/ml. See pictures of resistant seedlings and sensitive ones (Figure 3).

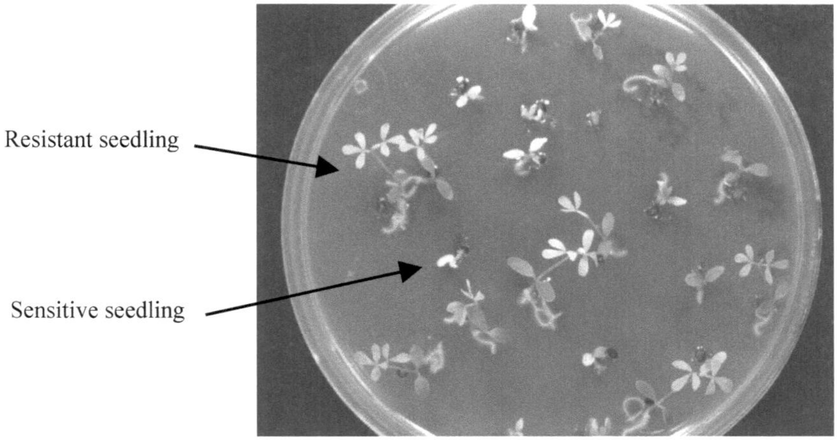

Figure 3. Lotus seedlings on selection media (Glufosinate 20μg/ml for BAR selection) 12 days after germination. Note that the root are growing and branching in the resistant seedlings.

CONCLUSION

Four hundred lines were transformed with the *Ds* trap constructs. *Lotus japonicus* lines containing single insertion of the *Ds* were generated from the primary transformants according to the scheme Figure 1. Our results show that the new *Ds* tagging system is active in *L japonicus*. The use of promoter acting late in plant development to drive the expression of the transposase gene would allow higher frequency of independent reinsertion of *Ds* element.

REFERENCES

Earp DJ, Lowe B, and Baker B. (1990) **Amplification of genomic sequences flanking transposable elements in host and heterologous plants: A tool for transposon tagging and genome characterization**. *Nucleic Acids Research* 18, 3271-3279.

Schauser L, Roussis A, Stiller J, and Stougaard J. (1999) **A plant regulator controlling development of symbiotic root nodules**. *Nature* 402, 191-195.

Suzuki Y, Uemar S, Saito Y, Murofushi N, Scmitz G, Theres K, and Yamaguchi I. (2001) **A novel transposon tagging element for obtaining gain-of-function mutants based on a self-stabilizing *Ac* derivative**. *Plant Molecular Biology* 45, 123-131.

Zhang S, Sandal N, Polowick PL, Stiller J, Stougaard J, and Fobert PR. (2003) **Proliferating floral organ (*Pfo*), a *Lotus japonicus* gene required for specifying meristem determinacy and organ identity, encodes an F-box protein**. *Plant Journal* 33, 607-619.

Chapter 5.5

MAPPING AND MAP-BASED CLONING

Niels Sandal[1,*], Katja Krause[2], Mirela Frantescu[1], Simona Radutoiu[1], Lene Krusell[2], Lene H Madsen[1], Norihito Kanamori[1], Esben Madsen[1], Leïla Tirichine[1], and Jens Stougaard[1]

[1]*Laboratory of Gene Expression, Department of Molecular Biology, University of Aarhus, Gustav Wieds Vej 10, DK-8000 Aarhus C, DENMARK;* [2]*Molecular Plant Nutrition Group, Max Planck Institute of Molecular Plant Physiology, Am Mühlenberg 1, D-14476 Golm, GERMANY; *Corresponding author*

Email: nns@mb.au.dk

Phone: +45 8942 5006 Fax: +45 8612 3178

Keywords: *Lotus japonicus*, *Lotus filicaulis*, map-based cloning, mapping, colormap, marker, *sym* gene, symbiosis

Development of a genetic system and genetic resources is one of the prerequisites for establishing Lotus japonicus *as a model legume. Successful identification and characterisation of several key plant genes controlling symbiosis with nitrogen fixing bacteria and mycorrhizal fungi demonstrate the potential of this approach. We provide here an overview of efforts to establish a genetic map and approaches using genetic marker information for positional cloning. In* L japonicus, *six independent mutant populations in the ecotype Gifu were obtained with EMS, fast neutrons, T-DNA, transposon, or tissue culture. More than 500 independent symbiotic mutants have been isolated and at least 24 independent loci found. General mapping populations are available from crosses between ecotype Gifu and ecotype MG-20 and between Gifu and* L filicaulis. *These parents are also used for establishing populations for map-based cloning. Five BAC and TAC libraries were constructed. The general strategy for map-based cloning in* L japonicus *is described and map positions of eleven* sym *genes are shown. Four* L japonicus sym *genes have been isolated using map-based cloning and two genes have been isolated with transposon tagging.*

INTRODUCTION

Part of the rationale for selecting *L japonicus* as a model legume (Handberg and Stougaard, 1992) was to develop a genetic system enabling characterisation of plant genes required for symbiosis with nitrogen fixing bacteria and mycorrhizal fungi. We have chosen the genetics of symbiosis for illustrating approaches and methodology used in mapping and map based cloning. In order to isolate symbiotic

A.J. Márquez (ed.), Lotus japonicus Handbook, 217-232.

mutants and genes from a homozygous genetic background the *L japonicus* ecotype Gifu was inbred for 9 generations to generate the B-129-S9 germplams (Stougaard and Beuselinck, 1996).

Six independent mutant populations have been obtained using chemical (EMS) mutagenesis (Perry *et al.*, 2003; Kawaguchi *et al.*, 2002; Szczyglowski *et al.*, 1998; Webb *et al.*, unpublished, Gresshoff *et al.*, unpublished), four using T-DNA or transposon insertion mutagenesis (Thykjær *et al.* 1995, Schauser *et al.* 1998; Webb *et al.*, 2000; Gresshoff *et al.*, unpublished), one using fast neutrons (Gresshoff *et al.*, unpublished), and several using tissue culture-derived mutants (Umehara and Kouchi, unpublished). Slightly different conditions were used for screening of mutant populations but in general the following phenotypes were scored: Absence of nodules (Nod^-), nodules not capable of nitrogen fixation (Fix^-), decreased or delayed nodulation (Nod^{+-}), increased or hypernodulation (Nod^{++}), Nodulation in the absence of *Rhizobium* (Nar) and absence of mycorrhization.

A summary of the type and number of symbiotic mutant lines so far identified within the *Lotus* community is presented in Table 1. In addition, other developmental or morphological mutants were found. Among these were dwarfs, leaf and root mutants, and flowering mutants, for example *pfo* characterised by excessive development of sepals and indeterminate growth of floral meristems (Zhang *et al.*, 2003).

Mutant phenotype	**Number of mutant lines**	**Mycor-rhization deficient**	**Number of loci**	**Alleles per locus**
Nonnodulation (Nod^-)	296	31	>10	1-12
Ineffective nodulation (Fix^-)	252	1	>11	1-3
Hypernodulation (Nod^{++})	14		3	1-6
Reduced or delayed nodulation (Nod^{+-})	61		ND	ND

Table 1. Summary of L japonicus *symbiotic mutants and alleles. Only few of the* Nod^- *and* Fix^- *mutants have been tested for mycorrhizal symbiosis and most of the mutants have not yet been assigned to complementation group.*

Diallelic crosses between all mutants with a similar phenotype to screen F1 plants for genetic complementation is usually next step following mutant isolation. Such an assignment to complementation groups identifies all alleles that contribute to phenotypic characterisation of mutants and genotyping of loci. Equally important it estimates the number of loci involved in a particular process and makes it possible to optimize use of resources required for map-based cloning of genes. Ideally, duplication of efforts between research groups working on *L japonicus* should be avoided and assignment of all available alleles to loci provides a common knowledge basis for project planning. However, diallelic crossing is a relatively slow process where progress is determined by generation time and slowed by continuously increasing number of individual crosses necessary to keep up with mutant isolation programs. Given the number of symbiotic mutant lines already available, this approach is unlikely to encompass all alleles in near future. Fortunately, genetic mapping has become easier and it is now possible to quickly define a rough map position for an allele using one or two microsatellite markers from each of the six linkage groups. Only one cross per mutant line is necessary to map a new allele relative to already known loci and together with TILLING of already cloned genes this could decrease the number of diallelic crosses needed to clean up mutant material.

At the end of a map-based cloning project, characterisation of several alleles provides very strong evidence that the correct gene has been identified at the locus. Depending on the strength of such evidence, this alone may serve to prove the identity of the gene underlying the phenotype or trait.

MAPPING AND CLONING IN LOTUS

Map-based cloning

The minimum requirements for accomplishing positional cloning successfully are:

- A mutation with good penetrance ensuring a mutant phenotype that is easily and reliably scored
- A cross to a polymorphic mapping parent producing sufficient seeds to establish a population of F2 plants segregating the mutation, a general genetic map with molecular markers
- A BAC (bacterial artificial chromosome)/TAC (transformation-competent artificial chromosome) library with large inserts
- Access to DNA sequence in the region delimited by recombination end-points
- Several mutant alleles and/or methods allowing complementation of mutants after transformation with the wild type gene.

Mapping partners

At present, two mapping partners, *Lotus filicaulis* (Sandal *et al.*, 2002) and *L japonicus* ecotype Miyakojima (MG-20) (Kawaguchi *et al.*, 2001), are available for establishing an F2 population segregating mutations obtained in a Gifu background. Each partner has advantages and disadvantages.

Lotus filicaulis

\+ High level of polymorphism; using a number of AFLP primer combinations, 40-50% of the bands were polymorphic

\+ Easy to map a gene

+ A genetic map has been developed (Sandal *et al.*, 2002).
– Low seed yield on F1 plant
– Relatively long generation time
– Several regions with distorted segregation especially middle part of chromosome 1
– Suppression of recombination on upper part of chromosome 3 and middle of long arm of chromosome 1

***Lotus japonicus*, MG-20**

+ High seed yield on F1 plant
+ Normal generation time, increasing number of microsatellite markers are developed from sequenced TACs and BACs
+ A genetic map has been developed (Hayashi *et al.*, 2002).
+ This ecotype is getting better and better as a mapping parent due to the genome sequencing of this ecotype at Kazusa DNA Research Institute. Upon completion of the complete MG-20 genome, SNP (single nucleotide polymorphisms) markers for Gifu/MG-20 will be developed on a genome-wide scale.
– Relatively low level of polymorphism; around 6% of AFLP bands polymorphic
– Compared to *L filicaulis*, more difficult to find linked markers, difficult to map a gene unless it is on a sequenced and mapped TAC clone
– Suppression of recombination on upper part of chromosome 1.

Crossing

To cross *L japonicus* ecotypes it is important to use female flowers that are at just the right stage of development. The pollen is then collected from a mature flower. Pictures of flowers where the pollen has not yet been liberated and the female parts are ready for fertilisation (Jiang and Gresshoff, 1997) are shown in Fig x. For crosses to *L filicaulis* (Grant *et al.,*1962), older flowers can be used since *L filicaulis* is less self fertile than *L japonicus,* at least in our greenhouse. The F1 plants can be discriminated from both parents based on the morphological characters. The F1 plant has broad leaves like the *L japonicus* parent and some of the flower characters are similar to *L filicaulis*. Alternatively, the F1 plant DNA can be analysed using a PCR marker giving a size difference to confirm cross-fertilisation.

Mapping

Efficient mapping relies on high-density genetic maps developed using molecular markers. Evolutionary distant relatives have higher levels of polymorphisms than found between closer relatives and provide a better source of DNA differences for molecular markers. However there are also disadvantages. Seed yield on F1 plants can be low and there will be regions with distorted segregation and often more inversions and translocations reflecting the evolutionary distance. For *L japonicus* two different maps balancing advantages and disadvantages were established, (Sandal *et al* 2002, Hayashi *et al* 2001). One was based on an interspecific cross between *L japonicus* and *L filicaulis,* the other on a cross between the *L japonicus* ecotypes Gifu and MG-20. The nomenclature of the linkage groups was agreed so that linkage group I corresponds to the largest *L japonicus* Gifu chromosome (chromosome 1) and the maps are shown with the short arm on the top (Sandal *et al.*, 2002, Hayashi *et al.*, 2002).

The alignment of the two maps and FISH analysis showed that there is a translocation between the top of chromosome 1 and the bottom of chromosome 2 between Gifu and MG-20 (Hayashi *et al.*, 2002). Furthermore, a large inversion of the short arm of chromosome 3 and a small inversion on the long arm of chromosome 1 were found between Gifu and *L filicaulis* (Pedrosa *et al.,* 2002). These chromosome rearrangements result in problems with genetic mapping in corresponding regions of relevant mapping populations. Alignment of two maps is very helpful in resolving these regional difficulties and in ordering markers. Additional small chromosome rearrangements can not be excluded and it may be important to do physical mapping with single copy DNA from the different chromosomes (and chromosome arms) as previously done by fluorescent *in situ* hybridisation (FISH) (Pedrosa *et al.*, 2002).

For the Gifu x *L filicaulis*, an F2 mapping population of 165 plants was established and using only 3 selected AFLP primer combinations it was possible to cover much of the *L japonicus* genome for fast mapping (Sandal *et al.*, 2002). The Gifu x MG-20 F2 population consists of 127 plants and the map established from this cross is currently used to support the genome sequencing at the Kazusa. Transformable artificial chromosomes (TACs) corresponding to markers and ESTs are selected for seed point sequencing. After sequencing a TAC clone (app. 100 kb), the sequence is scanned for the presence of microsatellite sequences like, for example, ATATATATAT or GTGTGTGT. This type of sequence changes rapidly during evolution, resulting in small insertion/deletions.

In many cases a size difference between the mapping parents is found that can be used as a microsatellite marker anchoring a sequenced TAC clone on the genetic map of Lotus (Sato *et al.*, 2001, Hayashi *et al.*, 2001, Nakamura *et al.*, 2002, Kaneko *et al.*, 2003, Asamizu *et al.*, 2003). For some of the SNPs found, it was possible to develop codominant derived cleaved amplified polymorphic sequences (dCAPS) (Neff *et al.*, 1998). For each TAC clone, around 10 genes have been sequenced and placed on the genetic map through microsatellite/dCAPS markers. Now, more than 500 TAC clones have been sequenced anchoring more than 5000 genes on the genetic map.

The microsatellite markers developed for the *L japonicus* Gifu x *L japonicus* MG-20 mapping population have also been tested for polymorphism between *L japonicus* Gifu and *L filicaulis* and 83% of the markers can be used in this cross either as dominant or codominant markers. For the remaining 17% of markers sequencing of the PCR products reveal SNPs for many of them. This efficient marker transfer provides an opportunity for direct and fast alignment of the two genetic maps.

After scoring genotypes of individual loci in mapping populations, genetic linkage maps were produced using Mapmaker and Joinmap programs. Usually a manual colormap (Kiss *et al.*, 1998) is also made to provide a visual overview of the linkage groups and to indicate quality of the map and markers. On average, only one recombination is seen per plant per chromosome in Lotus F2 populations. Apparent, close double recombination events in any one F2 plant will be obvious from the colormap. This might be caused by mistakes in the experiment or in the scoring and should therefore be checked again.

BAC libraries

For physical mapping and conversion of genetic linkage into physical distance, large insert libraries are needed. A number of *L japonicus* BAC libraries are available (Table 2). At Kazusa a number of *L japonicus* MG-20 BAC and TAC libraries were made from partial *Mbo*I or *Hin*dIII digestions of genomic DNA to support genome sequencing (Sato et al., 2001). More recent libraries from Kazusa were constructed with *Eco*RI or *Sau*3A. A BAC library of *L japonicus* Gifu carrying very large inserts was made by Kawasaki and Murakami, (2000). A transformation-competent Gifu BAC library was constructed by Men *et al.*, (2001) and we have constructed a *L japonicus* Gifu BAC library from a partial *Bam*HI digests (Sandal *et al.*, unpublished). This library consists of 30,000 clones with an average insert size of 75 kb. On average two copies of each single copy gene is found in this library. In general, it is a great advantage to have BAC and TAC libraries constructed with different restriction enzymes increasing chances of covering the whole genome. For fine mapping sequencing directly on BACs or TACs, is very important and access to BAC or TAC libraries for both mapping parents is an advantage.

Ecotype	Restriction enzyme	Vector	Ave. insert size	Number of clones	Screening
Gifu (a)	*Hin*dIII	PCLD04541	94 kb	33408	Hybrid-isation
Gifu (b)	*Bam*HI	pBAC/SacB I	75 kb	32000	Hybrid-isation + PCR
Gifu (c)			140 kb	30720	Hybrid-isation
MG-20 (d)	*Hin*dIII/*Mbo* I	pYLTAC7	87, 96, 105, 106 kb	7.7 genomes	PCR

Table 2. L japonicus *BAC libraries constructed in different laboratories: (a) Men et al., 2001; (b) Sandal et al., unpublished; (c) Kawasaki and Murakami, 2000; (d) Sato et al., 2001.*

Map-based cloning

Most of the *L japonicus* Gifu mutants originate from chemical mutagenesis programs so map-based cloning is needed to isolate the gene involved. To initiate a positional cloning project a mutant is crossed to MG-20 and/or *L filicaulis*. MG-20 appears to be more efficient as pollen recipient than the Gifu parent. Alternatively, since most mutations are recessive, the mutant can be chosen as pollen recipient and the F1 plant tested for wild type phenotype. In the next generation, F2 plants segregating mutants are grown under conditions where mutant phenotype can be distinguished from wild type. For recessive mutations, mutants are selected for genotyping because they will be homozygous for the mutant parent genotype (DNA polymorphisms) in the chromosomal region carrying the locus and gene of interest. For dominant mutations, plants with wild type phenotype are genotyped as they are homozygous for the recessive allele from the wt parent. With a mapping population of only 20-30 mutants, it is possible to place the locus on a chromosome and within a region of the linkage map. See Table 3 showing a colormap for 20 mutant plants from a cross of the mutant *har1* to MG-20. Using this approach, a number of symbiotic loci/genes have now a rough position and some were subsequently mapped and cloned (Figure 1).

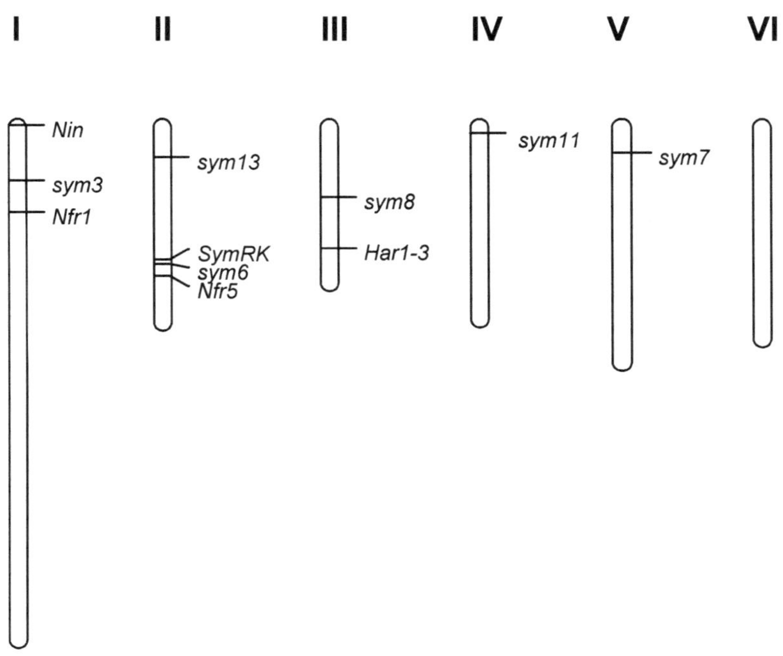

Figure 1. Map positions of 11 sym genes on the six linkage groups/chromosomes of L japonicus *based on the* L filicaulis *x* L japonicus *Gifu genetic map (Sandal et al., 2002). The map position of sym6 could also be above SymRK. nin, nfr1, nfr5 and sym3 have a Nod⁻ phenotype. sym3 does not have mycorrhizal symbiosis. sym6, 7, 8, 11 and 13 have a Fix⁻ phenotype. har1 has a hypernodulation phenotype. Where the gene has been isolated, the gene is indicated as the wild type gene (first letter capital).*

In practice, microsatellite markers are highly useful for this crude mapping. PCR primers are placed flanking the microsatellite sequence and it is possible to detect the PCR fragment size difference between the mapping parents and both fragments in heterozygotes. To resolve small differences, a PCR product of 150-200 bp is typically chosen. To work efficiently with the mapping population it is a great advantage to run the mapping by PCR in half or whole titertrays using multichannel pipette and running the products in gels with slots that fit a multichannel pipette. The products are separated on 2-4% agarose gels. Generally, 2-2.7% agarose can be boiled in a microwave oven while 4% gels are boiled in an autoclave to dissolve all agarose. For size differences of 1-3 bp it will be necessary to run polyacrylamide gels. Microsatellite markers are codominant which makes them very useful but single nucleotide polymorphisms (SNPs) found by sequencing can be developed into markers using parent specific or dCAPS primers (Neff *et al.*, 1998).

An alternative to microsatellite markers is to use AFLP (Vos *et al.*, 1995) and bulked segregant analysis (Michelmore *et al.*, 1991). For AFLP experiments, a cross to *L filicaulis* is often used because of the high level of polymorphism that leads to rapid identification of linked markers. A cross to MG-20 can also be used for AFLP analysis using High Efficiency Genome Scanning (HEGS) (Kawasaki and Murakami, 2000). This approach can provide new markers that might be closer than the already known microsatellite markers. A closely linked AFLP marker band can be extracted from the gel, re-amplified, and cloned. It can then be used to isolate BAC/TAC clones to build a contig of the region.

For regional mapping, it is often an advantage to arrange linkage information in colormaps (Kiss *et al.*, 1998). Different genotypes are indicated in Excel tables (see Table 3). Use the "conditional formatting" option of Excel (see manual for relevant Excel version) to colour the cells according to the genotype.

TM marker	Chr	Map pos	M1	M2	M3	M4	M5	M6	M7	M8	M9	M10
TM0181	1	0.0	B	h	h	h	h	a	h	B	h	h
TM0193	1	23.2	B	B	h	h	B	h	h	h	h	a
TM0122	1	80.2	B	h	a	h	a	h	h	a	h	a
TM0076	2	47.9	h	h	h	h	a	h	B	B	a	
TM0550	2	65.0	B		h	h	B	h		B		h
TM0080	3	10.6	a	B	B	h	B	h	h	h	h	a
TM0049	3	54.2	h	B	B	h	B	B	h	h	B	B
TM0203	3	64.4	B	B	B	h	B	B	B	h	B	B
TM0127	3	82.5	B	B	B	B	B	B	B	B	B	B
TM0256	4	5.3	a	B	a	B	h	h	h	h	a	B
TM0030	4	34.7	h	B	h	a	B	h	h	a	a	B
TM0095	5	18.6	h	h	B	a	B	h	h	B	a	a

TM marker	Chr	Map pos	M1	M2	M3	M4	M5	M6	M7	M8	M9	M10
TM0186	5	38.2	h	B	h	a	B	a	h	h	a	a
TM0362	5	53.9	h	B	h	h	B		h	h	h	h
TM0331	6	29.6	B	a	B	h	h	h	a	B	h	h
TM0367	6	48.7	h	a	h	B	a	h	a	B	h	h

TM marker	Chr	Map pos	M11	M12	M13	M14	M15	M16	M18	M19	M20	M21
TM0181	1	0.0				a						h
TM0193	1	23.2	h	B	a	a	B	a	B	a	B	B
TM0122	1	80.2	a		B	a	B	a	h	B	h	a
TM0076	2	47.9	a		h	h	a	h	h	h	h	h
TM0550	2	65.0	B	h	B	B	h	h	h	h	h	h
TM0080	3	10.6	h	B	B	B	h	h	B	h	h	B
TM0049	3	54.2	B	B	B	B	a	B	B	h	B	
TM0203	3	64.4	B	B	B	B	h	B	B	h	B	B
TM0127	3	82.5	B	B	B	B	B	B	h	B	B	B
TM0256	4	5.3	B	B	B	h	h	B	h	h	h	a
TM0030	4	34.7	B	B	h			B			a	
TM0095	5	18.6	B	h	a	h	h	h	h	h	h	a
TM0186	5	38.2	B	h	a	h	h	h	h	B	h	a
TM0362	5	53.9	h	h	a	h	h	h	h	a	h	a
TM0331	6	29.6	h	h	h	h	B	B	a	h	B	h
TM0367	6	48.7	h	h	a	h	h	B	a	h	B	a

Table 3. Colormap of the initial mapping of the har1 mutation. From left to right are the TM marker number, the chromosome number, and the cM position on the chromosome. a: homozygous L japonicus *MG-20, B: homozygous* L japonicus *Gifu (har1), h: heterozygous. The map shows 20 mutant plants tested with microsatellite markers from different chromosomal regions. At the location of the recessive mutant gene, all mutant plants will be homozygous* L japonicus *Gifu. It is clear that har1 is located on the lower end of chromosome 3. It can also be seen that the use of only one microsatellite marker from the middle of each of the chromosomes would have indicated the correct chromosome. Then different microsatellite markers from this chromosome could further pinpoint the position more precisely.*

For fine mapping, large mapping populations are needed (several hundred mutant plants). With additional mutants, the chance of finding a recombination very close to the gene increases. At the end of the process close recombination end-points on either side of the gene are needed. The closest markers are then used to screen BAC/TAC libraries to start building the physical map. See Figure 2 for an overview of the procedure. New markers developed from the BAC/TAC will show the direction towards the gene and are used as probes for isolation of additional clones walking towards the gene (see Figure 2 for details). For marker development, polymorphisms will be found primarily in non-coding regions.

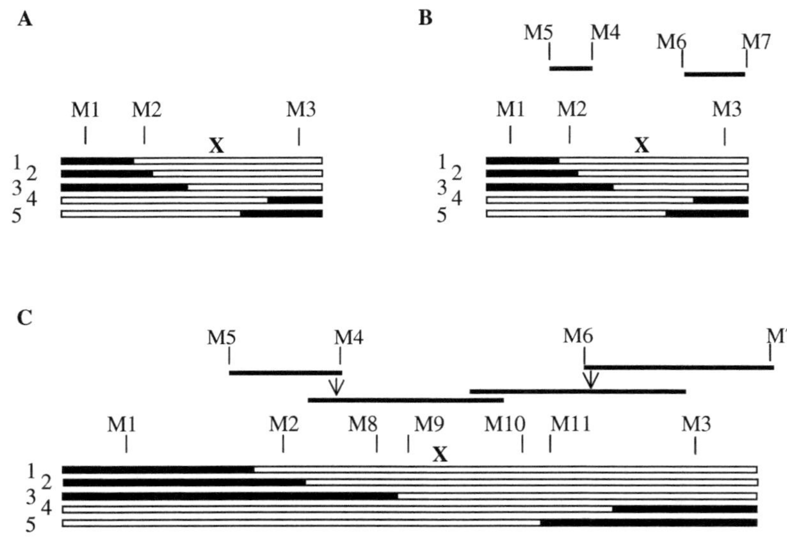

Figure 2. Overview of map-based cloning, fine mapping, and physical contig building.

A. Narrowing down the region containing the gene of interest using molecular markers. This illustrates the genetic composition of the chromosome containing the gene of interest from 5 mutant plants (1 to 5) from an F2 mapping population. In this example, the gene X is recessive, so that all mutant plants must be homozygous for the parental genotype of the mutant (in our case Gifu) in the region surrounding the gene. Black region indicates heterozygosity, white region indicates homozygosity in the region surrounding X. Three markers have been scored: Marker M1 shows heterozygocity in plants 1, 2, and 3, while marker M2 shows heterozygosity in plants 2 and 3 and homozygosity in plant 1, indicating that in plant 1 a breakpoint of recombination lies between M1 and M2. This also shows the direction towards the gene X, since the gene must be located in the region of homozygosity in all five plants. The direction is thus to the right of M1. It is also to the right of M2, since M2 is heterozygous in plants 2 and 3, showing that M2 is outside the region containing the gene. A similar rationale can be used for marker M3. This marker must be on the opposite side of gene X, since it detects recombination breakpoint in plants different from those detected with M1 and M2.

B. *The first step of physical contig building. A probe representing M2 is used to isolate a BAC or TAC clone(s). End sequences from this BAC are used to make primers to amplify fragments from the parents in which polymorphisms are used to develop new markers: M4 and M5. Checking the genotype of the five plants with M4 and M5 orients the BAC clone. Only plant 3 is heterozygous for M4 while plants 1, 2, and 3 show heterozygocity for M5 demonstrating that M4 is closer to gene X than M5 (and M2). A similar approach is taken for M3 to isolate BAC clone(s) on this side of the gene, giving markers M6 and M7.*

C. *Closing the contig. Using the relevant end sequences of the first BACs as probes, new BAC clones are isolated. The orientation of these is determined by doing PCR with end primers on the previous BACs. Primers from the end extending the contig will not be productive on the previous BAC. This is continued until the BACs meet to form a physical contig across the region. During this process, more markers are developed from (partial) BAC sequences to anchor and orient the BACs and to further delimit the region containing X. In this example, diagnostic markers are M8/M9 and M10/M11 defining the closest recombination breakpoints on the left and right side of X, respectively. When the marker information is exhausted, the region between the two closest markers must be sequenced and analysed for candidate genes.*

One of the pitfalls in contig building is the presence of repetitive or duplicated regions in the genome. Therefore, overlap between clones must be verified by regional sequence analysis (e.g. using the primers for BAC ends) to assure that the BAC clone contains sequences identical to the probe. It is often useful to Blast the sequence to the DNA databases to exclude repetitive Lotus sequences before use as a hybridization probe.

In the end, only very few mutant plants with recombination breakpoints delimiting the locus are of interest (plants 3 and 5, Figure 2). In order to define these few recombination end points more precisely the genotype of the relevant plants can be determined in polymorphic sites within the delimited region by sequencing of PCR products.

Once the region is narrowed down as much as possible, the sequence of the region between the closest markers must be determined and analysed for candidate genes using BlastN and BlastX searches in public databases, focusing on genes likely to have a function in the process of interest. Mutant alleles are then sequenced to identify premature stop codons, missense mutations, changed exon/intron splice sites, insertions/deletions, or even changes in regulatory elements of the promoter.

The final proof of the identity of a cloned candidate gene is complementation of mutants by the wild type gene. For most of the symbiotic mutants, this can be done by hairy root complementation via *Agrobacterium rhizogenes* (ibid). Some of the hypernodulation mutant phenotypes are determined by the shoot genotype (for example *har1*) and in this situation stable transformants are required.

General map-based cloning strategy

Select *L japonicus* Gifu mutant
↓
Cross to e.g. *L japonicus* Miyakojima
↓
Rough mapping
↓
Fine mapping
↓
BAC/TAC contig with flanking markers
↓
Candidate genes
↓
More alleles/completion

1. Mutant populations have been made in *L japonicus* Gifu. Decide which gene to clone. The mutant should have a phenotype that is easily scored. It should preferably be caused by a mutation in a single gene (3:1 segregation).
2. Cross to *L japonicus* MG-20 and/or *L filicaulis*. Collect seeds from the F1 plant. For recessive mutations screen out F2 mutant plants since they are homozygous for the mutant genotype at the locus of interest. For dominant mutations, wild type plants are used for mapping since they are homozygous for the wild type genotype at the locus of interest.
3. Rough map using microsatellite markers or AFLP and bulked segregant analysis. For recessive mutations, find the region with excess of homozygous *L japonicus* Gifu alleles. For dominant mutations, find the homozygous region with excess *L japonicus* MG-20/*L filicaulis* alleles. This will identify chromosome and chromosomal region.
4. Fine map with a larger population and more microsatellite and/or AFLP markers.
5. When the closest markers are less than 0.5-1 cM from the gene, start to isolate BAC/TAC clones using these flanking markers. Develop new markers from BAC/TAC ends to determine direction to the gene. Use these markers to isolate more BAC/TAC clones. Repeat this procedure until a contig containing markers on both sides of the gene of interest is completed. Determine the sequence of the BAC/TAC clone(s).
6. In this contig, look for candidate genes by BlastN/BlastX database searches in public databases. If a candidate gene is found, sequence the gene from a mutant. If a mutation is found, sequence all allelic mutants. If no mutation is found in the candidate gene, consider other candidate genes. All genes inside the flanking markers could eventually be sequenced or the mapping population could be increased to narrow down the region with closer recombinations.
7. Unequivocal evidence for correct gene assignment can be obtained by complementation of a mutant by *Agrobacterium rhizogenes* or *Agrobacterium tumefaciens* transformation.

From commencement of this undertaking in 1999 until end of 2003, two Lotus genes were cloned using transposon tagging and four using map-based cloning approaches (Table 4). Map-based cloning projects of around ten, mostly symbiotic, genes are close to completion and some will be published in 2004. This rapid progress demonstrates the power of genetics and shows that *L japonicus* is a prime model legume. With completion of the genome sequence in near future opening the possibility of high throughput SNP mapping positional cloning will be both easier and faster. This success in genetic analysis of primarily legume specific processes will no doubt create increased interest for Lotus as experimental system for scientific questions shared with other plants.

Gene	Type	Cloning method	Reference
Nin	Putative transcription factor	Transposon tagging	Schauser *et al.*, 1999
Pfo	F-box protein	Transposon tagging	Zhang *et al.*, 2003
SymRK	Receptor kinase	Map-based cloning	Stracke *et al.*, 2002
Har1	Receptor kinase	Map-based cloning	Krusell *et al.*, 2002 Nishimura *et al.*, 2002
Astray	Basic leucine zipper and RING-finger motif	Candidate approach	Nishimura *et al.*, 2002
Nfr1	Receptor kinase	Map-based cloning	Radutoiu *et al.*, 2003
Nfr5	Receptor kinase	Map-based cloning	Madsen *et al.*, 2003

Table 4. Genes cloned from Lotus mutants, map-based, Ac, or a candidate approach.

Mapping resources for the future - recombinant inbred lines

Stabilised mapping populations, recombinant inbred lines (RILs), are important resources for future mapping of single genes as well as quantitative trait loci (QTLs). Since our interspecific population represents alleles adapted to very different environmental conditions (Algeria and Japan) recombinant inbred lines have been developed by generations of selfing of our F2 plants in the *L filicaulis* x *L japonicus* Gifu mapping population. The advantage of recombinant inbred lines is that after 8 generations they are expected to be almost completely (99.22%) homozygous. Plants from seeds of a RIL will thus be phenotypically and genetically identical to the parent and therefore different research groups can use seeds of the same lines and use the same map. Accumulation of marker information will further allow genotyping of each plant. At present (November 2003), 78 lines have reached S8. A number of other lines have reached S5, S6, and S7 so that we expect soon to have around 100 RILs.

Because of the possibility of new recombination events in the heterozygotic regions, the map of the recombinant inbred lines will be expected to have the double length of the F2 map and thereby better resolution (Burr and Burr, 1991). Mapping on 44

RILS has now started with 50 microsatellite markers also used for the *L japonicus* Gifu x *L japonicus* MG-20 map. Therefore, the maps can be aligned. To obtain seeds of these RILs send an Email to sandal@biobase.dk.

149 RILs have also been developed for *L japonicus* Miyakojima MG-20 x *L japonicus* Gifu B-129: (www.kazusa.or.jp/lotus/RIline/ or http://cryo.naro.affrc.go.jp/sakumotu/mameka/lotus-e.htm). The RILs have been developed by Kazusa DNA Research Institute from F2 seeds provided by M Kawaguchi. The final number of these LjMG RILs will be 210. An initial map on this mapping population can be seen on www.kazusa.or.jp/lotus/RIline/RI_map.html The marker scoring can be downloaded as an Excel file from this website.

Finally, Peter Gresshoff's laboratory has developed RILs for *L japonicus* Gifu x *L japonicus* Funakura. These ecotypes are more closely related than Gifu and Miyakojima. These RILs have been used to examine glutamine synthetase function for plant biomass (Limani *et al.*, 1999). To obtain seeds of these RILs send an Email to p.gresshoff@botany.uq.edu.au.

For your level of interest, it is therefore worthwhile to analyze *L japonicus* ecotypes Gifu, Miyakojima, and Funakura and *L filicaulis* to look for differences and then map the character in the RILs for which the parents show the difference. As they are separate species *L filicaulis* and *L japonicus* would be expected to show more differences of biological (phenotypical) characters than the ecotypes and seeds from these RILs are available on a collaboration basis for analysis of different traits and subsequent mapping of single genes and QTLs.

ACKNOWLEDGEMENTS

We would like to thank Martin Parniske, Krzysztof Szczyglowski, Peter Gresshoff, Masayoshi Kawaguchi and Yosuke Umehara for making unpublished results available to us.

REFERENCES

Asamizu E, Kato T, Sato S, Nakamura Y, Kaneko T, and Tabata S. (2003). **Structural analysis of a *Lotus japonicus* genome. IV. Sequence features and mapping of seventy-three TAC clones which cover the 7.5 Mb regions of the genome**. *DNA Research* 10, 115-122.

Burr B, and Burr FA (1991). **Recombinant inbreds for molecular mapping in maize: theoretical and practical considerations**. *Trends in Genetics* 7, 55-60.

Grant WF, Bullen MR, and de Nettancourt D. (1962). **The cytogenetics of *Lotus*. I. Embryo-cultured interspecific diploid hybrids closely related to *L corniculatus* L** *Canadian Journal of Genetics and Cytology* 4, 105-128.

Handberg K, and Stougaard J. (1992). ***Lotus japonicus*, an autogamous, diplod legume species for classical and molecular genetics**. *The Plant Journal* 2, 487-496.

Hayashi M, Miyahara A, Sato S, Kato T, Yoshikawa M, Taketa M, Hayashi M, Pedrosa A, Onda R, Imaizumi-Anraku H, Bachmair A, Sandal N, Stougaard J, Murooka Y, Tabata S, Kawasaki S, Kawaguchi M, and Harada K. (2001). **Construction of a genetic linkage map of the model legume *Lotus japonicus* using an intraspecific F_2 population**. *DNA Research*

8, 301-310.

Jian, Q and Gresshoff PM. (1997). **Classical and molecular genetics of the model legume *Lotus japonicus*.** *Molecular Plant-Microbe Interactions* 10, 59-68.

Kaneko T, Asamizu E Kato T, Sato S, Nakamura Y, and Tabata S. (2003). **Structural analysis of a *Lotus japonicus* genome**. III. Sequence features and mapping of sixty-two TAC clones which cover the 6.7 Mb regions of the genome. *DNA Research* 10, 27-33.

Kawaguchi M, Motomura T, Imaizumi-Anraku H, Akao S and Kawasaki S. (2001). **Providing the basis for genomics in *Lotus japonicus*: the accessions Miyakojima and Gifu are appropriate crossing partners for genetic analyses.** *Molecular Genetics and Genomics* 266, 157-166.

Kawaguchi M, Imaizumi-Anraku H, Koiwa H, Niwa S, Ikuta A, Syono K and Akao S. (2002). **Root, root hair, and symbiotic mutants of the model legume *Lotus japonicus*.** *Molecular Plant-Microbe Interactions* 15, 17-26.

Kawasaki S and Murakami Y. (2000). **Genome analysis of *Lotus japonicus*.** *Journal of Plant Research* 113, 497-506.

Kiss G.B, Kereszt A, Kiss P and Endre G. (1998). **Colormapping: a non-mathematical procedure for genetic mapping.** *Acta Biologica Hungarica* 49, 125-142.

Krusell L, Madsen LH, Genua A, Szczyglowski K, Aubert G, Duc G, Sato S, Tabata S, Bruijn Fd, Pajuelo E, Sandal N, and Stougaard J. (2002). **A receptor-like kinase mediates shoot control of root development and nodulation.** *Nature* 420, 422-426.

Limami A, Phillipson B, Ameziane R, Pernollet N, Jiang Q, Roy R, Deleens E, Chaumont-Bonnet M, Gresshoff PM, and Hirel, B. (1999). **Does root glutamine synthetase control plant biomass production in *Lotus japonicus* L?** *Planta* 209, 495-502.

Madsen EB, Madsen LH, Radutoiu S, Olbryt M, Rakwalska M, Szczyglowski K, Sato S, Kaneko T, Tabata S, Sandal N, and Stougaard, J. (2003). **A receptor kinase of the LysM type is involved in legume perception of rhizobial signals.** *Nature* 425, 637-640.

Men AE, Meksem K, Kassem MA, Lohar D, Stiller J, Lightfoot D, and Gresshoff PM. (2000). **A bacterial artificial chromosome library of *Lotus japonicus* constructed in an *Agrobacterium tumefaciens*-transformable vector.** *Molecular Plant-Microbe Interactions* 14, 422-425.

Michelmore RW, Paran I, and Kesseli RV. (1991). **Identification of markers linked to disease resistance genes by bulked segregant analysis: a rapid method to detect markers in specific genomic regions using segregating populations.** *Proceedings of the National Academy of Sciences USA* 88, 9828-9832.

Nakamura Y, Kaneko T, Asamizu E, Kato T, Sato S, and Tabata S. (2002). **Structural analysis of a *Lotus japonicus* genome. II. Sequence features and mapping of sixty-five TAC clones which cover the 6.5 Mb regions of the genome.** *DNA Research* 9, 63-70.

Neff MM, Neff JD, Chory J, and Pepper AE. (1998). **dCAPS, a simple technique for the genetic analysis of single nucleotide polymorphisms: experimental applications in *Arabidopsis thaliana* genetics.** *The Plant Journal* 14, 387-392.

Nishimura R, Hayashi M, Wu G.J, Kouchi H, Imaizumi-Anraku H, Murakami Y, Kawasaki S, Akao S, Ohmori M, Nagasawa M, Harada K, and Kawaguchi M. (2002). **HAR1 mediates systemic regulation of symbiotic organ development.** *Nature* 420, 426-429.

Nishimura R, Ohmori M, Fujita H, and Kawaguchi, M. (2002). **A Lotus basic leucine zipper protein with a RING-finger motif negatively regulates the developmental program of nodulation.** *Proceedings of the National Academy of Sciences U S A* 99, 15206-15210.

Pedrosa A, Sandal N, Stougaard J, Schweizer D, and Bachmair A. (2002). **Chromosomal**

map of the model legume *Lotus japonicus*. *Genetics* 161, 1661-1672.

Perry JA, Wang TL, Welham TJ, Gardner S, Pike JM, Yoshida S, and Parniske M. (2003). **A TILLING reverse genetics tool and a web-accessible collection of mutants of the legume *Lotus japonicus*.** *Plant Physiology* 131, 866-871.

Radutoiu S, Madsen LH, Madsen EB, Felle HH, Umehara Y, Grønlund M, Sato S, Nakamura Y, Tabata S, Sandal N, and Stougaard J. (2003). **Plant recognition of symbiotic bacteria requires two LysM receptor-like kinases.** *Nature* 425, 585-592.

Sandal N, Krusell L, Radutoiu S, Olbryt M, Pedrosa A, Stracke S, Sato S, Kato T, Tabata S, Parniske M, Bachmair A, Ketelsen T, and Stougaard J. (2002). **A genetic linkage map of the model legume *Lotus japonicus* and strategies for fast mapping of new loci.** *Genetics* 161, 1673-1683.

Sato S, Kaneko T, Nakamura Y, Asamizu E, Kato T, and Tabata S. (2001). **Structural analysis of a *Lotus japonicus* genome. I. Sequence features and mapping of fifty-six TAC clones which cover the 5.4 Mb regions of the genome.** *DNA Research*, 8, 311-318.

Schauser L, Roussis A, Stiller J, and Stougaard J (1999). **A plant regulator controlling development of symbiotic root nodules.** *Nature* 402, 191-195.

Schauser L, Handberg K, Sandal N, Stiller J, Thykjær T, Pajuelo E, Nielsen A, and Stougaard J. (1998). **Symbiotic mutants deficient in nodule establishment identified after T-DNA transformation of *Lotus japonicus*.** *Molecular and General Genetics* 259, 414-423.

Searle IR, Men AE, Laniya TS, Buzas DM, Iturbe-Ormaetxe I, Carroll BJ, and Gresshoff PM. (2003). **Long-distance signaling in nodulation directed by a CLAVATA1-like receptor kinase.** *Science* 299, 109-112.

Stougaard J, and Beuselinck PR. (1996). **Registration of GIFU B-129-S9 *Lotus japonicus* germplasm.** *Crop Science* 36, 476.

Stracke S, Kistner C, Yoshida S, Mulder L, Sato S, Kaneko T, Tabata S, Sandal N, Stougaard J, Szczyglowski K, and Parniske M. (2002). **A plant receptor-like kinase required for both bacterial and fungal symbiosis.** *Nature* 417, 959-962.

Szczyglowski K, Shaw RS, Wopereis J, Copeland S, Hamburger D, Kasiborski B, Dazzo, FB, and de Bruijn, FJ. (1998). **Nodule organogenesis and symbiotic mutants of the model legume *Lotus japonicus*.** *Molecular Plant-Microbe Interactions* 11, 684-697.

Thykjær T, Stiller J, Handberg K, Jones J and Stougaard J. (1995). **The maize transposable element *Ac* is mobile in the legume *Lotus japonicus*.** *Plant Molecular Biology* 27, 981-993.

Vos P, Hogers R, Bleeker M, Reijans M, van de Lee T, Hornes M, Frijters A, Pot J, Peleman J, Kuiper M, and Zabeau M. (1995). **AFLP: a new technique for DNA fingerprinting.** *Nucleic Acids Research* 23, 4407-4414.

Webb KJ, Skøt L, Nicholson MN, Jørgensen B, and Mizen S (2000). ***Mesorhizobium loti* increases root-specific expression of a calcium-binding protein homologue identified by promoter tagging in *Lotus japonicus*.** *Molecular Plant-Microbe Interactions* 13, 606-616.

Zhang S, Sandal N, Polowick PL, Stiller J, Stougaard J, and Fobert PR. (2003). **Proliferating floral organs (PFO), a *Lotus japonicus* gene required for specifying floral meristem determinacy and organ identity, encodes an F-box protein.** *The Plant Journal* 33, 607-619.

Chapter 5.6

AN *IN SILICO* STRATEGY TOWARDS THE DEVELOPMENT OF LEGUME GENOME ANCHOR MARKERS USING COMPARATIVE SEQUENCE ANALYSIS

Leif Schauser[1*], Sreenath Subrahmanyam[1], Lene Heegaard Madsen[2], Niels Sandal[2], and Jens Stougaard[2]

[1]*Bioinformatics Research Center, University of Aarhus, Hoegh Guldbergs Gade 10, 8000 Aarhus* DENMARK; [2]*Laboratory of Gene expression, Department of Molecular Biology, Aarhus University, Gustav Wieds Vej 10, DK-8000 Aarhus C,* DENMARK; [*]*Corresponding author*

Email: schauser@daimi.au.dk

Phone: +45 8942 3239 Fax: +45 8942 3077

Keywords: Comparative genome analysis, evolutionarily conserved sequences, molecular marker development

Development of molecular markers and the transfer of marker information from one species to another are limiting steps in the assembly of genetic maps and the use of map information in breeding programs. To identify potential marker sequences more efficiently, we have established procedures combining multi-species EST and genome sequence data for a genome-wide, in silico *identification of molecular markers. Taking advantage of information from a few related species, comparative EST sequence analysis identifies evolutionarily conserved sequences (ECSs) that with high probability are conserved in less characterised species in the same family. The chance of observing variation between any two mapping parents is increased by selecting ECS that are interrupted by introns in corresponding genomic regions. Our procedure simultaneously optimizes (1) primer selection for stable performance of PCR across species by choosing ECS as the target sequences for priming, (2) the likelihood of polymorphism discovery by selecting intron-containing ECSs, (3) marker transfer between species, and (4) information content by counting copy numbers of homologous sequences in* Arabidopsis. *We illustrate our procedure in legumes, where model plant genome and EST sequence data have great potential re influencing crop legume breeding programs.*

A.J. Márquez (ed.), Lotus japonicus Handbook, 233-242.

INTRODUCTION

Two different genetic maps were recently established for *Lotus japonicus* and used for positional cloning of genes (Sandal et al., 2002; Hayashi et al., 2001; Pedrosa et al., 2002). One is based on the ecotypes GIFU and MG-20, the other on the interspecific cross between GIFU and *L filicaulis*. Both maps were originally assembled from anonymous AFLP markers and a few gene-specific markers (Sandal et al., 2002; Hayashi et al., 2001). More recently, seed point sequencing of the *Lotus* genome has provided numerous microsatellite markers that has increased the marker density of both maps substantially. Unfortunately, markers identified by these methods do not transfer well to crop legumes.

A bottleneck in establishing high-density genetic maps for crop legumes is the identification of useful polymorphic markers. The most common variations found between any two mapping parents are single nucleotide polymorphisms (SNPs) and small insertions/deletions (INDELS). Potentially rich resources for SNPs are genome projects and expressed sequence tag (EST) collections. This is only true, however, if the collection of individuals used for these large-scale sequencing projects were polymorphic. While this is the case for the human genome project (Sachidanandam et al, 2001), it is not for most other genome projects, such as the *Arabidopsis thaliana* or the rice genome projects (Arabidopsis Genome Initiative, 2000; Yu et al., 2002; Goff et al., 2002).

In Arabidopsis, SNPs/INDELS were obtained later by sampling random sequences from a polymorphic ecotype (Landsberg *erecta*) and comparison to the Columbia genome sequence (Jander et al., 2002). This brute force approach will eventually unravel all differences between these two ecotypes. In Arabidopsis, this was feasible due to its compact and gene-dense genome of 125 Mbp, which contains unusually low amounts of repetitive DNA. However, such an approach is expensive and specialised, and not immediately applicable in other species; in other plants, the cost per informative polymorphism would be higher or prohibitively expensive as large proportions of the genomes consist of highly repetitive non-informative DNA. For example in legumes, the smallest genomes studied (*Lotus japonicus* and *Medicago truncatula*) are estimated to be 420- 470 Mbp in size (Young et aL 2003), and most of the approximate 300 Mbp difference to the Arabidopsis genome is thought to be attributable to the accumulation of repetitive DNA, mainly retrotransposons.

In order to circumvent problems associated with repetitive DNA, methods focusing on unique sequences need to be developed. Towards this goal, we employ a strategy based on identification of single or low copy number evolutionarily conserved sequences (ECSs) within legume transcriptomes. These sequences are used as PCR primer annealing sites for subsequent polymorphism discovery. This approach ensures that gene-rich regions of the genome are the primary targets of this effort. Our bioinformatic approach is based on differences in the evolutionary rate of DNA changes in a genome.

During evolution, many functional sequences, such as coding regions and regulatory elements, are under strong purifying selection. In contrast, intron sequences are less constrained and will display a higher degree of mutational variation between any two ecotypes / varieties. Since the nature of PCR limits the detection of polymorphisms to the region between the conserved primer anchoring sequences, we attempt to maximize the chance of including polymorphisms in such a region. This is done by considering the evidence for two ECS flanking one or more introns in the orthologous gene of the Arabidopsis genome. Although the evolutionary constraints on the exact sequence of the intron are relaxed, the position and approximate length of the intron is usually conserved, even over long evolutionary distances (Roy et al., 2003). We then rank the ECSs according to criteria regarding intron length in order to obtain high quality candidate sequences for primer design for PCR amplification. The output of our algorithm will be an ordered list of primers pairs for the amplification of sequences that have a high probability of capturing polymorphisms between varieties of any legume. Subsequent sequencing of intron spanning PCR products will reveal the presence of any polymorphism between varieties of a legume species in these regions.

METHODS

The EST clusters used for this analysis were retrieved from the Institute of Genome Research (TIGR). We downloaded the gene indices (clustered EST collections, Quackenbush et al., 2000) for *Medicago truncatula* and *Glycine max*. We constructed our own gene index for *Lotus japonicus* by clustering 22.000 EST sequences deposited at NCBI with additional 44.500 EST sequences from the Kazusa DNA Research Institute (Asamizu and Tabata, unpublished) using TIGRs TGICL software (Pertea *et al.* 2003). The Arabidopsis proteome and coding sequences (CDS) sequences were downloaded from the TIGR FTP site. The Lotus genomic sequences were retrieved using NCBIs ENTREZ.

The Blast package (Altschuler et al 1995) was obtained from the NCBI. For comparison of nucleotide sequences, we used the megablast program with a word size of 20 and cut off e-value of 2e-40. For DNA-protein comparisons, we used the BlastX program (e-value 10e10-6). A series of Perl scripts were generated to parse the Blast outputs and assemble sequence collections of ECS.

PCR amplification and sequencing of amplicons was performed using standard laboratory protocols.

RESULTS

The algorithm designed to identify molecular markers is divided into five operational steps and is best illustrated as a pipeline adding consecutive comparative selection criteria (see Figure 1).

Step 1: Identifying expressed evolutionary conserved sequences (ECS) from different legume species

Regions of strong homology between collections of ESTs from different legume species were identified. We have here used the ready clustered gene indices as input, but any assembled collection of ESTs could serve as entry points. Expressed ECS, as defined by stringent criteria (word size 20, e-value 2e-40), were passed on to the second step. The principle behind step one is that ECSs represent exons that have been under strong purifying selection during evolution i.e. they display a less-than-average diversity between legume species.

Step 2: Counting copy numbers in the Arabidopsis proteome

In order to get a score for the information content of an ECS, we counted the number of highly homologous sequences encoded by the Lotus representative of the ECS in the Arabidopsis proteome. Repeated sequences are not useful for mapping purposes, since polymorphisms might reflect paralogous origin rather than allelic variation. Furthermore, allelic variation at a candidate marker locus can be partially or completely masked by the presence of paralogues, reducing the information content of this marker. Several rounds of genome duplication and gene family amplification have occurred before the split between Leguminosae (Rosid I) and Brassicacae (Rosid II). The Arabidopsis genome has been subject to at least one round of duplication since the divide. The diploid legume species Lotus and Medicago do not seem to have undergone a similar duplication (Bowers et al, 2003). Therefore, counts of Arabidopsis genes can be taken as a conservative estimate (i.e. overestimate) of the legume count.

Step 3: ECS-genome alignment.

In order to maximize chances of detecting polymorphisms, three criteria are set that characterize the alignment of ECSs with genomic sequences. The first criterion is the presence of introns in the corresponding genomic region. Here we only make use of the *Lotus* genome sequence, but this could easily be extended to, Medicago and Arabidopsis genomic sequences. This is informative, since the presence, location, and approximate length of introns are highly conserved features, even among distant species. The presence of introns enhances the chance of encountering polymorphic sites due to the accumulation of mutations in regions under low selection pressure. We also score the length of introns. This quantity is of interest for two reasons: Firstly, short introns are less likely to be polymorphic than longer ones, and thus longer introns are of interest. Secondly, the final PCR reaction using degenerate primers is limited to the maximum amplicon size of ~3 kb using standard polymerases.

Step 4: Ranking ECSs

The ECSs are now ranked according to the previous steps in order to maximise the likelihood of detecting polymorphisms. In this study, we only consider ECS with an Arabidopsis gene count of 1 or 2 and containing at least one intron in the Lotus genome. We then rank the ECS according to the overall length of the alignment and the length of introns.

Step 5: Primer design

Multiple alignments of ECSs with indication of intron position are generated and primers are designed using this alignment as a guide. This step is currently done manually. Again, it can easily be automated by applying software such as PRIMO (Li et al., 1997).

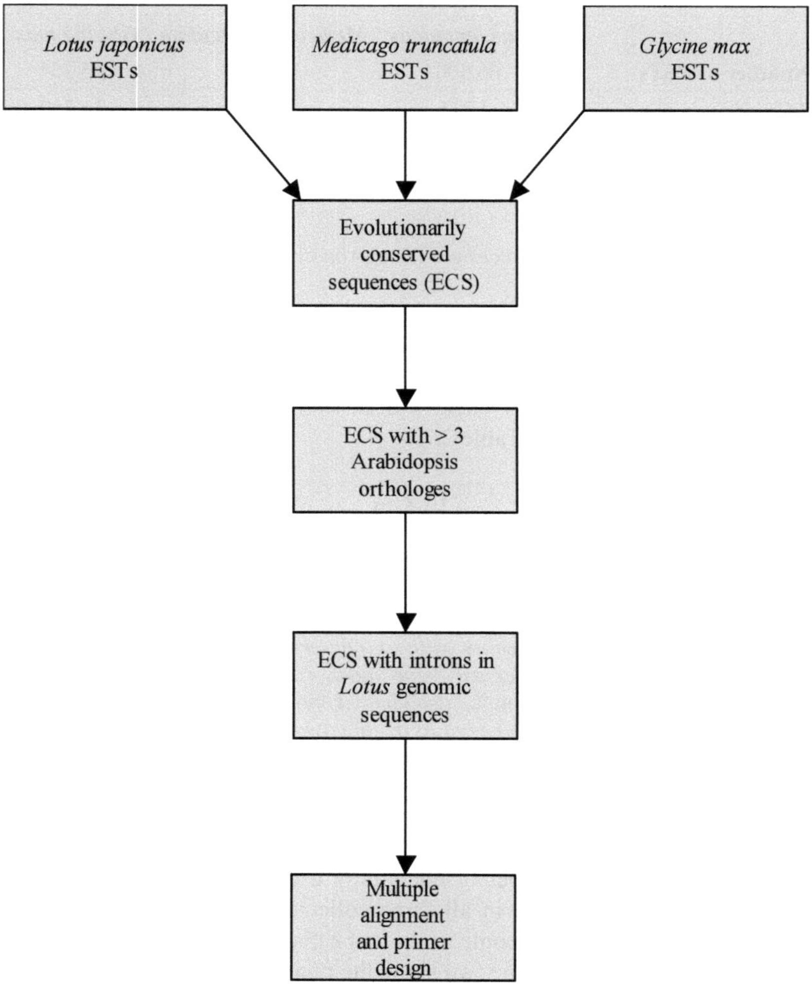

Figure 1. Pipeline of the marker candidate algorithm. In the first step, EST collections of selected species are compared. Evolutionary conserved sequences are passed on to the next step. Here the number of sequences with homology to the Arabidopsis *reference proteome is estimated. In this study, we only considered sequences genes with one or two hits in the* Arabidopsis *proteome. ECSs passing this criterion are compared to Lotus genomic sequences and ranked according to overall length of the ECS and optimal length of introns. The ECSs are multiply aligned and primers are designed using this alignment as input.*

We will demonstrate the flow of the pipeline with a preliminary run of legume sequences. Starting material for our analysis were the gene indices generated at TIGR (*Medicago*, *Glycine*) and Lotus indices generated by us from a subset of Lotus ESTs. Gene indices (GIs) are clusters of ESTs using TIGRS Assembler software. GIs consists of tentative EST clusters (TCs) and singleton ESTs. The number of ESTs and GIs generated and used for this study are shown in Table 1.

	Lotus japonicus	***Medicago truncatula***	***Glycine max***
Number of ESTs	66,500	170,676	285,351
TC	4,715	16,086	24,750
Number of GIs	17,057	33,765	55,990
Number of nuclear GIs	16,951	33,706	55,943

Table 1: Numbers of EST sequences used in this analysis.

Because chloroplast genomes are not a target for mapping purposes, ESTs derived from this organelle were removed, leaving nuclear-encoded GIs. By repeated application of step one in the pipeline, two pairwise comparisons were merged into a three-way comparison of the EST clusters from Lotus, *Medicago* and *Glycine*. The resulting numbers are shown in Table 2.

	Lj-Mt	**Lj-Gm**	**Mt-Gm**	**Lj-Mt-Gm**
Number of ECSs	6,261	8,652	10,655	4,463

Table 2. Numbers of sequences with a conserved counterpart in pairwise comparisons with the other legume gene index collections. Lj denotes Lotus japonicus; *Mt,* Medicago truncatula; *and Gm,* Glycine max. *Sequences common to all pairwise comparisons are referred to as three-way ECSs (Lj-Mt-Gm). For the purpose of this illustration, we limited further analysis to the orthologue set of 4,463 ECSs with representatives in all three EST collections.*

In step two of the pipeline, where the assembled ECS are compared to the Arabidopsis proteome, the number of potentially useful ECSs was further reduced. Of 4,463 sequences represented in all three collections, 867 also had exactly one match in the Arabidopsis proteome. As an estimate of homologous sequence numbers in a diploid plant genome, we show the number of significant hits obtained by comparing the 4463 ECS with the proteome of Arabidopsis (Figure 2).

ECSs with one hit in the Arabidopsis proteome were used for further analysis. The approximately 200 ECSs with no hits may have potential as general legume marker candidates, but many of them correspond to ribosomal rRNA molecules that are not represented in the proteome, and we did not pursue any of these candidates. In total, we identified 867 ECSs with one orthologous sequences in the Arabidopsis proteome and 812 ECSs with two homologous sequences in the Arabidopsis proteome.

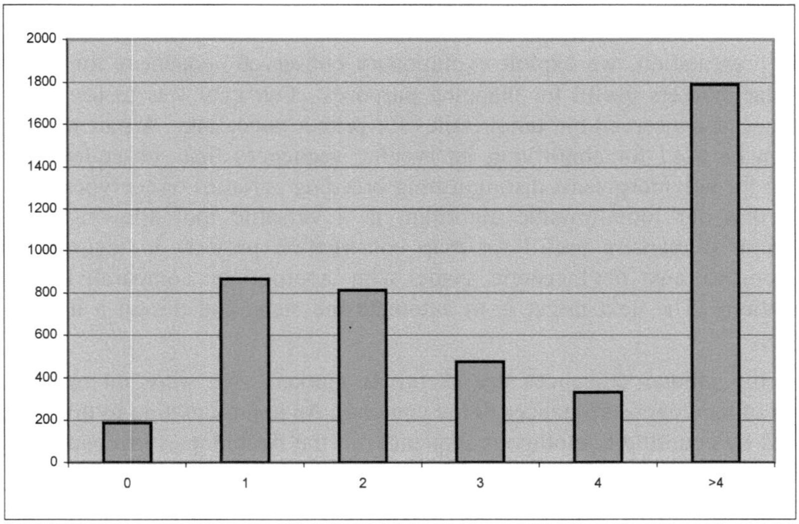

Figure 2. Distribution of ECS according to their number of homologous sequences in the Arabidopsis proteome, as defined by significant BlastX hits. Those in class 0 are either legume-specific ECS or non-coding RNAs (not represented in the proteome). For this study, we only considered the approximately 1,600 ECSs with one hit in the Arabidopsis proteome.

Applying step three and four, where ECSs are compared with genomic sequences and scanned for introns, the variation potentials of the ESCs were assessed. For this analysis, we have initially chosen to compare the 867 Lotus EST clusters with one Arabidopsis orthologue from the previous step with the published genomic sequences from Lotus. Introns were inspected for suitable length and for total ECS length. From this 49 ECS were selected. Repeating this step with the 812 ECSs with two Arabidopsis homologues resulted in an additional 47 ECS sequences.

Step 5 is not yet automated and included in the algorithm and this step is currently operated manually. For this purpose the multiple sequence alignment of ECSs from *Glycine max, Lotus japonicus*, *Medicago truncatula* together with the Arabidopsis CDS was conducted using standard DNA comparison programs software, and primer design was done by manual inspection of this alignment.

Preliminary experimental evidence demonstrates the efficiency and robust performance of the bioinformatic approach. Out of the first 19 EST alignments based on ECS sequences with one hit in the Arabidopsis proteome and available genome information from Lotus, primers with only limited degeneracy could easily be designed for 18. In a pilot experiment testing this first set of degenerate primers on the parents of a mapping population of distantly related *Arachis hypogaea* (peanut), 50% of the primers were productive under our standard low stringency PCR conditions (annealing at 50°C). Sequence determination showed that most primers amplified the expected genomic fragment and that ~90% of these sequences contained polymorphisms (SNPs, indels, or size polymorphisms) between the peanut mapping parents which could be exploited for mapping purposes.

DISCUSSION

In this presentation, we exploit evolutionary conserved sequences for developing molecular markers useful for mapping purposes. Our goal was to use these ECS sequences as conserved but unique sites for primer annealing. A pair of such sites can then be used for amplifying intervening sequences that subsequently can be scanned for polymorphisms distinguishing breeding varieties or ecotypes. We have shown that our bioinformatic algorithm is a versatile tool allowing the quick generation of markers useful for map construction projects in legumes and by extension, to any phylogenetic clade with appropriate comparative sequence information. The next target is to automate the steps and design a user-friendly interface.

Since only unique sequences are useful as markers, we were interested in the number of paralogous sequences in the genome. An approximation to this number is obtained by counting homologous sequences in the Arabidopsis proteome. Strictly speaking, we are not able to discern between orthologous and paralogous origin of homologous sequences using this approach. However, for those sequences with only one homologue in the Arabidopsis genome, we can reasonably assume orthology. The ECS with no homologous proteins in Arabidopsis could represent legume specific genes. However, they could equally well represent conserved UTRs or non-coding RNAs, and would hence not be represented in the proteome. It might be meaningful to count the number of homologous DNA sequences in the Arabidopsis genome instead, since these ECSs also might be of value for marker development purposes. Our analysis found 4% of such ECS. Blast searches revealed their identity as non-translated ribosomal rRNA molecules deriving from repetitive regions of the genome and hence useless as marker candidates. Sequences with no hit in the Arabidopsis genome are likely to be either newly evolved sequences, arisen since the divide between legumes and Arabidopsis, or sequences of ancient origin, that have since become deleted form the Arabidopsis lineage. Among the ECS, we only find 4 % with no Arabidopsis homologue. This means, not surprisingly, that strong sequence conservation among legumes is predictive of the presence of homologous sequences in Arabidopsis.

A particular use of this algorithm is the transfer of genome information from model species to crop plants. In plant breeding programs, traits of economic importance are screened out of large populations. Breeders introduce variation through crosses between varieties and in some cases wild relatives but there is rarely a simple method for following the segregation of the trait or allele of interest. Instead of screening for the traits *per se*, which can be difficult to score due to environmental conditions, late onset or small contributions to the phenotype, breeders often use linked markers as indicators of inheritance. For this purpose, molecular markers are best suited, since they are co-dominant, cheap and readily scored. Dense genetic maps spanning all linkage groups are of invaluable help for breeding purposes. The success of applying our pipeline depends on the variation between any two varieties used for mapping.

In a current study, we are using this pipeline in an attempt to identify polymorphisms in diploid peanut varieties used as crossing parents Preliminary results suggest a success rate of PCR amplification in peanut of 50% when all possible primers are considered. Given the large number of marker candidates generated once the pipeline is fully implemented, a more stringent selection of primer sets to include only those with the highest ranking could probably lead to more efficient amplification in distantly related legumes. On the other hand, for more closely related legumes it will be of value to develop the full set of ECS based primer sets as a tool to bridge the genetic maps of model and crop legumes. When looking for introns, we have here only exploited the genomic sequence information of *Lotus japonicus*, but clearly, we could also make use of the Medicago genomic sequences. Given the observed conservation of intron positions it should even be possible to use the Arabidopsis genome as a reference. Thus, we should be able to generate several thousands of marker candidates for any legume cross in the near future.

The comparative approach described here is broadly applicable to all EST resources collected form species with appropriate phylogenetic distance and reference genome information. We are currently evaluating the feasibility of this pipeline to marker development in monocotyledons.

ACKNOWLEDGEMENTS

We would like to thank E Asamizu and S Tabata for sharing unpublished EST sequence information. This work was funded by the Danish Agricultural and Veterinary Research Council (grant 23-01-0110) to JS. LS is supported by SNF grant 21-01-0329.

REFERENCES

Altschul SF, Gish W, Miller W, Myers E, and Lipman DJ. (1990) **Basic local alignment search tool.** *Journal of Molecular Biology* 215, 403-410.

Arabidopsis Genome Inititative (2000) **Analysis of the genome sequence of the flowering plant *Arabidopsis thaliana*.** *Nature* 408, 796-815.

Bowers JE, Chapman BA, Rong J, Paterson AH. (2003) **Unravelling angiosperm genome evolution by phylogenetic analysis of chromosomal duplication events.** *Nature* 422, 433-438.

Chiapello H, Lisacek F, Caboche M, Henaut A. (1998) **Codon usage and gene function are related in sequences of *Arabidopsis thaliana*.** *Gene* 209, 1-38.

Fulton TM, Van der Hoeven R, Eannetta NT, Tanksley SD. (2002) **Identification, analysis, and utilization of conserved ortholog set markers for comparative genomics in higher plants.** *Plant Cell* 14, 1457-67.

Goff SA, Ricke D, Lan TH, Presting G, Wang R, Dunn M, Glazebrook J, Sessions A, Oeller P, Varma H, Hadley D, Hutchison D, Martin C, Katagiri F, Lange BM, Moughamer T, Xia Y, Budworth P, Zhong J, Miguel T, Paszkowski U, Zhang S, Colbert M, Sun WL, Chen L, Cooper B, Park S, Wood TC, Mao L, Quail P, Wing R, Dean R, Yu Y, Zharkikh A, Shen R, Sahasrabudhe S, Thomas A, Cannings R, Gutin A, Pruss D, Reid J, Tavtigian S, Mitchell J, Eldredge G, Scholl T, Miller RM, Bhatnagar S, Adey N, Rubano T, Tusneem N, Robinson R, Feldhaus J, Macalma T, Oliphant A, and Briggs S. **A draft sequence of the rice genome (*Oryza sativa* L ssp. *japonica*).** *Science* 296:92-100.

Hayashi M, Miyahara A, Sato S, Kato T, Yoshikawa M, Taketa M, Hayashi M, Pedrosa A, Onda R, Imaizumi-Anraku H, Bachmair A, Sandal N, Stougaard J, Murooka Y, Tabata S, Kawasaki S, Kawaguchi M, Harada K. (2001) **Construction of a genetic linkage map of the model legume *Lotus japonicus* using an intraspecific F2 population**. *DNA Research* 8, 301-310.

Jander G, Norris SR, Rounsley SD, Bush DF, Levin IM, Last RL (2002) **Arabidopsis map-based cloning in the post-genome era**. *Plant Physiology* 129, 440-450.

Li P, Kupfer KC, Davies CJ, Burbee D, Evans GA, Garner HR. (1997) **PRIMO: A Primer Design Program That Applies Base Quality Statistics for Automated Large-Scale DNA Sequencing**. *Genomics* 40, 476-485.

Pedrosa A, Sandal N, Stougaard J, Schweitzer D, and Bachmair A. (2002) **Chromosomal map of the model legume *Lotus japonicus***. *Genetics* 161, 1661-1672.

Pertea G, Huang X, Liang F, Antonescu V, Sultana R, Karamycheva S, Lee Y, White J, Cheung F, Parvizi B, Tsai J, and Quackenbush J. (2003). **TIGR Gene Indices clustering tools (TGICL): a software system for fast clustering of large EST datasets**. *Bioinformatics* 19, 651-652.

Quackenbush J, Liang F, Holt I, Pertea G, and Upton J. (2000) **The TIGR Gene Indices: reconstruction and representation of expressed gene sequences**. *Nucleic Acids Research* 28, 141-145.

Roy SW, Fedorov A, Gilbert W. (2003) **Large-scale comparison of intron positions in mammalian genes shows intron loss but no gain**. *Proceedings of the National Academy of Sciences USA* 100, 7158-7162.

Sachidanandam R, Weissman D, Schmidt SC, Kakol JM, Stein LD, Marth G, Sherry S, Mullikin JC, Mortimore BJ, Willey DL, Hunt SE, Cole CG, Coggill PC, Rice CM, Ning Z, Rogers J, Bentley DR, Kwok PY, Mardis ER, Yeh RT, Schultz B, Cook L, Davenport R, Dante M, Fulton L, Hillier L, Waterston RH, McPherson JD, Gilman B, Schaffner S, Van Etten WJ, Reich D, Higgins J, Daly MJ, Blumenstiel B, Baldwin J, Stange-Thomann N, Zody MC, Linton L, Lander ES, Altshuler D; International SNP Map Working Group (2001) **A map of human genome sequence variation containing 1.42 million single nucleotide polymorphisms**. *Nature* 409, 928-933.

Sandal N, Krusell L, Radutoiu S, Olbryt M, Pedrosa A, Stracke S, Parniske M, Bachmaier A, Sato S, Tabata S, Ketelsen T, and Stougaard J. (2002) **A genetic linkage map of the model legume *Lotus japonicus* and strategies for fast mapping of new loci**. *Genetics* 161, 1673-1683.

Young ND, Mudge J, and Ellis TH. (2003) **Legume genomes: more than peas in a pod**. *Current Opinion in Plant Biology* 6, 199-204.

Yu J, Hu S, Wang J, Wong GK, Li S, Liu B, Deng Y, Dai L, Zhou Y, Zhang X, Cao M, Liu J, Sun J, Tang J, Chen Y, Huang X, Lin W, Ye C, Tong W, Cong L, Geng J, Han Y, Li L, Li W, Hu G, Huang X, Li W, Li J, Liu Z, Li L, Liu J, Qi Q, Liu J, Li L, Li T, Wang X, Lu H, Wu T, Zhu M, Ni P, Han H, Dong W, Ren X, Feng X, Cui P, Li X, Wang H, Xu X, Zhai W, Xu Z, Zhang J, He S, Zhang J, Xu J, Zhang K, Zheng X, Dong J, Zeng W, Tao L, Ye J, Tan J, Ren X, Chen X, He J, Liu D, Tian W, Tian C, Xia H, Bao Q, Li G, Gao H, Cao T, Wang J, Zhao W, Li P, Chen W, Wang X, Zhang Y, Hu J, Wang J, Liu S, Yang J, Zhang G, Xiong Y, Li Z, Mao L, Zhou C, Zhu Z, Chen R, Hao B, Zheng W, Chen S, Guo W, Li G, Liu S, Tao M, Wang J, Zhu L, Yuan L, and Yang H. (2002) **A draft sequence of the rice genome (*Oryza sativa* L ssp. *indica*)**. *Science* 296, 79-92.

Chapter 5.7

INFORMATION TRANSFER: MAPPING AND CLONING IN OTHER LEGUMES

Lene Heegaard Madsen and Jens Stougaard

Laboratory of Gene Expression, Department of Molecular Biology, University of Aarhus, DENMARK.

Email: lhm@mb.au.dk

Phone: +45 8942 5011 Fax: +45 8620 1222

Keywords: Information transfer, legume, genetic markers

One of the big advantages of a model legume is the relative ease with which information from gene isolation and characterization of gene function can be transferred to other plants, including crop plants such as pea, bean, soybean, and peanut. This transfer of information includes both the cloning of small specific DNA regions of known map position in the model, for positioning on the crop genetic map, and the cloning of entire genes encoding functions of scientific or more applied interest. One important achievement from this type of work will be transfer of genetic markers from the model's high-density map to linkage maps of crop legumes. This will ease genetic map building, improve the density of the crop genetic map, and help in identification of molecular markers for breeding purposes (see also Chapter 5.6 of this Handbook). Taking advantage of the fast progress of Lotus genome sequencing, a limited number of genetic markers can provide invaluable information on genes located between the same markers in syntenic regions within crop genomes.

INTRODUCTION

Information transfer work involving comparative analysis of important control genes such as *Nin, SymRK, Nfr1*, and *Nfr5* from different legumes with different nodulation types and specificities towards diverse rhizobial strains and lipochitin-oligosaccharide Nod-factors may answer many questions. For instance, this approach may give insight into genome evolution, levels of general or regional syntony between model and crop genomes, and the biodiversity of symbiosis (or of any other field of interest). Here, information transfer from a model plant to a crop plant is illustrated by describing the isolation of a pea gene using information from Lotus. A number of different approaches are described and resources available in public domain databases included.

A.J. Márquez (ed.), Lotus japonicus Handbook, 243-248.

GENE CLONING

Once a gene with an interesting and important function is cloned in *Lotus*, one of the obvious applications is comparative analysis of structure and function after isolation of the corresponding gene from another legume of interest, in our case often pea. The choice of pea is based on several criteria. Pea belongs to a different cross-inoculation group than Lotus. The microsymbiont of Lotus is *Mesorhizobium loti* and Lotus forms determinate nodules. In contrast pea develops indeterminate nodules in symbiosis with *Rhizobium leguminosarum* bv. *viciae*. This biological diversity clearly adds to the potential information value of analysing the same nodulation gene in two different legumes. Furthermore, a very large collection of pea mutants exist and many of these are available from the pea germplasm collection at JIC Norwich, UK (www.jic.bbsrc.ac.uk/germplas/pisum/index.htm) or from researchers in the field. These mutants are invaluable when it comes to proving that a gene isolated from pea is actually the orthologue of the Lotus gene. If a pea mutant with a phenotype comparable to that of the Lotus mutant from which the gene was originally isolated is available, formal proof of functional homology of the genes from Lotus and pea can be obtained by showing that the pea mutants carry mutations in this particular gene. Another possibility is complementation of Lotus mutant(s) with the pea gene. This approach will however only succeed if the gene function is well conserved.

Direct hybridization strategy for gene isolation

If genomic libraries or cDNA libraries from appropriate tissue is available for the legume of interest, a direct hybridization approach using a probe derived from the Lotus sequence can be used. However, some complications caused by non-specific hybridization signals may be encountered because lowered stringency of hybridization would be necessary to ensure detection of signals from target DNA. Furthermore, the gene region used as a probe should be chosen with care, especially if the gene in question contains regions coding for well-conserved domains. For example, kinase domains or homeodomains, which are very well conserved between proteins of quite different function, should be avoided. If possible a region encoding an amino acid segment specific for this particular gene should be chosen and regions encoding e.g. signal peptides or transit peptide should be avoided since these are not nearly as well conserved as the rest of the coding sequence.

If sequence information is available only from the model, a good test for hybridization specificity and the stringency conditions that allow detection of target without cross hybridization to more distantly related sequences is to perform a genomic Southern blot using the model probe on a number of different digests of target plant genomic DNA. Enzymes that give good results are standard 6 base pair cutters like *Hind*III, *Eco*RI, *Eco*RV, *Xba*I, *Dra*I, to mention a few. Hybridization at lowered temperature (e.g. 55°C) could be tried as a first approach, however the exact conditions must be determined experimentally and depends on the relatedness of the sequences in question. When using a *Lotus* probe to detect target sequences from pea we most often use hybridization at normal stringency (65°C, standard hybridization buffer, (see below)) followed by washes using 2xSSC, 0.1% SDS and then, if filters have high background when checked with a monitor continued washing using 1xSSC, 0.1% SDS at 65°C.

After exposure, complexity of the hybridization signal will show whether the probe is specific for the desired target or whether conditions more stringent (lower salt, higher temperature) are required. If this hybridization test shows a low complexity of the signal, one may choose to continue using the heterologous probe for cDNA or gene isolation. However, if the pattern is complex, or if difficulties from cross hybridizing related sequences such as members of gene families must be avoided, an alternative strategy is to develop PCR primers to amplify a fragment of the homologous gene from the DNA of the target organism. This will then allow hybridization to be carried out at high stringency.

PCR strategy to isolate a gene fragment from the DNA of the target plant

Comparison of Lotus gene sequences to available gene or EST (expressed sequence tag) sequences from different legumes or even Arabidopsis (Figure 1) often leads to identification of regions of sufficient DNA sequence conservation for design of only slightly degenerate PCR primers. Such primers can then be used in PCR reactions with low annealing temperature to amplify similar sequences from the target DNA. If cDNA sequences are available from other legumes, these are most conveniently presented in the TIGR databases (The Institute for Genomic Research, TIGR: www.tigr.org/tdb/tgi/). In this database, sequences are searchable on an organism basis, and the result often quickly gives a good overview of the complexity of a gene family (and therefore the risk of amplifying the "wrong" homolog), and gives information about the tissues from which the relevant ESTs were isolated. This may also give a hint whether the sequence is relevant or not, although expression in unexpected tissues is not *per se* a reason to exclude a sequence from analysis.

However, mere sequence conservation between expressed sequences such as ESTs is not sufficient information to design functional primers. Another parameter is the position of introns, which might interrupt the conserved regions making them unsuitable for primer design. One such example is shown in Figure 2. By simple comparison of three legume ESTs for which no gene structure information was available seven well-conserved regions for primer design were identified. However, by superimposing the intron/exon structure of the corresponding Arabidopsis gene only two of the primers remained, while the others were likely to span intron/exon junctions. One might argue than the intron/exon structure may not be conserved between legumes and Arabidopsis, however in our experience this is very often the case, and therefore comparison to Arabidopsis gene structure is a valid tool in this kind of analysis.

Similarity region I	**Identities = 30/33 (90%)**
Nin	692 TGGATCTTTA**GCCCTTCCTGTCTTCGAAAGAGG** 724
at4g35270	795 TGGATCATTAGCCCTTCCTGTGTTTGAAAGAGG 827
Similarity region II	**Identities = 36/40 (90%)**
Nin	1275 TTTGTTTT**GGAGTTTTTCCTTCCAAAAGATTGCC**ATGACA 1314
at4g35270	1426 TTTGTGTTGGAGTTTTTCTTTCCAAAAGCTTGCCTTGACA 1465
Similarity region III	**Identities = 52/57 (91%)**
Nin	1843 AAGTTCTAA**GACAGTACTTTGCAGGAAGC**...
at4g35270	1988 AAGTTCTTCGACAATACTTTGCAGGAAGC...
Nin	...**CT**AAAAGATGCAGCAAAGAGCATTGGTG 1899
at4g35270	...CTCAAAGATGCAGCCAAGAGCATTGGTG 2044

Figure 1. Regional matches between Lotus Nin *mRNA and* Arabidopsis *at4g35270 mRNA. Although the overall similarity between the two sequences is low (less than 40%), three regions with high nucleotide similarity between* Lotus Nin *mRNA and the* Arabidopsis *homolog at4g25270 were detected in a blastn search of the NCBI non-redundant database (www.ncbi.nlm.nih.gov/blast) using the* Lotus Nin *mRNA as the query. These regions had sufficient sequence conservation to design slightly degenerate primers based on the sequences shown in bold. The forward primer 5′-GCCCTTCCTGTSTTYGAAAGAGG-3′ derived from region I, and the reverse primer 5′- AGGCTTCCTGCAAAGTAYTGTC-3′ derived from region III can be used in a PCR reaction at ~50°C to produce a ~1500 bp fragment of the pea Nin homolog, while combination of the same forward primer and the reverse primer 5′-GGCAAKCTTTTGGAARGAAAAACTCC-3′ derived from region II gives an ~850 bp fragment (Borisov* et al. *2003).*

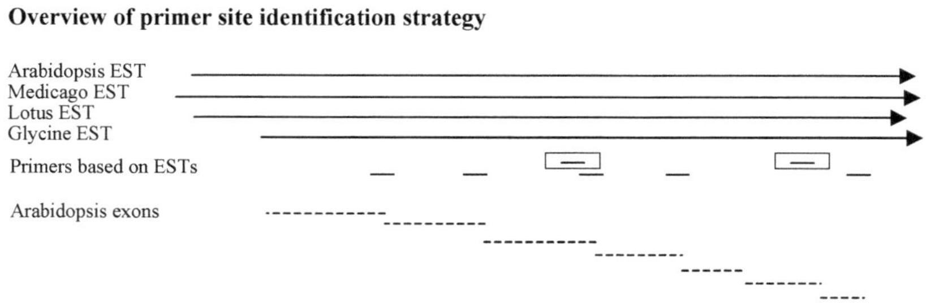

Figure 2. Simple sequence alignment of homologous EST sequences from Medicago, Lotus, *and* Glycine *identified 7 well conserved regions suitable for primer design. After superimposing the exon/intron structure of the corresponding* Arabidopsis *gene, only two remained (boxed)*

5′-RACE and 3′-RACE

If no libraries are available for the organism of interest (or if the desired clone is not found) 5′- and 3′-RACE (rapid amplification of cDNA ends) and subsequent cloning and sequencing is a good alternative to obtain the full coding sequence of the gene. To optimize this approach it is important to know the expression pattern of the gene in order to isolate RNA in which the transcript is present. This is often known from studies performed in the model plant. In our experience, the SMART-RACE kit from Clontech (K1811-1) is very efficient for both 5′- and 3′-RACE and this has given the best results for genes of low expression level. However, also the procedure described in the 5′- 3′- RACE kit from Roche works quite well (Cat. No. 1 734 792).

After cDNA isolation and sequence determination, the sequence of the corresponding gene can be determined using primers for the cDNA to amplify the genomic region. Primers that accidentally span exon-exon junctions will not be productive, but if introns are not too long, amplification of the whole coding region of the gene is a relatively straightforward task and the exon/intron structure can then be derived. To obtain information of non-transcribed regions such as the promoter, isolation of a genomic clone is required. Alternatively, more sophisticated PCR methods, such as Inverse PCR (Sambrook et al. 1989) or TAIL PCR (Liu and Whittier 1995) can be applied.

Identification of homologous mutants

To provide formal proof that the gene isolated from the crop plants is the functional homolog of the gene from the model, several different approaches can be envisioned. One is the functional complementation of mutants identified in the model by transfer of the crop plant gene. This approach is limited to genes encoding proteins in which all functional domains are well conserved. Another approach involves the identification, in the crop plant, of mutants with the same phenotype as the mutants defining the model plant gene and subsequent sequencing of the crop plant mutant alleles.

Transfer of molecular markers for genetic map anchor points

One of the long term applied goals of mapping efforts in model plants is to unify the genetic systems of the models with that of the crops. To this end, we are currently trying to correlate the genetic maps of Lotus and Arachis (peanut) and we hope to extend this analysis to other crop legumes such as bean (see chapter 5.6 of this Handbook). In short, our aim is to transfer map position information of so-called anchor markers between the two maps. Assuming a limited number of rearrangements during legume genome evolution, we expect to find colinearity/synteny between marker positions on corresponding chromosomes or chromosome regions of the two maps. Exploiting the wealth of information accumulating from the Lotus sequencing project it will be possible to make predictions about gene content and gene identity between anchor markers in the crop. Furthermore, if a gene or trait of interest is pinpointed between two markers in the crop it may be possible to derive new, closer markers and finally identify candidate genes simply by inspecting the corresponding genomic region of Lotus, as has been elegantly demonstrated between barley and the monocot model rice (Collins et al. 2003).

REFERENCES

Borisov AY, Madsen LH, Tsyganov VE, Umehara Y, Voroshilova VA, Batagov AO, Sandal N, Mortensen A, Schauser L, Ellis N, Tikhonovich IA, and Stougaard J. (2003) **The *Sym35* Gene Required for Root Nodule Development in Pea Is an Ortholog of *Nin* from *Lotus japonicus*1**. *Plant Physiology* 131, 1009-1017.

Collins NC, Thordal-Christensen H, Lipka V, Bau S, Kombrink E, Qiu JL, Huckelhoven R, Stein M, Freialdenhoven A, Somerville SC, and Schulze-Lefert P. (2003) **SNARE-protein-mediated disease resistance at the plant cell wall**. *Nature* 425, 973-977.

Liu YG, and Whittier RF. (1995) **Thermal asymmetric interlaced PCR: automatable amplification and sequencing of insert end fragments from P1 and YAC clones for chromosome walking**. *Genome* 25, 674-681.

Sambrook J, Frisch EF, and Maniatis T. (1989) **Molecular Cloning, a laboratory manual**. Second edition, Cold Spring Harbor Laboratory Press.

SECTION 6

TRANSFORMATION AND REVERSE GENETIC ANALYSIS

Chapter 6.1

AGROBACTERIUM-MEDIATED *IN VITRO* TRANSFORMATION

Patrizia Lombari, Emanuela Ercolano, Hicham El Alaoui, and Maurizio Chiurazzi*

*International Institute of Genetics and Biophysics; Via Marconi 12; 80125, Napoli; ITALY; *Corresponding author.*

Email: chiurazz@igb.cnr.it

Phone: +39 081 725 7256 Fax: +39 081 593 6123

Keywords: organogenesis, *in vitro*, Agrobacterium, transformation, regeneration

A major obstacle for understanding the molecular basis of root nodule morphogenesis has been for many years the lack of a well-developed genetic system in a legume that is amenable to the advanced technologies developed for model plants such as Arabidopsis thaliana. *The diploid self-fertile plant* Lotus japonicus *has been identified and exploited as model legume by the research community and is now widely used in genetic and molecular studies. Methodologies have been developed over the last nine years allowing the routinely production of transformants. Since in planta transformation procedures have not been described yet in* Lotus japonicus, in vitro *procedures still represent the only available tools to obtain transgenic Lotus plants. The aim of this article is to review several protocols for both Agrobacterium tumefaciens and rhizogenes mediated in vitro transformation.*

A. *TUMEFACIENS*-MEDIATED *IN VITRO* TRANSFORMATION

Introduction

The essential requirements in a gene transfer system for production of transgenic plants are: (1) availability of a target tissue including cells competent for plant regeneration, (2) a method to introduce DNA into those regenerable cells, and (3) a procedure to select and regenerate transformed plants at a satisfactory frequency. Regeneration procedures can be based on direct or indirect somatic embryogenesis. In *L. japonicus*, the attempting to obtain transgenic plants via direct somatic embryogenesis have been unsuccessful and therefore several procedures involving first a dedifferentiation or callus induction step followed by the organogenesis phases have been set up (Handberg and Stougaard 1992; Stiller *et al.* 1997; Martirani *et al.* 1999; Lohar *et al.* 2001). Organogenesis *in vitro* depends on the

A.J. Márquez (ed.), Lotus japonicus Handbook, 251-259.

application of exogenous phytohormones, in particular auxin and cytokinin, and on the ability of the tissue to respond to these phytohormones changes during culture. In general, three phases of organogenesis are recognisable, based on temporal requirement for a specific balance of phytohormones in the control of organogenesis. In the first phase, cells in the explants acquire "competence" which is defined as the ability to respond to hormonal signals of organ induction. The process of acquisition of organogenic competence is referred to as "dedifferentiation." The competent cells in cultured explants are canalised and determined for specific organ formation under the influence of the phytohormone balance through the second phase. Then the morphogenesis proceeds independently of the exogenously supplied phytohormones during the third phase.

The practical requirements to be satisfied in a transformation regeneration procedure are:

- Ready availability of the target tissue. Possibility to maintain a continuous supply of explants as starting material for transformation
- High efficiency, economy and reproducibility, to readily produce many independent transformants for testing
- Technical simplicity, involving a minimum of demanding or inherently variable manipulations
- Unequivocal selection or efficient screening to recover transgenic plants from transformed cells
- Minimum time in tissue culture, to reduce associated costs and avoid undesired somatoclonal variation
- Stable, uniform (nonchimeric) transformants for vegetatively propagated species
- Simple integration pattern and low copy number of introduced genes, to minimise the probability of undesired gene disruption at insertion sites, or multicopy-associated transgene silencing
- Stable expression of introduced genes in the pattern expected from the chosen control sequences.

In this article, we'll review several transformation procedures that satisfy this list of requirements.

Root explants as starting material

Materials

All regeneration media are based on Gamborg's B5 medium (Gamborg, 1970) and contain B5 basal salts (Sigma, cat. # G-5768), B5 vitamins (Sigma, cat. # G-1019), 3% sucrose, pH 5.7 and 1 % agar (Sigma, cat. # A-7921) if not otherwise specified. Media differ in hormone and antibiotics composition:

- Callus inducing medium (CIM): 3 mg/L^{-1} indoleacetic acid (IAA); 0.15 mg/L^{-1} 2,4-dichlorophenoxyacetic acid (2,4-D); 0.6 mg/L^{-1} benzyladenine

(BA); 0.3 mg/L^{-1} isopentenyladenine (IPA).

- Shoot inducing medium (SIM): 0.2 mg/L^{-1} benzyladenine (BA) plus 10 mM $NH_4^+SO_4$.
- Shoot elongation medium (SEM): B5 medium, 1% Sucrose.
- Root inducing medium (RIM): B5/2 medium; 0.1 mg/L^{-1} naphthaleneacetic acid (NAA).

Preparation of hormone and antibiotics stock solutions followed manufacturers instructions. Petri dishes and plant containers (Sigma, cat. # P1552) were used for cultivation of explants.

21 days transfer intervals are used during selection of transformed tissue.

Transformation/regeneration procedure

The description of this procedure has been adapted from Lombari *et al.* 2002. *Lotus japonicus*, GIFU ecotype F9 seeds are sterilised (20 min in 25% commercial bleach [1 % hypochlorite], 0.1 % Triton) and washed 6 times in sterile water. After resting overnight at 4°C, seeds are spread on the surface of 1% agar plates and left o/n, cap side down at 4°C. Plants are then transferred in the culture cabinet (23°C, 16 h photoperiod). Young seedlings (6 days old) are transferred on B5 standard medium plates (15/140 mm Petri dishes) placed vertically. The plants are allowed to grow for 3 weeks. The freshly harvested roots are spread evenly on CIM such that the entire root system is in contact with the medium. During this incubation, a visible increase of the root thickness is observed. Usually roots from one plate of plants are divided onto two CIM plates. Plants can be also stored at 4°C for two months and pre-incubated at 23°C for two days before to proceed with incubation of the excised roots on CIM medium.

For co-cultivation, the roots, which have been incubated on CIM for five days, are dipped in the appropriate Agrobacterium culture (about 5 X 10^9 cells/ml) and wounded by cutting into 0.5-cm pieces and by squeezing firmly with forceps (about 16 explants per plant and 30 explants/100 mm Petri dish). The root pieces are then blotted quickly on sterile paper and transferred on fresh CIM medium plates (about 40/100 mm Petri dishes). After two days of co-cultivation, the bacterial mass is removed by rinsing the explants in fresh B5 medium, blotting on sterile paper and transfer on CIM medium plus 200 mg/L^{-1} cefotaxime.

After two days, the explants are transferred onto CIM medium containing 200 mg/L^{-1} cefotaxime and 15 mg/L^{-1} hygromycin (4 days after infection).

After three to four weeks, the putative transformants are scored for resistance and green sectors (0.2-0.4 cm) are isolated and transferred on SIM containing 200 mg/L^{-1} cefotaxime and 15 mg/L^{-1} hygromycin plus 10mM $NH_4^+SO_4$. Green calli are compact and easily separable from root segments (Figure 1). The average number of Hyg^R calli obtained per explant is 1.23 and 70-90% of segments provided transgenic calli (about 22 per plant). The time of exposure of the explants to the CIM could be either further shortened by moving the explants to the shoot inducing medium (SIM) before the appearing of the green calli (3 weeks). This short incubation does not affect the frequency of callus formation and the Hpt^R green calli showing up on the

SIM maintain a good cell division capacity by reaching a considerable size before to start shoot differentiation.

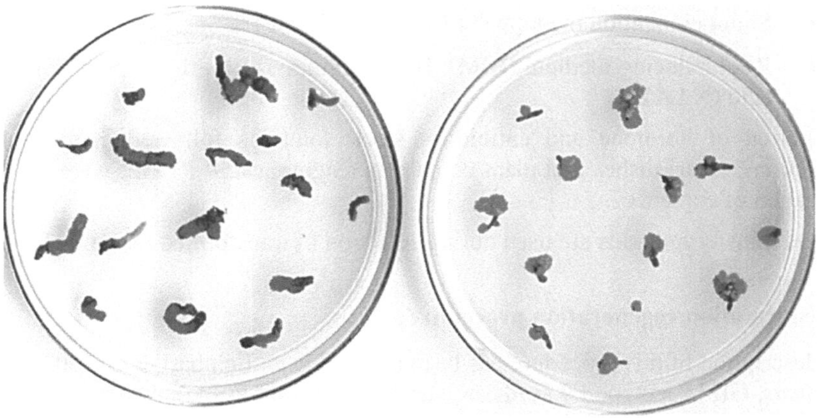

Figure 1. On the left, root explants not infected with Agrobacterium after an incubation of five weeks on CIM plus 15 mg/L^{-1} hygromycin. On the right, root explants with Hyg^{R} green calli obtained after an incubation of four weeks on CIM plus 15 mg/L^{-1} hygromycin.

Rooted plants stored at 4°C

↓

Root explant incubation on CIM

↓ 5 days

Co-cultivation

↓ 2 days

Pre-incubation on CIM plus Cef

↓ 2 days

Selection on CIM plus Hyg

↓ 21 days

Selection on SIM plus Hyg

↓ 10-45 days

Incubation on SEM

↓ 20 days

Transfer of shoots to RIM

↓ 10 days

Incubation on REM

↓ 15 days

Transfer to pots

Figure 2. Flowchart of the root explant transformation regeneration procedure.

Calli with small shoots emerging are then transferred to SEM. Further elongation is achieved after transfer the shoots on B5 medium without hormones and $NH_4^+SO_4$. Root induction is achieved on 2-4 cm long shoots by a ten days subculture on RIM and further subculture on B5/2 medium without hormones. The rooted T_1 plants are healthy and produced normal flowers and seeds. A schematic description of the hypocotyl procedure is shown in Figure 2.

Hypocotyl as starting material

Materials

All regeneration media are based on Gamborg's B5 medium and contain B5 basal salts (Sigma, cat. # G-5768), B5 vitamins (Sigma, cat. # G-1019), 2% sucrose, 20 mg/L^{-1} Hygromycin, 5 mg/L^{-1} G418, 15 mg/L^{-1} PPT, and gelrite 0.2% (unless otherwise specified), pH 5.7. Media differ in hormone and antibiotics composition.

- **Co-cultivation medium**: 10 fold diluted B5 liquid medium without sucrose, BA 0.5 mg/L^{-1}, NAA 0.05 mg/L^{-1}
- **Callus-inducing medium**: (CIM): BA 0.5 mg/L^{-1}, NAA 0.05 mg/L^{-1} cefotaxime 300 mg/L^{-1}
- **Shoot-inducing medium**: (SIM): BA 0.2 mg/L^{-1}, 10mM NH_4SO_4, cefotaxime 300 mg/L^{-1}
- **Shoot-growth medium**: (SGM): BA 0.2 mg/L^{-1}, cefotaxime 300 mg/L^{-1}
- **Shoot-elongation medium**: (SEM): no hormones
- **Root-inducing medium**: (RIM): B5/2 medium, NAA 0.5 mg/L^{-1}, 0.4% Gelrite
- **Root-elongation medium**: (REM): B5/2 medium, 0.4% Gelrite.

Transformation-regeneration procedure

The description of this procedure has been adapted from Handberg and Stougaard 1992; Stiller *et al.* 1997. Sterilised seeds (20 min. in 2% hypochlorite, 0.01% Tween 20, several washing with sterile water) are germinated 7 days on water-soaked filter paper in darkness under a 23°/20°C (16/8 hours) temperature regime. Sterile hypocotyls are cut transversally or longitudinally in 2-3 pieces on filter paper soaked with culture of Agrobacterium. After cutting, explants are transferred immediately onto a pile of filter paper soaked with co-cultivation medium for 7 days in the dark. To provide mixture, additional medium is added next to the filter paper. After 7 days, the top filter paper with the hypocotyl pieces is moved to CIM plus cefotaxime.

Selection of transformed cells is started 5 days later by moving individual hypocotyl pieces on CIM plus G418, hygromycin or PPT. Green calli are obtained in average after a 4-week incubation. Transformed tissues are maintained on selective plates until the appropriate size for transfer to shoot induction plates is reached. Shoot induction is achieved in 3-6 weeks. Explants are transferred to fresh medium every week during selection, shoot induction and elongation stages. Well-grown shoots are cut off and vertically inserted into RIM and incubated for a week. Then, root elongation is performed on REM for 3-4 weeks. A schematic description of the hypocotyl procedure is shown in Figure 3.

Hypocotyl co-cultivation
↓ 7 days
Regeneration medium
↓ 5 days
Selection medium
↓ 2-5 weeks
Shoot induction
↓ 3-6 weeks
Shoot elongation
↓ 4-6 weeks
Root induction
↓ 1 week
Root elongation
↓ 3-4 weeks
Transfer to pots

Figure 3. Flowchart of the hypocotyl explant transformation regeneration procedure.

A RHIZOGENES-MEDIATED TRANSFORMATION AND NODULATION OF COMPOSITE PLANTS

Introduction

The establishment of hairy root transformation and nodulation procedures in several legumes, including *Lotus corniculatus*, *Lotus japonicus*, *Vicia hirsuta*, and soybean, has enabled the studies on regulation and function of genes involved in nodule organogenesis and functioning (nodulins) in different genetic backgrounds. This experimental system represents the so-called short-cut system that allows the analysis of the promoter activity of nodulins as well as other genes involved in different organogenesis programs that take place in the root system. The transgenic root system obtained after the A. rhizogenes infection is suitable for the analysis of the promoter activity by exploiting appropriate reporter fusions and can be obtained in about one month hence avoiding the time consuming procedure for plant regeneration.

Hypocotyl infection

This procedure has been adapted from Stiller *et al.,* 1997. Sterilised seeds are germinated on a wet filter paper in the dark, 24°C for 2 days. Germinated seeds are transferred on the top of agar slopes in square Petri dishes (1/2 X B5 medium, no sucrose, 1.2% agar) with root tips dipped into the medium. The seedlings are incubated in a growth chamber at vertical position (16/8 hrs day/night cycle). Three days later plantlets with unfolded cotyledons are inoculated in the hypocotyl area with a syringe and 1-2 drops of *A. rhizogenes* inoculum (late logarithmic phase; growth in YEB medium) is applied at the wound. Approximately 3 days post infection wounded hypocotyl areas become swollen and 2-3 days later, small callus starts to grow from the wound. The first hairy roots appear about 14 days post-infection and 1-3 cm long hairy roots develop in 3-4 weeks.

Composite plants (wild type shoot over a transgenic root system) with 1-2 cm long hairy roots are transferred on 1/2 X B5 medium to promote seedling growth and the primary root is cut off. Four days later plants are transferred on no Nitrogen source medium slopes (B&D, Jenssen) and inoculated by applying a drop of *M. loti* inoculum on newly grown root tips.

Root infection

This procedure has been adapted from Martirani *et al.,* 1999. Surface sterilised seeds are germinated and grown vertically on 1% agar plates. Six days-old seedlings are transferred on NLN medium (Duchefa) containing 1% sucrose. After two days, the primary roots of the seedlings can be cut at different distance from the root tip. The freshly cut surface is inoculated with *A. rhizogenes* strains grown overnight in LB medium. After two days of co-cultivation, the seedlings are washed and blotted on sterile paper before to be transferred on B5 medium supplemented with 1% sucrose and 200 mg/L cefotaxime. A callus is visible at the wound site 5 days after infection. After 10 days, microcalli with emerging hairy roots appear clearly at the wound sites (Figure 4).

Figure 4. Plant infected with A. rhizogenes *showing a well-developed transgenic root system.*

The composite plants with hairy roots are maintained on B5 medium for at least two weeks until a massive root system is obtained. During this incubation, the roots that arise above the wounded site are eliminated with scissors. Analysis of the root system obtained after infection with a *A. rhizogenes* strain equipped with a T-DNA vector carrying a 35S-*gus*A cassette indicates that in average 80% of the plants are transformed and about 70% of the roots show GUS activity (Figure 5), indicating a high frequency of co-transformation of the *gusA* and pRi-born T-DNAs.

Figure 5. GUS staining obtained in hairy roots co-transformed with the T-DNA vector carrying the 35S-gusAint cassette. See colour plate 4 (G) for this figure in colour.

Composite plants are transferred to slanted fresh NLN medium 1% agar. Sterile filter paper is first applied over the taller part of the slope where the root system will be located. This is to avoid growth of the roots into the agar. Petri dishes are located vertically in a growth cabinet at 22°C with a 16-hrs photoperiod. The roots are maintained in the dark by using aluminium foil. Each primary meristem is inoculated with a drop of *Mesorhizobium loti* NZP2235 suspension containing about 10^6 bacteria.

COMMENTS

In this article, we describe several Agrobacterium-mediated transformation procedures adapted from different papers. These protocols differ for some features such as the target tissue used as starting material, the infection procedures, and regeneration media. In particular, the efficiency of transformation in the *A. tumefaciens*-mediated procedure is comparable (about 1.5 transformants per explant) whereas the incubation time on CIM medium differs significantly (one month shorter in the root explant procedure). On the other hand, the use of the mutagen 2,4-D is not provided in the CIM medium of the hypocotyl transformation procedure. Furthermore, an important point of the root explant procedure is the possibility to use plants stored in the cold room for long time as source of the explant hence providing a possibility to have a continuous supply of starting material for transformation.

In planta transformation procedures like those published for *A. thaliana* (Bechtold *et al* 1993; Clough and Bent 1998) and, more recently, for *Medicago truncatula* (Trieu *et al.* 2000) represent the next challenge for the *L. japonicus* community. However, the techniques described here ensure high reproducibility and efficiency with reasonably quick regeneration times (about 4-5 months to obtain primary transformants to be transferred in soil for *A. tumefaciens*).

REFERENCES

Bechtold N, Ellis J, and Pelletier G. (1993) **In planta *Agrobacterium*-mediated gene transfer by infiltration of adult *Arabidopsis thaliana* plants**. *Comptes Rendus de l'Academie des Sciences, Serie III-Sciences de la Vie* 316, 1194-1199.

Clough SJ and Bent AF. (1998) **Floral dip. a simplified method for *Agrobacterium*-mediated transformation of *Arabidopsis thaliana***. *The Plant Journal* 16, 735-743.

Lohar D, Schuller K, Buzas DM, Gresshoff PM, and Stiller J. (2001) **Transformation of *Lotus japonicus* using the herbicide resistance bar gene as selectable marker**. *Journal of Experimental Botany* 52, 1697-1702.

Martirani L, Stiller J, Mirabella R. Alfano F, Lamberti A, Radutoiu SE, Iaccarino M, Gressohff PM, and Chiurazzi M. (1999) **Establishment of a T-DNA Tagging Program in the Model Legume *Lotus Japonicus***. Expression patterns, activation frequencies and potential for insertional mutagenesis. *Molecular Plant-Microbe Interactions* 12, 275-284.

Handberg K and Stougaard J. (1992) ***Lotus japonicus*, an autogamous, diploid legume species for classical and molecular genetics**. *The Plant Journal* 2, 487-496.

Stiller J, Martirani L, Tuppale S, Chian R, Chiurazzi M, and Gresshoff PM. (1997) **High frequency transformation and regeneration of transgenic plants in the model legume *Lotus japonicus***. *Journal of Experimental Botany* 48, 1357-1365.

Trieu AT, Burleigh SH, Cardailsky IV, Maldonado-Mendoza IE, Versaw WK, Bleylock LA, Shin H, Chiou TJ, Katagi H, Dewbre GR, Weigel D, and Harrison MJ. (2000) **Transformation of *Medicago truncatula* via infiltration of seedlings or flowering plants with *Agrobacterium***. *The Plant Journal* 22, 531-541.

Chapter 6.2

INDUCTION OF HAIRY ROOTS FOR SYMBIOTIC GENE EXPRESSION STUDIES

Clara L Díaz, Mette Grønlund[1], Helmi RM Schlaman, and Herman P Spaink*

*Institute of Biology, Leiden University, Wassenaarseweg 64, 2311 AL Leiden, THE NETHERLANDS; [1]Current address: Laboratory of Gene Expression, Department of Molecular Biology, University of Aarhus, Gustav Wiedsvej 10, DK-800 Aarhus, DENMARK; *Corresponding author*

Email: spaink@rulbim.leidenuniv.nl
Phone: +31 715 275 063; Fax: +31 715 275 088

Keywords: Agrobacterium rhizogenes, Rhizobium, nodulation

The model legume Lotus japonicus *can be transformed and regenerated efficiently with* Agrobacterium tumefaciens *or* A. rhizogenes. *However, it takes between 8 to 12 months to obtain seeds of transgenic plants. We therefore developed a rapid and efficient transformation protocol using* A. rhizogenes *to induce transgenic hairy roots that can be inoculated with* Mesorhizobium loti *2 weeks after transformation. The first nodules emerge 8 to 10 days after inoculation, as on the roots of wild type Lotus plants and expression of plant genes involved in any step of nodulation can be completed within two months after the start of a transformation-nodulation experiment. A large number of seedlings can be tranformed in one experiment, allowing addressing of a number of variables in one single tranformation-nodulation experiment.*

INTRODUCTION

Lotus japonicus (Regel) Larsen has been proposed as a model plant to study the symbiosis between leguminous plants and rhizobia using classical and molecular genetics because of its small genome size, short life cycle and high seed yield (Handberg and Stougaard 1992, Jiang and Gresshoff 1997). Moreover, *L japonicus* can be regenerated after transformation of hypocotyls segments by *Agrobacterium tumefaciens* (Handberg and Stougaard 1992, Handberg et al 1994, Oger et al 1996, Stiller et al 1997, Quaedvlieg et al 1998) transverse cotyledon, or root segments (Aoki et al 2002; Lombari et al 2003). Highly efficient transformation and

A.J. Márquez (ed.), Lotus japonicus Handbook, 261-277.

regeneration of *L japonicus* plants using *A. rhizogenes* has also been obtained (Stiller et al 1997; Martirani et al 1999). In these protocols, tissue culture periods ranging from 4 to 9 months before transfer of the transgenic plantlets to soil are reported, with seed setting beginning 3-4 months later.

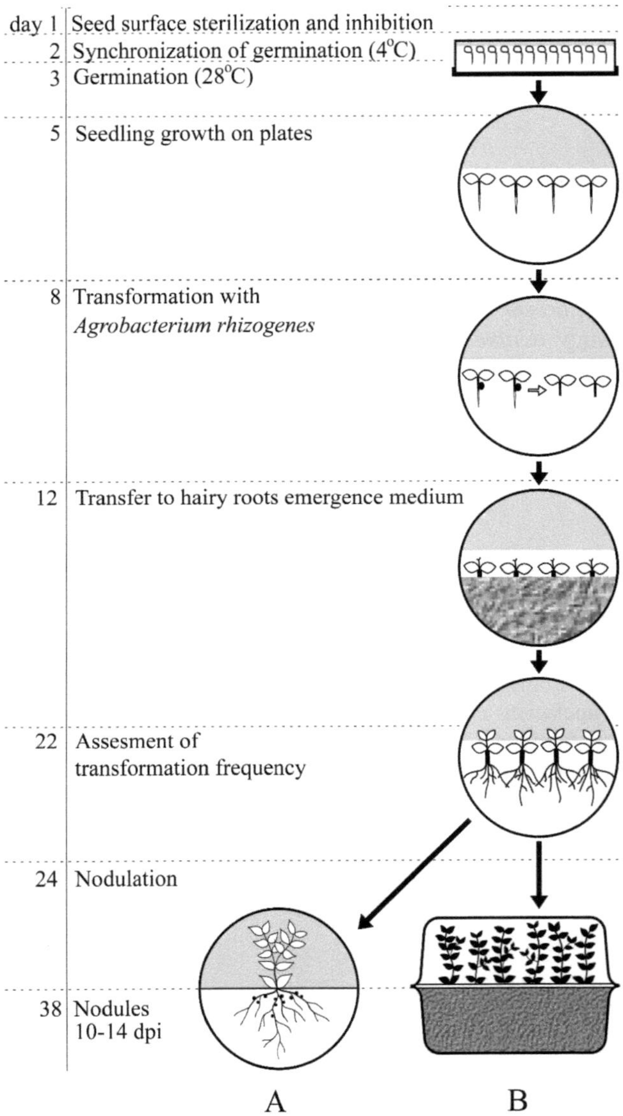

Figure 1. Flow chart of a hairy roots induction-nodulation experiment on L. japonicus *seedlings. The left columm gives the timing where each step is carried out starting from seeds to the observation of the first root nodules by eye inspection on plants grown on agar-plates (A) or in nodulation boxes (B).*

Although useful for the production of transgenic plants, *A. tumefaciens*-based transformation-regeneration protocols are not practical when the aim is to rapidly test gene expression in Lotus roots during nodulation. Therefore we set up a faster and more efficient transformation system for *L japonicus* ecotype 'Gifu' using *A. rhizogenes* carrying the genes of interest in binary vectors. Nodulation assays on transformed hairy roots can be set up 2 weeks after transformation and nodulation kinetics proceed as in wild type plants. Nodules are pink and able to fix nitrogen, as shown by acetylene reduction assay. Because large number of plants can be transformed (per experiment, 200 to 300 seedlings can be comfortably transformed by one experienced worker), statistical analysis of the expression of the gene or genes of interest can be easily achieved.

The protocol we describe in this chapter is based on the induction of hairy roots on white clover seedlings (Díaz et al., 1989) and can be divided into four steps: seed surface sterilization and germination, co-cultivation and transformation, hairy root emergence, and nodulation of transformed plants. The approximate timing of each step appears in the flow chart depicted in Figure 1.

MATERIALS AND INSTRUMENTATION

Solutions for seed surface sterilization

Sterilised, deionized water suitable for tissue culture (Millipore) is used in all solutions. For surface sterilisation, we use a 4-5% natrium hypochlorite solution that can be purchase in local supermarkets. It does not contain detergents.

Bacterial Media

Solid medium for Agrobacterium rhizogenes

- Bactotryptone (DIFCO) 50.0 g
- Yeast extract (DIFCO)" 5.0 g
- NaCl 8.0 g
- $MgSO_4 \cdot 7H_2O$ 12.3 g
- 1 M Tris 1 mL

Adjust pH to 6.4 to 6.8, bring to 1 L and add Select Agar (Gibco BRL, Difco Lab. Paisley, Scotland) to 1.8%. Sterilize for 20 min at 120°C.

Solid medium for *Mesorhizobium loti* (YMB)

- Mannitol 10.0 g
- Yeast extract (DIFCO) 0.4 g
- $MgSO_4 \cdot 7H_20$ 0.2 g
- NaCl 0.1 g

- $K_2HPO_4 \cdot 3H_2O$ (or K_2HPO_4) 0.65 g (or 0.5 g)

Adjust pH to 7 using 2 M HCL, bring to 1 L, and add Select Agar (Gibco BRL, Difco Lab. Paisley, Scotland) to 1.8% and sterilize for 20 min at 120°C.

Medium for germination of *L japonicus* seeds (Jensen)

- $CaHPO_4$ 1.0 g
- K_2HPO_4 (or $K_2HPO_4 \cdot 3H_20$) 0.2 g (0.26 g)
- $MgSO_4 \cdot 7H_20$ 0.2 g
- NaCl0 2 g
- $FeCl_3 \cdot H_20$ 0.1 g
- Trace elements solution 12.5 mL

Bring to 1 L. To solidify, add 9 g Daishin agar (Daiichir Seed Company, Japan). Sterilize for 30 min at 110°C. (Liquid Jensen medium can be sterilized for 20 min at 120°C and stored at room temperature for up to 5 months. Formation of a precipitate is a normal feature of this medium (Van Brussel et al., 1982)).

Trace elements solution

- $CuSO_4 \cdot 5\ H_20$ 35.4 mg
- $MnSO_4 \cdot 4H_20$ (or $MnSO_4 \cdot H_20$) 609 mg (or 462 mg)
- Zn $SO_4 \cdot 7H_20$ 97.4 mg
- H_3BO_3 1269 mg
- $NaMoO_4 \cdot 2H_2O$ 398 mg

Bring to 1 L Do not sterilize this solution. Keep in frozen in aliquots.

Medium for co-cultivation and nodulation of transformed *L japonicus* seedlings

Prepare solidified Jensen as described. Allow to cool down to 65°C and add (per L) 1.5 mL of a sterilized 1 M NH_4NO_3 solution.

Medium for hairy roots emergence (HRE)

This medium is based on SH medium (Schenk and Hildebrandt, 1971) and contains 300 μg cefotaxime per mL.

- 20X concentrated SH-A salts
- 20X concentrated UM-C vitamins
- 10 g sucrose
- 3 mL 1 M MES

To prepare a litre of medium, combine 700 mL deionized, tissue culture grade water with the above components. Rinse the tubes containing the stocks 2-3 times with water and add to the cylinder. Adjust pH to 5.8 using 1 M KOH. Adjust volume to1 litre. For solidified medium, add Daichin agar to 0.9%. Sterilize for 30 min at 110°C. Add 3mL cefotaxime stock solution after cooling the medium to 65°C and pour plates immediately. Allow the plates to dry for 45 minutes before storage. Liquid HRE sterilized medium can be kept frozen in aliquots for 3 months. Allow the medium to come to RT before adding cefotaxime (see below) and use immediately.

Concentrated (20X) SH-A salts

- KNO_3 50 g
- $MgSO_4.7H_2O$ 8 g
- $NH_4H_2PO_4$ 6 g
- $CaCl_2.2H_2O$ 4 g
- $MnSO_4.4H_2O$ 0.2 g
- H_3BO_3 0.1 g
- $ZnSO_4.7H_2O$ 0.02 g
- KI 0.02 g
- $CuSO_4.5H_2O$ 0.004 g
- $NaMoO_4.2H_2O$ 0.002 g
- $CoCl_2.6H_2O$ 0.002 g
- $FeSO_4.7H_2O$ 0.3 g
- NaEDTA(TiTriplex) 0.4 g

Use deionized tissue culture grade water to bring to a final volume of 1 L For $FeSO_47H_2O$, prepare a solution in 100 mL warm water. For NaEDTA (TiTriplex), prepare a solution in 100 mL warm water. Dissolve the other salts in 700 mL water. Add the $FeSO_4.7H_2O$ and NaEDTA solutions and adjust the volume to 1 L Store 50 mL aliquots at -20°C until use.

Concentrated (20X) UM-C vitamins

- Myoinositol 2.0 gL^{-1}
- nicotinic acid 0.1 gL^{-1}
- Pyridoxine HCl (Vitamin B6) 0.2 gL^{-1}
- thiamine HCl 0.2 gL^{-1}
- Glycine 0.04 gL^{-1}

Use deionized tissue culture grade water. Dissolve in 900 ml H_2O, adjust volume to 1L, divide in aliquots of 50 mL and keep at –20°C until use.

MES (2-[N-Morpholino]ethane-sulfonic acid), 1M stock solution

To prepare 150 mL of a 1 M solution weight 29.28 g MES (Sigma, Cell Culture) and add to 125 mL water. Adjust to pH 5.8 with a 10 M KOH solution. Adjust MES volume to 150 mL, divide in aliquots of 6 mL and store at –20°C until use. Before adding to HRE medium, warm the MES solution to avoid precipitates.

Cefotaxime sodium (Duchefa, Haarlem, The Netherlands)

For a 10 mL stock solution containing 100 mg mL^{-1}: dissolve 1 g cefotaxime in 9 mL water and filter-sterilize it. Keep at 4°C in the dark, use within 2 months.

Preparation of growth plates

Germination plates

Germination plates are 100X25 mm (Labtec, Nalge Nunc Int. Aurora, Ohio, USA, article LX4031). They contain 25 mL Jensen medium, solidified with 0.9% Daichin agar. After pouring and drying for 45 min, two filter papers (cut following the contour of the plate and sterilized as described in *VI*.5) are placed on the inner surface of the lids. Closed plates, wrapped in plastic, can be stored for 2-3 months at room temperature.

Transformation plates

Transformation plates (94X16 mm, C.A. Greiner & Shöne Ges. m.b. H., Kremsmünster, Austria, article 633102) contain 30-35 mL solidified Jensen medium (0.9% Daichin agar) supplemented with 1.5 mM NH_4NO_3. After drying for 45 min., semicircles of sterilized filter paper, 5.5 cm in diameter (see Preparation of growth plates, Filter paper), are placed on top of the agar, to cover half the plate. Closed plates, wrapped in plastic, can be stored (filter side up) up to 2 months at RT.

HRE (hairy roots emergence) plates

HRE (hairy roots emergence) plates (94X16 mm, C.A. Greiner & Shöne Ges. m.b. H., Kremsmünster, Austria, article 633102) contain solidified HRE medium supplemented with 300 μg mL^{-1} cefotaxime. Plates can be stored (medium side up) up to one week wrapped in plastic in the dark at 4°C.

Nodulation plates

Nodulation plates, (150X20 mm, from Sarstedt Inc., Newton, Ohio, USA, article 82.1184), contain 60-70mL Jensen supplemented with 1.5 mM NH_4NO_3. After drying for 45 min, a semicircle of sterilized filter paper 8.5 cm in diameter (see Preparation of growth plates, Filter paper) is placed on top of the agar, as to cover half a plate. Closed plates, wrapped in plastic, can be stored (filter side up) for 2 months at RT.

Filter paper

Filter paper (Schleicher &Schuell sheets, article 334580) is used to maintain a thin layer of nutrient solution in direct contact with roots. Semicircles (5.5 cm in diameter for transformation and root hair emergence plates, and 8.5 cm for nodulation plates, respectively) are cut following the contour of the base of the plates and spread as the tiles of a roof on a sheet of aluminium foil, in such a way that they can be individually pick up with a forceps without compromising sterility. Filter papers, wrapped in aluminium foil are sterilized for one h at 120°C.

Plastic strips

Plastic strips (PVC) used in nodulation plates are 140X18 mm, 1 mm in width. They have a split 7 mm in length and 1.5 mm in width located at the center of the strip. The split should be wide enough to let the stem of a Lotus plant go through it without damage. The strip is inserted into the agar so far that when the lid is placed and the plate stands in vertical position, it prevents direct light reaching the roots. Plastic strips are kept in 70% ethanol and are air-dry (down flow cabinet) before placing them in nodulation plates.

Parafilm

It is important to use Parafilm™ correctly in order to avoid leaking of liquid contaminated with Agrobacterium. To seal transformation and HRE plates unroll Parafilm and cut 7X5 cm strips. For nodulation plates, obtain 14X5 cm strips of Parafilm. Plates are sealed by pressing longitudinally stretched Parafilm strips on ¾ of each plate, allowing gases to diffuse above cotyledonary and real leaves. To avoid leakage due to tearing, Parafilm strips should be slightly stretched, just to obtain firm attachment to the plates. Placing a piece of black cardboard between the plates also helps to prevent tearing.

Nodulation boxes

The bottom part of tissue culture boxes (bottom RA60, lids RDA145/60, sold by ECOLINE BVBA, Pierkenstraat 82, B-9620 Zottegem, Belgium, teL +32 09 360 97 10, Fax +32 054 300 696) are filled with wet clay pellets to 4-5 cm in height. Clay pellets (the type used in hydroponic plant culture) are 2-4 mm in diameter and can be purchased from Hydro Jongkind BV, Alsmeer, Holland, htpp://www.jongkind-grond.nl/ or from Handelsvertreter Deutschland und Österreich, Herget's Blumenzwiebel, Rosenstrasse 28, D-82024 Taufkirchen, Germany; Tel. +49 896 124 706, Fax +49 896 125 128. To sterilize, pellets are placed in pyrex oven dishes, rinsed several times with deionized water, and baked overnight in open dishes in an oven at 180°C. The lids of the pyrex dishes (baked separately) are placed as soon as the glass is cold and the oven is opened. Securing the lids with tape and wrapping the dishes in aluminum foil allows storage for long periods.

To prepare nodulation boxes, dried pellets are submerged in sterilized Jensen medium containing 1.5 mM NH_4NO_3 for 10 min. Excess of liquid is discarded and the bottom of the boxes is filled wit soaked clay particles using a large, previuously sterilized spoon. Closed, filled containers can be prepared advance and kept at RT for 2 weeks.

Flow Cabinets

Laminar-flow cabinets

Laminar-flow cabinets are used to prepare plates and nodulation boxes, to surface sterilize seeds, and to spread germinated seedlings on transformation plates.

Down-flow cabinets

Down-flow cabinets are used for transformation, transfer to HRE plates and nodulation plates, inoculation with rhizobia and addition of liquid nutrient media, if required.

The working surface in down-flow cabinets is protected from contamination with Agrobacterium by opening plates on a 1 L beaker placed on a piece of highly absorbent paper towel (25x15 cm). Towels are placed on aluminium foil and sterilized for 1 h at 120°C before use. If required, pieces of absorbent paper towel (2x2 cm), sterilized as above described, can be used to dry the edges of Petri dishes.

Items required for working in flow cabinets, such as glassware, forceps (straight, thin, light in weight and 24 cm long, Duchefa, Haarlem, The Netherlands), pipettes, spoons, etc., are previously sterilized before use.

Growth Conditions

Temperature, photoperiod, and light sources

Growth of seedlings before transformation, co-cultivation, hairy root emergence, and nodulation takes place at 20°C, with 16 h light and 8 h darkness. Transformation frequencies varying between 60 and 100% have been obtained by growing the plants with Philips TLD 30W/95, TLD 30W/33 and TLD30W/84 tubes at light intensities varying between 2000 to 6000 Lux.

Stress during tissue culture conditions

In all experiments, we have observed that some seedlings can show signs of stress, which is apparent one to two weeks after growth in HRE plates. These signs are: anthocyanin formation in stems, yellowing of real leaves, retardation or failure to grow and production of new leaves, appearance of brown patches (possible tannins) on the root surface and retardation of root growth and lateral root formation. Plants that do not seem to be affected by stress (about 80% for plants grown under Philips TLD 30W/95 and 40-60% for plants grown under the other two light sources) are green, grow vigorously, and have whitish roots. Apparently, stress is not due to the composition of the growth media, or to transformation, because non-transformed control plants grown on different plant media and under the same light conditions as transformed plants, can show the same signs of stress. After a period of stress (about 10 days) 20-30% of the transformed stressed plants grown in HRE medium seem able to recover spontaneously. Stress can also be alleviated by shielding the roots from direct light and by transferring to fresh HRE medium.

Stress during nodulation

Nodulation in agar plates also induces stress in transformed plants, even if the roots are shielded from direct exposure to light. Nodulation in boxes, with roots growing in the dark in clay pellets, does not seem to induce stress. Moreover, stressed plants seem to recover quickly when transferred to these boxes.

EXPERIMENTAL PROCEDURES

Seed surface sterilization and germination

- Weigh seeds. Each seed is about 1 mg. An experienced operator can transform 300 to 400 seedlings in one experiment. Transfer to a 100 mL sterilized Erlenmeyer.
- Add enough concentrated H_2SO_4 to cover de seeds, let stand for 2 min.
- Add 50 mL water, mix quickly, and pour out immediately. Add more water, swirl seeds, and pour out. Repeat this step 5 times.
- Add a 4-5% sodium hypochlorite solution, just to cover the seeds, incubate for 10 min.
- Rinse as described in the third step.
- For imbibition: add enough water to obtain a 2-3 mm layer above the seeds and keep at room temperature in the dark. Seeds swell rapidly in the next 6 h, after this they can be placed at 4°C. Alternatively, let imbibition proceed overnight at room temperature.
- Seeds are rinsed twice with water before pouring them on germination plates (see Preparation of growth plates, Germination plates). Place two sterilized filter papers on the inner surface of the lid of the plate and add water to wet them well, but do not overflow. Set imbibed seeds on the agar surface, close about ¾ of the plate with Parafilm, or completely around with leucopore.
- Synchronization of germination is achieved by placing the seeds at 4°C in the dark for 24-48 h. Set the plates with seeds resting on the surface of the agar.
- For germination, plates are placed up-side down at 28°C in the dark for 2 days, as to the roots will grow toward the wetted filter paper. A very high percentage of germination is obtained with this method. Roots show many root hairs and are uniformly about 1 cm in length.

Infection and co-cultivation with Agrobacterium rhizogenes

- Pipette 1.5-2 mL Jensen medium containing 1.5 mM NH_4NO_3 onto transformation plates (see Preparation of growth plates, Transformation plates). Take care to use only clear Jensen solution, as the precipitate formed in the medium will interfere with microscopy.
- Exposure of the germination plates (up-side down) to light for 4-6 h triggers expansion of the cotyledons and facilitates removal of the seed coat. After this period, germination plates are turned up side up and 5 mL Jensen is pipetted onto the seedlings to protect them against desiccation and to make seed coat removal an easier manipulation. In addition, small groups of seedlings can be easily separated and individual seedlings can be recovered without damage. Seedlings with roots similar in length are transferred to transformation plates. To synchronize the expansion of the cotyledons and to ensure straight-growing hypocotyls, the seed coat is removed by applying a slight pressure over the cotyledons. Seed coat removal results in vigorously growing seedlings that seem to help dealing with the stress caused by sectioning of the root and transformation. However, successful transformation with *A. rhizogenes* can be carried out without removal of the seed coat. Five or six undamaged seedlings are chosen and placed in a row, with the cotyledons on the filter paper, about 3 mm from the edge of the paper. Parafilm is used to seal about ¾ of the plates, leaving the part above the epicotyls open. Plates are placed vertically in a tray and transferred to the growth cabinet for 2 days.
- Have ready: plates of fully-grown *A. rhizogenes* LBA 1334 (Offringa et al., 1986) carrying the binary plasmids of interest (2 plates per strain, grown for 2-3 days at 28°C on fresh selection plates, see Bacterial media, Solid medium for Agrobacterium). We use LBA 1334 because it transforms Lotus so efficiently, as we report in the comments.
- Infection: carefully, open transformation plates and streak *A. rhizogenes* on the hypocotyls, about 2-3 mm below the point of insertion of the cotyledons. The bacterium is taken directly from a selection plate with the closed tip of a glass Pasteur pipette. Antibiotics such as rifampicin, kanamycin, and spectinomycin do not seem to affect the plant. To synchronize the emergence of hairy roots, hypocotyls are sectioned with a scalpel in the middle of the zone streaked with Agrobacterium. Roots are discarded and if needed, Jensen is pipetted to keep a high degree of humidity into the plates. Sectioned plants will remain attached to the filter paper if cotyledons are placed on the paper' surface. Plates are sealed with Parafilm (as described in Preparation of growth plates, Parafilm) and placed in a growth tray. Note that no incubation time is required between application of Agrobacterium and sectioning of the roots. However, to increase the efficiency of this procedure, seedlings contained in 25-30 plates are usually infected before proceeding with sectioning.
- Co-cultivation: trays are placed in the growth cabinet for 5 days. At the

end of this period, many seedlings present swellings at the wound site. In some cases, hairy tumours and emerging roots can be observed. It is very important to keep the plates very wet in order to allow this rapid response (when standing vertically, a little bit of liquid Jensen should be visible on the bottom of the plate). In some cases, emergence of the first real leaf could cause the seedling to fall down. Apparently, this does not affect transformation.

Hairy root emergence

To obtain a high degree of humidity, 1.5-2 mL liquid HRE medium (hairy roots emergence Mat. and Inst. *V*) containing 300 mg L^{-1} cefotaxime is added onto emergence plates (Mat. and Inst. *VI*.3).

On the transformation plates, seedlings that have fallen to the bottom of the plate are replaced in the row. Carefully, a new semicircle of filter paper is placed on the seedlings, in such a way as to cover the area of response to *Agrobacterium* infection. Sliding the bottom filter towards the centre of the plate facilitates the deposition of the cover filter. Cotyledons and apical meristems should not be covered with filter paper. The two filters, carrying the seedlings in between, are transferred to HRE plates. Plates, sealed with Parafilm as described, are placed in growth trays and are taken to the growth cabinet. Take care not to transfer seed coats or cut roots, as they will contribute to the development of infections.

To control seedling development and humidity of the plate during hairy root emergence, add liquid HRE medium containing cefotaxime if necessary. One week after transfer, about 80% of the seedlings will present more than two roots and 10-15% will have one or two roots. In addition, the first two real leaves will emerge and unfold. Seedlings will grow vigorously and new roots will emerge in the following days.

Transformation frequencies (for example, GUS activity) can be assayed 10-14 days after transfer to HRE plates.

If seedlings are going to be kept in tissue culture conditions for longer than 21 days, transfer transformed seedlings to fresh HRE plates. Pour 150X20 mm plates with HRE medium (follow instructions described in Preparation of growth plates, Nodulation plates). Place 4 or 5 seedlings per plate, cover roots with filter paper, and pipette 5-7 mL liquid HRE containing cefotaxime to keep humidity high. These larger plates allow space for seedling growth and development.

Nodulation of transgenic hairy roots

Nodulation on agar plates: a transformed seedling is transferred to a nodulation plate (see Preparation of growth plates, Nodulation plates). The hairy roots tumor and emerged roots should be in contact with the filter paper. Rhizobia are applied to root tips and onto the hairy roots tumor directly from a plate, with the closed tip of a Pasteur pipette. Between 5-7 mL Jensen medium containing 1.5 mM NH_4NO_3 are added and a new filter paper is placed as to cover the roots and the tumor. To avoid direct exposure of the roots to the light, a plastic strip is inserted into the agar above the edges of the filter papers. About ¾ of the plate is sealed with a piece of Parafilm. A second piece of Parafilm is placed on top of the first one, to avoid leakage. The plates are placed vertically in a tray, between pieces of black cardboard (12X15 cm) and transferred to the growth cabinet.

Nodulation in tissue culture boxes (see Preparation of growth plates, Nodulation boxes) is achieved by planting inoculated roots in clay pellets contained in tissue culture boxes. 100% of innoculated plants nodulate and the average number of nodules counted 6 weeks after inoculation with *M. loti* strain R7A is 14 (most confined to one well-developed transformed root); wild type plants present 6-8 nodules at this time point. The roots of transformed seedlings on the HRE plates are inoculated by applying rhizobia directly from a plate, as already described. A hole in the clay pellets is made with a tweezer and an inoculated plant is carefully transferred to the container. Clay pellets are then displaced to completely cover the roots. This procedude is facilitated if seedlings are inoculated 10 to 20 days after transfer to HRE plates, when the longest roots are between 3-5 cm in length. Usually, 6 to 8 plants can be placed in one container. The container is closed and transferred to the growth room.

COMMENTS

Flow chart

In this protocol, we have given the minimal time required for each step to be accomplished with success (70-100% transformation frequencies using different binary vectors, see table 1 below for details). The percent of roots showing GUS activity is consistent with the expected frequency of roots co-transformed with the gusA and pRi born T-DNAs. However, there are margins of variation in the timing of each step that do not affect transformation frequencies and facilitate carrying out some steps. For instance, imbibed seeds resting on the agar surface of the plate can be kept at 4°C as long as the seeds do not dry out (about one month), as drying out dramatically decreases germination. In addition, germination proceeds well, although slower, at 25°C with a 16h light/8h darkness photoperiod. Infection with *A. rhizogenes* can be performed within 2 to 5 days after transfer of seedlings to transformation plates, without compromising transformation frequencies. Note that transfer to plates containing HRE medium on time (4-6 days after infection with Agrobacterium) speeds up root emergency and decreases stress.

Agrobacterium rhizogenes *LBA1334*

As can be seen in Table 1 (next section), this strain is very efficient in the transfer of DNA carried by the binary plasmids we tested into *L japonicus* cells. In addition, LBA1334 induces a very small tumour from which multiple cells are able to rapidly differentiate into roots (Figure 2), which can effectively be used for nodulation studies. This strain, a derivative of *A. tumefaciens* strain C58C9 with a chromosomal marker for rifampicin (Rif), carries the agropine-type pRi1855 with a spectinomycin resistance marker (Spc) in a segment that is not involved in virulence (Offringa et al., 1986). Binary plasmids can be efficiently crossed by triparental mating (Ditta et al, 1980) or electroporation (den Dulk-Ras et al., 1995). To prepare selection plates use LC medium (see Bacterial media, Solid media for Agrobacterium) containing Rif 20μgmL^{-1}, Spc 250μgmL^{-1} and the bacterial selection marker encoded in the binary plasmid.

We have also found that LBA1334 is resistant to chloramphenicol. Binary vectors from the pPZP family (Hajdukiewicz et al., 1994) can be selected using kanamycin 100μgmL^{-1} since neomycin phosphotransferase II is expressed under the control of 35S CaMV promoter, which is active in this strain.

We never keep Agrobacterium strains on plates; instead, we prepare a back-up stock (kept at -80°C) and a working stock (kept at -80 or -20°C) that is used to inoculate freshly poured plates containing the required antibiotics 4 or 5 days before transformation is planned. LBA1334 from stocks grows well within 48 h at 26-28°C. A second set of freshly poured plates containing the required antibiotics is then used to grow Agrobacterium for transformation. Stocks are obtained from fully grown plates prepared as described above by suspending the bacteria in 3 mL previously sterilized 0.6% peptone-15% glycerol before freezing.

Frequency of transformation

The following table summarizes transformation frequencies of various binary vectors, based on histological staining of the CaMV35S-*gus:intron* reporter gene using 5-bromo-4chloro-3-indolyl-β-glucoronic acid as substrate, performed 7 to 20 days after transfer to HRE medium. Seedlings can present 2 to 10 rapidly growing roots, and many loci of transformed cells that are not yet differentiated into (growing) roots (see also Figure 2).

Binary vector	**Percent of plants with transformed roots**[a]	**Percent of GUS-expressing roots**[1]
pBin19-gusAINT[b]	85±15	62±8
pPZP-based[c]	90±10	75±5
pCAMBIA1301[d]	78±5	50±10

Table 1. Transformation frequencies of L japonicus *by LBA 1334-carrying various binary vectors with the gus:intron reporter gene driven by the CaMV 35S promoter. Data are the mean of at least three independent transformation experiments. GUS assay was performed in lots of 20 to 50 plants. [a]Only emerged roots were considered when calculating the % of transformed roots at the time of staining. [b]Vancanneyt et al (1990). [c]Hajdukiewicz et al (1994). [d]Robert s et al (1997).*

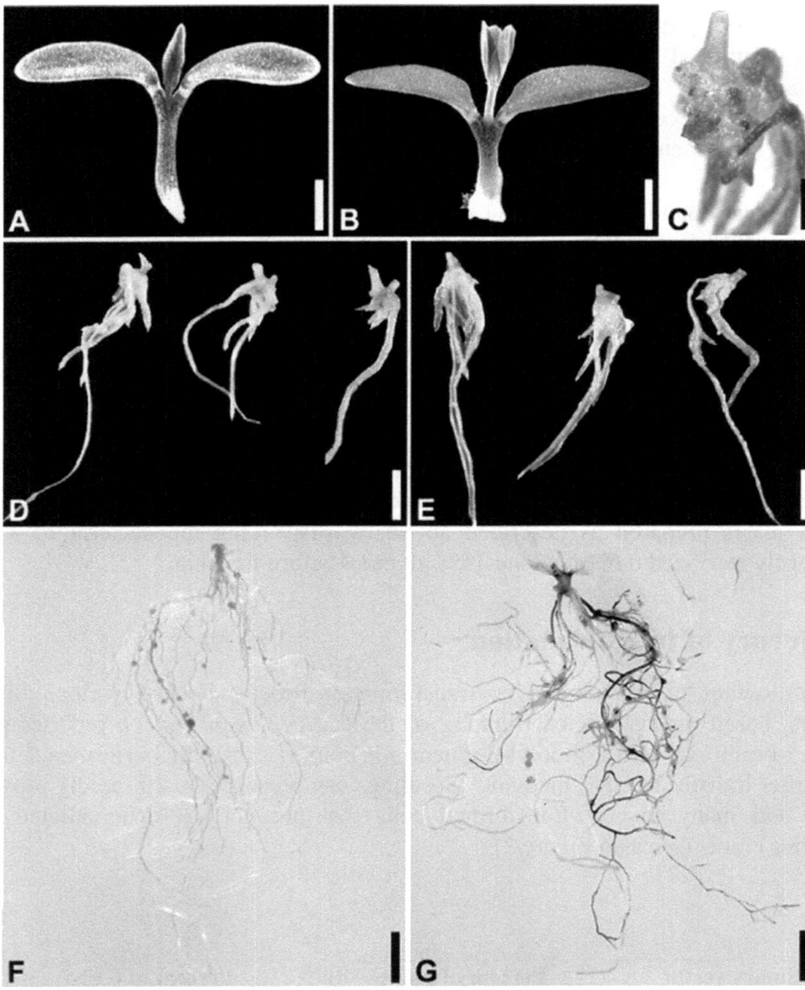

Figure 2. Hairy roots induced by A. rhizogenes *LBA 1334 in* L japonicus *seedlings. Panel A: there is no swelling observed at the base of the hypocotyl of a Lotus seedling that has not been infected with LBA 1334 4 days after sectioning of the main root. In contrast, a slight swelling and root hairs appears in the area of wounding in a seedling 4 days after sectioning of the main root at the base of the hypocotyl and subsequent infection with LBA 1334 (panel B). Panels C, E, and G show GUS expression as blue precipitate using 5-bromo-4chloro-3-indolyl-β-glucoronic acid as substrate in roots transformed with CaMV35S-gus:intron cloned in various binary vectors. GUS histological staining was performed as described in Scarpella* et al. *(2000). Panel C shows that all emerging hairy roots of a Lotus seedling were tranformed with LBA1334 carrying GUS as reporter gene cloned in pCAMBIA1301. Note that GUS is also expressed in loci that have not yet differentiated into hairy roots. Depending on the binary vector used, 5-10% of the transformed plants will show GUS expression in all emerging roots, the rest of the transformed plants wil present a combination of GUS-stained and non-stained roots*

(see Table 1 for more details). Staining was performed 7 days after transfer of the seedling to HRE medium. Panel E shows blue roots of seedlings transformed with LBA1334 carrying gus in a pPZP derivative and panel D, roots emerging in control seedlings, infected with LBA 1334, that do not show blue staining. GUS histological staining was performed 10 days after transfer to HRE. As many as 20 roots can be initiated, however only a few (2-10 roots) grow rapidly and produce lateral roots at this stage of development. Panel F shows the results of GUS staining of a control plant tranformed with LBA 1334 carrying pBin19 and panel G, the staining of a plant tranformed with LBA 1334 carrying pBin19 CaMV35S-gus:intron and showing GUS activity in nodulated hairy roots. Note that in both cases nodules are concentrated on the most developed root, which in 60-70 % of the cases, is transformed. Predominance of 1 or 2 roots is apparent after 2-3 weeks growth in HRE, clay particles or Jensen-agar plates. Staining of plants in panels F and G was performed 5 weeks after inoculation with M. loti *strain R7A and growth of roots in clay particles. Bars in panels A, B, C= 1mm; in panels D, E= 4.5 mm; in panels F, G= 10 mm. See colour plate 4 (A-G) for this figure in colour.*

In vitro culture of GUS-expressing hairy roots and Southern blot analysis

To analyse the number of T-DNA inserts of transformed roots, 4-6 root tips (approximately 0.3 mm) were cut from lateral roots of individually sectioned nodulated hairy roots before staining for GUS activity. Root tips were placed on solidified half strength Murashiga and Skoog (MS) medium containing 3% sucrose, 300 μgmL^{-1} cefotaxime, 100 μgmL^{-1}vancomycin and 50 μgmL^{-1} G418, using 94x16 mm plates (CA Greiner & Shöne Ges. m.b. H, Kremsmünster, Austria, article 633102). Root tips were covered with a drop of half strength MS medium containing the mentioned antibiotics, to avoid bacterial overgrowth. After overnight incubation at 20°C in the dark, excess of liquid was removed, and root tips were allowed to dry. Root tips corresponding to positive GUS staining roots were grown for 2-3 months in the dark, with transfer to fresh medium every 2 weeks. Bacterial growth was not observed after 4-6 transfers. G418 seems to retard root growth, and in some experiments, it was omitted or substituted by timentin 100 μgmL^{-1}.

DNA was isolated from 0.1-0.2 g root tissue as described by van Slogteren et al. (1987). This procedure yielded approximately 2μg DNA, of which 1μg was used for Southern blot analysis. Other DNA extraction procedures we tried yielded DNA of poor quality that either could not be digested by restriction enzymes or resolved properly by electrophoresis.

Southern analysis of 12 independent transformation events obtained with the pPZP111-derivative, some of which derived from the same composite plant, resulted in 75% roots containing one T-DNA integration; 17%, two T-DNA integrations and 8% more than two T-DNA integrations. These results show that pPZP vectors carried by LBA 1334 are excellent tools to deliver the DNA of interest into hairy roots of *L japonicus*.

ACKNOWLEDGEMENTS

The authors want to thank Peter Hock for drawing figure 1 and Adrie 'tHooft for mounting figure 2. CL Díaz and HP Spaink were supported by the EU International Cooperation Project ICA4-CT-2001-10056, M Grønlund was supported by the EU Research Training Network HPRN-CT-2000-00086.

REFERENCES

Aoki T, Kamizawa A, and Ayabe S. (2002) **Efficient *Agrobacterium*-mediated transformation of *Lotus japonicus* with reliable antibiotic selection**. *Plant Cell Reports* 21, 238-243.

Den Dulk-Ras A, and Hooykaas PJJ. (1995) **Electrophoration of Agrobacterium tumefaciens**. In: *Plant Cell Electrophoration and Electrofusion Protocols. Methods in Molecular Biology* (Nickoloff JA Ed.) Humana Press Inc, Totowa, N J, pp. 55, 63-72.

Díaz CL, Melchers LS, Hooykaas PJJ, Lugtenberg BJJ, and Kijne JW. (1989) **Root lectin as a determinant of host-plant specificity in the Rihizobium-legume symbiosis**. *Nature* 338, 579-581.

Ditta G, Stanfield S, Corbin D, and Helinski DR. (1980) **Broad host range cloning system for gram negative bacteria: construction of a gene bank of *Rhizobium meliloti*.** *Proceedings of the National Academy of Sciences USA* 77, 7347-7351.

Jiang Q, and Gresshoff PM. (1997) **Classical and molecular genetics of the model legume *Lotus japonicus***. *Molecular Plant-Microbe Interactions* 10, 559-568.

Hajdukiewicz P, Svab Z, and Maliga P. (1994) **The small, versatile pPZP family of *Agrobacterium* binary vectors for plant transformation**. *Plant Molecular Biology* 25, 989-994.

Handberg K, and Stougaard J. (1992). ***Lotus japonicus*, an autogamous, diploid legume species for classical and molecular genetics**. *The Plant Journal* 2, 487-496.

Handbeg K, Stiller J Thykjaer T, and Stougaard J. (1994) **Transgenic plants: *Agrobacterium*-mediated transformation of diploid legume *Lotus japonicus***. In: *Cell Biology: a laboratory handbook.* (Celis JE, Ed.), Academic Press, New York, pp. 119-128.

Lombari P, Ercolano E, El Alaoui H. and Chiurazzi M. (2003) **A new transformation-regeneration procedure in the model legume Lotus japonicus: root explants as a source of large numbers of cells susceptible to Agrobacterium-mediated transformation**. *Plant Cell Reports* 21, 771-778.

Martirani L, Stiller J, Mirabella R, Alfano F, Lamberti A, Radutoiu SE, Iaccarino M, Gresshoff P, and Chiurazzi M. (1999) **T-DNA Tagging of Nodulation- and Root-Related genes in *Lotus japonicus*: Expression Patterns and Potencial for Promoter trapping and Insertional Mutagenesis**. *Molecular Plant-Microbe Interactions* 12, 275-284.

Offringa IA, Melchers LS, Regensburg-Tuink AJG, Constantino P, Schilperoort RA, and Hooykaas PJJ. (1986) **Complementation of *Agrobacterium tumefaciens* tumor inducing *aux* mutants by genes from the T_R-region of the Ri plasmid of *Agrobacterium rhizogenes*.** *Proceedings of the National Academy of Sciences USA* 83, 6935-6939.

Oger P, Peit A, and Dessaux, Y. (1996) **A simple technique for direct transformation and regeneration of the diploid legume species *Lotus japonicus***. *Plant Science* 116, 159-168.

Quaedvlieg NEM, Schlaman HRM, Admiraal PC, Wijting S, Stougaard J, and Spaink HP.

(1998) **Fusions between the green fluorescent protein and β-glucoronidase as sensitive and vital bifunctional reporters in plants.** *Plant Molecular Biology* 37, 715-727.

Roberts, C, Rajagopal, S, Smith, LM, Nguyen, TA, Yang, W, Nugrohu, S, Ravi, KS, Vijayachandra, K, Harcourt, RL, Dransfield, L, Desamero, N, Slamet, I, Hadjukiewicz, P, Svab, Z, Maliga, P, Mayer, JE, Keese, PK, Kilian, A, and Jefferson, RA. (1997) **A comprehensive set of modular vectors for advanced manipulations and efficient transformation of plants.** *Rockefeller Foundation Meeting of the International Program on Rice Biotechnology,* September 15-19, Malacca, Malaysia.

Scarpella E, Rueb S, Boot KJM, Hoge H and Meijer A. (2000) **A role for the rice homebox gene *Oshox1* in provascular cell fate commitment.** *Development* 127, 3655-3669.

Schenk RU, and Hildebrandt AC. (1971) **Medium and techniques for induction and growth of monocotyledonous and dycotiledonous plant cell cultures.** *Canadian Journal of Botany* 50, 199-204.

Stiller J, Martinari L, Tuppale S, Chian R-J, Chiurazzi M, and Gresshoff PM. (1997) **High frequency transformation and regeneration of transgenic plants in the model legume *Lotus japonicus*.** *Journal of Experimental Botany* 48, 1357-1365.

Van Brussel AAN, Tak T, Wetselaar A, Pees E, and Wijffelman CA. (1982) **Small leguminosae as test plants for nodulaiton of *Rhizobium leguminosarum* and other *Rhizobia* and *Agrobacterium* harbouring a leguminosarum plasmid.** *Plant Science Letters* 27, 317-325.

Vancanneyt G, Schmidt R, O'Connor-Sanchez A, Willmitzer L, and Rocha Sosa M. (1990) **Construction of an intron-containing marker gene: splicing of the intron in transgenic plants and its use in monitoring early events in *Agrobacterium*-mediated plant transformation.** *Molecular and General Genetics* 220, 245-250.

Van Slogteren GMS, Hoge JHC, Hooykaas PJJ, and Schilperoort RA. (1987) **Clonal analysis of hetrogeneous crown gall tumor tissues induced by wild-type and shooter mutant strains of *Agrobacterium tumefaciens*-expression of T-DNA genes.** *Plant Molecular Biology* 2, 321-333.

Chapter 6.3

TRANSFORMATION-REGENERATION PROCEDURE FOR *LOTUS JAPONICUS*

Leila Tirichine, José A. Herrera-Cervera, and Jens Stougaard*

Gene Expression Laboratory, Department of Molecular and Structural Biology, Gustav wieds vej 10 8000 Aarhus DENMARK; **Corresponding author*

Email: stougaard@mb.au.dk

Phone: +45 8942 5011 Fax: +45 8620 1222

Keywords: Lotus, transformation, regeneration

Lotus japonicus *has gained considerable attention because of its increasing use as a model legume. It is thus important to have an effective transformation and regeneration procedure to allow the transfer of foreign genes into Lotus for complementation experiments and the studies of regulation of nodulation and developmental processes.* Agrobacterium tumefaciens *mediated transformation is preferred because of its simplicity and efficiency in providing stable integration of transferred DNA into the plant genome.* In vitro *shoot regeneration of* Lotus japonicus *has been reported from hypocotyl explants (Thykjaer* et al., *1998). The objective of this work was to develop an improved* in vitro *regeneration and an* Agrobacterium tumefaciens *transformation method for* Lotus japonicus *hypocotyl explants.*

EXPERIMENTAL PROCEDURES

Plant material and transformation

Germination

1. Transfer the seeds to be germinated in a container resistant to sulfuric acid.
2. Cover the seeds with sulfuric acid (95-97%) and let them scarify for 2 to 3 min.
3. Wash the seeds 4 times with distilled water
4. Cover the seeds with commercial bleach (8-12% active chloride) and let them set for 20 minutes. The commercial bleach is diluted 20 times (1ml in 20 ml of distilled water plus Tween20 (5 μl))
5. Wash the seeds 5 times with sterile water.
6. In a sterile Petri dish, put a few layers of sterile filter paper.

A.J. Márquez (ed.), Lotus japonicus Handbook, 279-284.

7. Wet the filter paper with sterile water and leave a film of water at the bottom.
8. Put the seeds on the top of the filter paper and spread them.
9. Let the seeds germinate at 26°C in continuous light for 3 days or until roots are about 0.5 cm long.

Geminate seeds for 1 week in continuous light at 26 °C (see surface sterilisation protocol). Streak the *Agrobacterium* strains on fresh selective plates. Start 2 to 3 days before transformation (it depends how fast the *Agrobacterium* strain grows). Prepare 2 to 3 Petri dishes for each strain to get enough cells to resuspend in YMB medium. Scrape 2-day-old *Agrobacterium* colonies off selective plates with a sterile toothpick or sterile spatula, and resuspend each strain to a density of approximately 5 x 10^9 per millilitre in 5 ml of YMB medium by whirling. Leave bacteria as milky cultures at room temperature while preparing the co-cultivation plates.

Prepare co-cultivation plates by placing 0.5 cm-high platforms of sterile filter papers in Petri dishes. Add co-cultivation medium until free liquid appears (12-15 ml). Prepare a new set of Petri dishes with two layers of sterile filter paper. Soak filter papers with 4 ml of *Agrobacterium* YMB suspension or 4 ml of YMB for control plates. Free liquid should appear on the surface of filter paper.

Transfer some seedlings to the *Agrobacterium* filters and using scalpel and forceps cut the seedlings below the shoot primordia and at the stem base. Once cut transversely, hypocotyls are cut longitudinally into two pieces to have more surface contact with *Agrobacterium* and more room for calli to form. Leave the cut hypocotyls soaking in *Agrobacterium* solution into 5 ml containers for approximately 30 min. Transfer the 3 mm hypocotyls pieces into the co-cultivation plates. Seal the Petri dishes with Parafilm but remember to perform some opening within the Parafilm with a sterile scalpel.

Incubate co-cultivation plates for 7 days at 21°C in the dark.

Regeneration procedure

Plant cells are totipotent, i.e., a single cell can give rise to an entire organism provided it is supplied with appropriate nutrients and stimulated with proper combinations of hormones to grow and differentiate. Thus, it is important to ensure the right media and concentration of growth factors.

To eliminate *Agrobacterium*, transfer explants with the top filter paper from the co-cultivation plates to regeneration medium with 300µg/m of cefotaxime. Leave for 5 days to allow the regeneration of the explants.

Start the selection of transgenic calli by including one of the following antibiotics: G418 (25 µg/ml), hygromycin (15 µg/ml). Transfer the explants with the filter paper to selective medium containing 300 µg/ml of cefotaxime. Incubate for 7 days at 26°C in continuous light. (If you have the kanamycin resistance gene in your construct, use Geniticin (G418) instead of kanamycin to select transformed calli, because it is more effective in Lotus. If you can't get calli with 25 µg/ml Geniticin, lower the concentration to 15 µg/ml. In either case, use a kill curve to determine which concentration is effective.)

Continue selection of transgenic calli by transferring explants, without filter paper to selective medium containing 300 µg/ml of cefotaxime and the appropriate antibiotic.

Continue selection and propagation of transgenic calli on callus medium for 5 weeks. Transgenic green calli are separated from explants as soon as possible and propagated in one piece. Keep transferring the calli to a new media each week.

Start shoot induction by transferring green transgenic calli to shoot induction medium with 300 µg/ml cefotaxime. Shoot structures will develop in the dark green zone appearing at the perimeter of the calli. Keep calli undivided; remove brown and light yellow tissue.

Continue propagation of transgenic calli on shoot induction medium for 3 weeks. Maintain counter selection of *Agrobacterium* with 300 µg/ml cefotaxime until the transfer to the shoot elongation medium. Keep transferring the calli to new shoot induction media each week.

Transfer calli with emerging shoot structures to shoot growth medium for 2 weeks. Keep transferring emerging shoots to new growth media each week.

Transfer calli with emerging shoot structures to shoot elongation medium for 2 weeks. Keep transferring emerging shoots to new elongation media each week.

Cut off individual shoots 2-4 cm long and move to root induction medium for 5 days or until you see the swelling of the base of the shoot (5 to 12 days) at 21°C with a photoperiod of 8/16 hours. Alternatively and based on the observation that cut Lotus shoots easily develop roots in a humid atmosphere, we strongly recommend moving directly the cut shoots to the greenhouse. Put them in very wet soil (without any fertilisers), spray them with water and put them in the minicaphouse. Remember to cover to ensure high level of humidity.

Move shoots to root elongation medium and incubate for 2 weeks or longer at 21°C until the root system is developed enough to be transferred to the greenhouse. Move the shoots with roots into pots containing soil, cover with a plastic bag or use the minicaphouse. Remove the plastic bag gradually, so plants adapt to the new surroundings.

Plant media

Co-cultivation medium

- 1/10 x Gamborg's B5 medium without sucrose
- 5 mM MES, pH 5.2
- 1/10 x B5 vitamins
- NAA (0.05 µg/ml)
- BAP (0.5 µg/ml)

Mix 387 mg ready-made B5 in 1 litre of water and autoclave. Before use, add 5 ml of autoclaved 1M MES, pH 5.2 stock solution, 100 µl of the vitamin stock for B5 medium, 100 µl of NAA, and 1 ml of BAP.

Regeneration medium

- MS with 2% sucrose
- 0.2% GelRite
- 0.05 μg/ml NAA
- 0.5 μg/ml BAP
- 300 μg/ml cefotaxime

For 1 litre, add 3.87 g of ready-made B5, 20 g of sucrose, and 2 g of GelRite. Adjust pH to 5.5 and autoclave. This is called LO medium. Add 1 ml of the vitamins for B5 medium, 1 ml of BAP, and 100 μl of NAA before pouring cooled medium into 9 cm Petri dishes.

Selection medium

Add the appropriate antibiotic to regeneration medium.

Shoot induction medium

- MS with 2% sucrose
- 0.2% GelRite
- 10 mM NH_4^+
- BAP at 0.5 μg/ml

Prepare 1 litre LO medium as above, then add 2.5 ml of autoclaved 2M $(NH_4)_2SO_4$ to hot medium. Then add 1 ml of BAP stock solution and 1 ml of the vitamin stock for B5 medium to cooled medium. Pour into Petri dishes.

Shoot growth medium*

- B5 with 2% sucrose
- 0.2% GelRite
- BAP at 0.2 μg/ml

Prepare 1 litre LO as above, then add 400 μl of BAP stock solution and 1 ml of the vitamin stock for B5 medium to cooled medium. Pour into plant containers.

Shoot elongation medium*

- B5 with 2% sucrose
- 0.2% GelRite
- BAP at 0.05 μg/ml

Prepare 1 litre LO as above, add 1 ml of the vitamin stock solution for B5 medium and 80 μl BAP and pour into plant containers. (If your shooting plants are looking fine, you can simply use half-strength B5 with only vitamins and no BAP.)

Note: **For shoot growth and elongation medium, we replace MS with B5. On B5, shooting improves and calli make proper shoots, not just leaves.*

Root induction medium

- Half-strength B5 with 1% sucrose
- 0.4% GelRite
- 0.5 to 1 µg/ml NAA

For 1 litre, add 1.94 of B5, 10 g sucrose, and 4 g of GelRite and autoclave. This is called L medium. Before use, add 500 µl of vitamin stock solution and the right amount of NAA depending in the concentration. Pour into plant containers.

Root elongation medium

- Half-strength B56 medium with 1% sucrose and 0.4% GelRite.

Use 1 litre L medium with 500 µl of the vitamin stock solution for B5 medium and pour into plant containers.

Agrobacterium media

Solid agar plates

- 0.5% yeast extract
- 1% bacto-tryptone
- 0.5% Nacl
- 1.4% agar, pH 7

For 1 litre, add 5 g of yeast extract, 10 g of bacto-tryptone, 5 g of NaCl, and 14 g of agar. Adjust pH to 7 and autoclave. Stored medium is remelted in a microwave oven and cooled. Antibiotics for selecting appropriate binary T-DNA vectors are added before plates are poured.

YMB liquid growth medium

- 0.2% Mannitol
- 0.04% Yeast extract
- 0.02% $MgSO_47H_2O$
- 0. 01% NaCl

For 1 litre, dissolve 2 g mannitol, 0.4 g yeast extract, 0.2 g $MgSO_47H_2O$ and 0.1 g NaCl and autoclave. Autoclave a 6.55% $K_2HPO_43H_2O$, pH 6.8 solution separately, add 10 ml per litre before use. This media is used to resuspend *Agrobacterium* for transformation.

Seedling media

Solutions for sterilisation and germination of seeds

- 2% hypochlorite
- 0.02% Tween 20

For 20 ml, mix 2.7 ml of a 15% sodium hypochlorite solution with autoclaved water in a sterile universal container. Add 4 μl Tween20, shake gently, and use the same day.

Support material in liquid cultures

Squares of filter paper (6 x 6 cm) are placed in aluminium foil and autoclaved.

Solutions and media for co-cultivation, selection of transgenic calli, and plant regeneration

Plant hormone stock solutions

- BAP at 0.5 mg/ml: Dissolve 25 mg BAP in 1 ml 1M NaOH and add water to 50 ml
- NAA at 0.5 mg/ml: Dissolve 25 mg NAA in 24 ml 96% ethanol and add water to 50 ml. All solutions are filter sterilised into 2 ml aliquots and stored at -20°C. Upon use, you can heat gently in a microwave oven.

Antibiotic stock solutions

- Kanamycin at 100 mg/ml: Dissolve 2.5 g kanamycin in 25 ml H2O, filter sterilise into 5 ml aliquots and store at -20°C.
- G418 (Geniticin) at 50 mg/ml: Dissolve 1.25 g G418 in 25 ml water. Filter sterilise into 5 ml aliquots and store at -20°C.
- Cefotaxime at 30 mg/ml: For 1 litre of medium, dissolve 300 mg cefotaxime (claforan) in 10 ml H_2O, filter-sterilise and use the same day or you can prepare a stock and keep at -20°C in 10 ml containers.

Other solutions

- 1M MES: Dissolve 4.88 g in 25 ml of water, adjust pH to 5.2 with KOH. Autoclave before use.
- 2M$(NH_4)_2SO_4$: Dissolve 198.24 g in 250 ml. Autoclave before use.

REFERENCES

Thykjaer T, Schauser L, Danielsen D, Finneman J, and Stougaard J. (1998) **Transgenic plants: Agrobacterium-mediated transformation of the diploid legume *Lotus japonicus*.** In: *Cell Biology: A laboratory handbook.* Second edition vol 3. Academic Press.

Chapter 6.4

AGROBACTERIUM RHIZOGENES PRI TL-DNA INTEGRATION SYSTEM: A GENE VECTOR FOR *LOTUS JAPONICUS* TRANSFORMATION

Simona Radutoiu, Lene H Madsen, Esben B Madsen, Anna Marie Nielsen and Jens Stougaard*

*Laboratory of Gene Expression, Department of Molecular Biology, University of Aarhus, DK-8200, DENMARK; *Corresponding author.*

Email: stougaard@mb.au.dk

Phone: +45 8942 5011 Fax: +45 8620 1222

Keywords: Agrobacterium, co-transformation, Lotus

Transfer of T-DNA from binary vectors to transgenic "hairy roots" rely on co-transformation of the T-DNA from the binary vector and the T-DNA from the Ri plasmid into the same plant cell. The frequency of co-transformation varies, so for some experiments an alternative method integrating the gene construct under study diretly into the Ri T-DNA segment may be useful.

METHOD

This co-transformation methodology has been developed using the *A. rhizogenes* pRi15834 TL-DNA segment. Three *A. rhizogenes* strains, AR1193, AR12 and AR14 containing pBR322 plasmid sequences within the TL-DNA serve as acceptor strains for pBR322-derived plasmids unable to replicate in Agrobacterium. Integration into the TL-DNA acceptor region occurs by homologous recombination through the pBR322 sequences (Stougaard et al., 1987). In addition, AR12 contains a 35S::GUS reporter gene and AR14 a 35S::CAT reporter gene (Hansen et al., 1989).

For conjugational transfer and integration of gene constructs into AR1193, AR12, and AR14, the pBR322 derived cloning vector, pIV10, can be used. The pIV10 plasmid is 4.1 kb, contains the *Eco*RI – *Hind*III polylinker cloning sites from pUC19, and for selection of integration events in Agrobacterium, the ampicillin resistance gene from pBR322 plasmid and spectinomycin/streptomycin resistance gene from R702 plasmid. Up to 15 kb of DNA has beeen cloned in pIV10, transferred to the acceptor strains by recombination and transformed into *Lotus*

A.J. Márquez (ed.), Lotus japonicus Handbook, 285-287.

japonicus hairy roots via the hypocotyl transformation procedure (Stougaard, J. 1995).

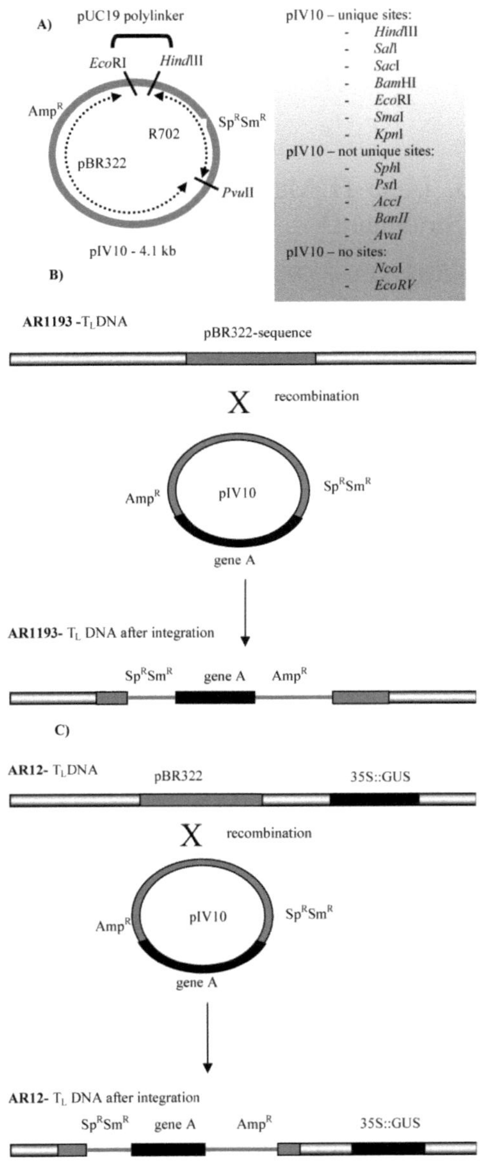

Figure1. Gene integration system into A rhizogenes *TL-DNA. (A) pIV10 vector for gene cloning and integration into* A rhizogenes *strains; (B) Structure of the TL segment before and after integration into AR1193 strain; (C) Structure of the TL segment before and after integration into AR12 strain.*

Selection of strains and plasmids

AR1193- Rifampicin 100 µg/ml

AR12 – Rifampicin 100 µg/ml

pIV10 – Ampicillin 100 µg/ml , Spectinomycin/Streptomycin 100 µg/ml

REFERENCES

Stougaard J, Abildsten D, and Marcker KA. (1987) **The *Agrobacterium rhizogenes* pRi TL-DNA segment as a gene vector system for transformation of plants**. *Molecular General Genetics* 207, 251-255.

Hansen J, Jørgensen J-E, Stougaard J, and Marcker KA. (1989) **Hairy roots -a shortcut to transgenic root nodules**. *Plant Cell Reports* 8, 12-15.

Stougaard J. (1995) ***Agrobacterium rhizogenes* as a vector for transforming higher plants**. *Methods in Molecular Biology* 49, 49-61.

Chapter 6.5

VECTORS FOR REVERSE GENETICS AND EXPRESSION ANALYSIS

Stig Uggerhøj Andersen, Cristina Cvitanich, Mette Grønlund, Hanne Busk, Dorthe Bødker Jensen, and Erik Østergaard Jensen*

*Laboratory of Gene Expression, Department of Molecular Biology, University of Aarhus, Gustav Wieds Vej 10, DK-8000 Aarhus C., DENMARK.; *Corresponding author.*

Email: eoj@mb.au.dk

Phone: +45 8942 5014   Fax: +45 8942 5012

Keywords: Vectors, transformation, reverse genetics, expression analysis, protein localization

The binary Agrobacterium *pPZP211 vector is stable and fully sequenced. We have produced derivatives of this vector that can be used for general over-expression, nodule-specific expression, expression pattern analysis, and protein localization studies.*

INTRODUCTION

The pPZP211 (Hajdukiewicz et al., 1994) derivatives are divided into two groups, pZ and pZN. The pZ vector series is based on the original pPZP211 vector, in which various constructs have been inserted into the pUC18 multiple cloning site (MCS). The presence of the strong cauliflower mosaic virus 35S (CaMV35S) enhancer-promoter (ep35S) (Guilley et al., 1982) in the pZ vector series could cause problems by overruling the specificity of a promoter under investigation, as observed with a similar vector (Hansen et al., 1999). To avoid this effect, ep35S was replaced with the *Agrobacterium tumefaciens* nopaline synthase promoter (pNos) in the pZN vector series.

RESULTS

pZ-35S

The pZ-35S vector is designed for over-expression of a gene of interest (GOI) from the CaMV35S promoter (p35S) (Figure 1A). The MCS allows cloning of a GOI

A.J. Márquez (ed.), Lotus japonicus Handbook, 289-292.

between p35S and the *A. tumefaciens* nopaline synthase terminator (pAnos). The pZ-35S vector has successfully been used to over-express intrinsic factor in *Arabidopsis thaliana* (Fedosov et al., 2003).

pZ-35S-GFP

The pZ-35S-GFP vector (Figure 1B) is designed for protein localization studies. An Xba I restriction site allows in frame fusion to Green Fluorescent Protein (GFP). p35S drives expression of the fusion protein. The pZ-35S-GFP vector has been used to determine the localization of CPP1 and NDX proteins by transient expression in onion epidermal cells and *Cantharantus roseus* cells, respectively (Cvitanich et al., 2000; Grønlund et al., 2003).

pZ-LB

The pZ-LB vector (Figure 1C) is designed for nodule specific expression from the *Glycine max* leghemoglobin *lbc3* promoter (plbc3), which is active only in the central infected tissue of the root nodule (Stougaard et al., 1987). The MCS allows cloning of a GOI between plbc3 and pAnos. The pZ-LB vector has been used to express *gus* in nodules (Cvitanich et al., 2000).

pZ-Enod12

The pZ-Enod12 vector (Figure 1D) is designed for nodule specific expression from the *Pisum sativum enod12* promoter (penod12). This promoter is active in nodule primordia and young nodules. Expression from the promoter has also been observed in pollen (Grønlund et al., 2003). The MCS allows cloning of a GOI between penod12 and pAnos.

pZN

This vector (Figure 1E) is designed to avoid ep35S influence on promoters used for tissue specific expression or expression pattern analysis. To this end, ep35S, used to control NptII expression in pPZP211, was replaced by pNos.

pZN-LB

This vector (figure 1F) is the same as pZ-LB but ep35S was replaced by pNos.

pZN-Enod12

This vector (Figure 1G) is the same as pZ-Enod12 but ep35S was replaced by pNos.

pZN-GUS

The pZN-GUS vector (Figure 1H) was designed for expression pattern analysis using an intron-containing β-glucuronidase reporter gene (GUSint). The MCS allows cloning of a promoter of choice in front of the promoterless GUSint. The GUSint gene is interrupted by the second intron (IV2) of the potato ST-LSI gene (Vancanneyt et al., 1990) to avoid expression in bacteria.

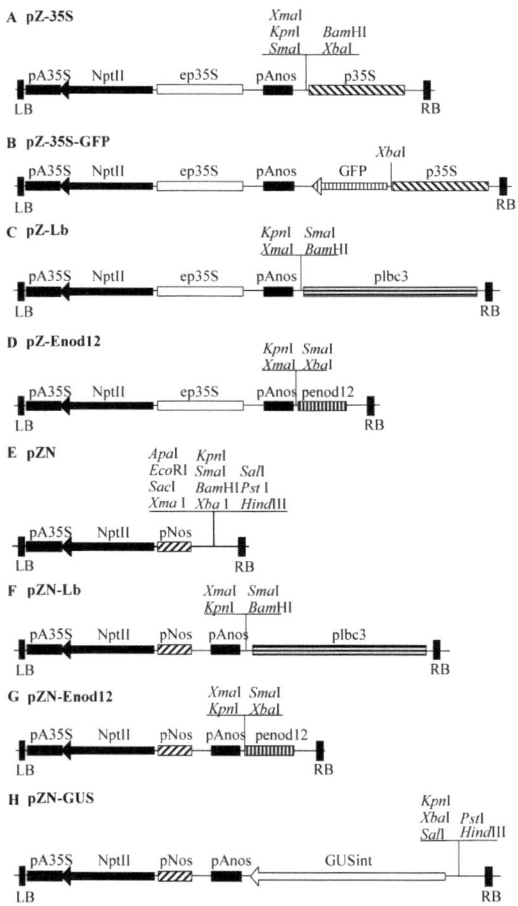

Figure 1. Vectors (pPZP211 derivatives) for reverse genetics and expression analysis. Only the region between left border (LB) and right border (RB) is shown. Unique restriction sites are indicated by restriction enzyme names. All vectors confer bacterial resistance to spectinomycin. pA35S: CaMV35S terminator. NptII: neomycin phosphotransferase II gene. ep35S: CaMV35S enhancer-promoter. p35S: CaMV35S promoter. pAnos: A. tumefaciens *nopaline synthase terminator. pNos:* A. tumefaciens *nopaline synthase promoter. GFP: Green Fluorescent Protein. plbc3:* Glycine max *leghemoglobin lbc3 promoter. penod12:* Pisum sativum *enod12 promoter. GUSint: intron-containing β-glucuronidase reporter gene.*

REFERENCES

Cvitanich C, Pallisgaard N, Nielsen KA, Hansen AC, Larsen K, Pihakaski-Maunsbach K, Marcker KA, and Jensen EO. (2000) **CPP1, a DNA-binding protein involved in the expression of a soybean leghemoglobin c3 gene**. *Proceedings of the National Academy of Sciences USA* 97, 8163-8168.

Fedosov SN, Laursen NB, Nexo E, Moestrup SK, Petersen TE, Jensen EO, and Berglund L. (2003) **Human intrinsic factor expressed in the plant *Arabidopsis thaliana***. *European Journal of Biochemistry* 270, 3362-3367.

Grønlund M, Gustafsen C, Roussis A, Jensen D, Nielsen LP, Marcker KA, and Jensen,EO. (2003) **The *Lotus japonicus* ndx gene family is involved in nodule function and maintenance**. *Plant Molecular Biology* 52, 303-316.

Guilley H, Dudley RK, Jonard G, Balazs E, and Richards KE. (1982) **Transcription of Cauliflower mosaic virus DNA: detection of promoter sequences, and characterization of transcripts**. *Cell* 30, 763-773.

Hajdukiewicz P, Svab Z, and Maliga P. (1994) **The small, versatile pPZP family of Agrobacterium binary vectors for plant transformation**. *Plant Molecular Biology* 25, 989-994.

Hansen AC, Busk H, Marcker A, Marcker KA, and Jensen EO. (1999) **VsENBP1 regulates the expression of the early nodulin PsENOD12B**. *Plant Molecular Biology* 40, 495-506.

Stougaard J, Sandal N, Grøn A, Kühle A, and Marcker KA. (1987) **5' Analysis of the soybean leghaemoglobin *lbc3* gene: regulatory elements required for promoter activity and organ specificity**. *EMBO Journal* 6, 3565-3569.

Vancanneyt G, Schmidt R, O'Connor-Sanchez A, Willmitzer L, and Rocha-Sosa M. (1990) **Construction of an intron-containing marker gene: splicing of the intron in transgenic plants and its use in monitoring early events in Agrobacterium-mediated plant transformation**. *Molecular General Genetics* 220, 245-250.

SECTION 7

PRIMARY AND SECONDARY METABOLISM

Chapter 7.1

NITRATE ASSIMILATION: INFLUENCE OF NITROGEN SUPPLY

Alicia Orea[1,2], Purificación Pajuelo[1*], José M Romero[2], and Antonio J Márquez[1†]

*[1]Departamento de Bioquímica Vegetal y Biología Molecular, Facultad de Química, Universidad de Sevilla, Apartado 553, 41080-Sevilla, SPAIN. [2]Instituto de Bioquímica Vegetal y Fotosíntesis, Centro de Investigaciones Científicas Isla de la Cartuja, Avda. Américo Vespucio, 49, 41092-Sevilla, SPAIN. *Present address: Escuela Universitaria de Ingeniería Técnica Agrícola; Carretera de Utrera, Km. 1; 41012-Sevilla, SPAIN. [†]Corresponding author.*

Email: cabeza@us.es
Phone: +34 *954 557* 145; Fax: +34 *954 626* 853

Keywords: roots, shoots, nitrate reductase, nitrite reductase, nitrate content

We have studied the effect of nitrogen supply on growth as well as nitrate assimilation in Lotus japonicus *plants. High nitrate concentrations (higher than 20 mM) were toxic for* L japonicus, *inhibiting growth, with the optimum nitrate concentration ranging between 4 and 8 mM. The maximum growth and levels of nitrate reductase (NR) and nitrite reductase (NiR) activity as well as nitrate accumulation was obtained in plants grown simultaneously on nitrate plus ammonium.* L japonicus *could also grow on ammonium as the sole nitrogen source, without a supplement of organic acids like succinate. Low external nitrate concentrations induced a higher growth of roots, while higher external nitrate concentration inhibited the growth of roots and favoured shoots growth. NR and NiR activities were always higher in roots than in shoots of* L japonicus *plants. An increase in external nitrate concentrations did not alter significantly the plant partitioning of nitrate assimilation among roots and shoots, being predominantly higher in the roots than in stems and leaves. The kinetics of NR activity induction by nitrate showed a different pattern in roots and in leaves, being this induction in roots independent on external nitrate concentration in the short-term. In contrast, the induction kinetics of NiR activity and nitrate accumulation showed the same pattern in roots and leaves, being dependent on the external nitrate concentration. In all cases the levels of NR and NiR activity and nitrate accumulation were higher in roots than in leaves. Significant levels of NR protein but no NR activity were detected in* NH_4^+*-grown plants.*

A.J. Márquez (ed.), Lotus japonicus Handbook, 295-313.

INTRODUCTION

Nitrate is the main nitrogen source for most higher plants. About 99% of the organic nitrogen in the biosphere is derived from nitrate assimilation (Heldt, 1997). Nitrate is transported into root cells where it is stored in the vacuole, reduced in the cytoplasm or transported to the leaves, where it can be stored or reduced (Crawford, 1995). Nitrate is reduced to ammonium by the nitrate assimilatory pathway (for review see: Hoff et al., 1994; Crawford, 1995; Daniel-Vedele and Caboche, 1996; Forde and Clarkson, 1999). This reduction takes place in two successive enzymatic steps. Nitrate is first reduced to nitrite by nitrate reductase, a cytoplasmic enzyme, using pyridine nucleotide as the source of reductant (NR, NADH- nitrate oxidoreductase, EC 1.6.6.1 and NAD(P)H-nitrate oxidoreductase EC 1.6.6.2). Nitrite is then translocated to the plastids where it is reduced to ammonium by nitrite reductase (NiR, EC 1.7.7.1.). Finally, ammonium is incorporated into amino acids by the glutamine synthetase (GS, EC 6.3.1.2) and glutamate synthase (GOGAT, EC 1.4.7.1 or EC 1.4.1.14) pathway (Ireland and Lea, 1999; Hirel and Lea, 2002).

NR is a highly regulated enzyme throughout a hierarchy of transcriptional and post-translational controls (Crawford, 1995; Daniel-Vedele and Caboche, 1996; Campbell, 2002). Regulation of NR responds to different factors, such as light, chloroplastic factors, reduced nitrogen compounds, circadian rhythms and cytokinins. However, nitrate is the main signal that induces the accumulation of NR (Stitt, 1999).

Nitrate assimilation takes place both in leaves and in roots (Heldt, 1997). The distribution of nitrate assimilation throughout the plants depends on the plant species, age and environmental factors, being the nitrate supply one of the most important factors (Smirnoff and Stewart, 1985; Wallace, 1986; Andrews,1986; Caba *et al.*, 1995; Woodall and Forde, 1996). Andrews (1986) established a classification of legume and non-legume plants in accordance with the partitioning of nitrate assimilation and its possible correlation to nitrate supply and the climatic origin of the plant.

Most of the studies performed in *Lotus japonicus* in relation to nitrogen metabolism have been focused on nitrogen fixation (Szczyglowski *et al.*, 1997; Colebatch *et al.*, 2002; Radutolu *et al.* 2003). Little work has been published with regard to nitrate assimilation. *Lotus japonicus* belongs to the temperate legume group that assimilates nitrate mainly in roots (Andrews, 1986; Woodall and Forde, 1996). We have recently characterized NiR and NR from roots of *L japonicus* (Orea *et al.*, 2001; Pajuelo *et al.*, 2002). In the present work, we show the effect of the nitrogen source on growth and distribution of nitrate assimilation in different organs of the plant, as well as the response of NR and NiR to nitrate supply.

RESULTS

Effect of the nitrogen source on growth

The growth of *L japonicus* plants, estimated as the number of fully developed trifolii, was comparatively determined 35 days after sowing at different nitrate concentrations.

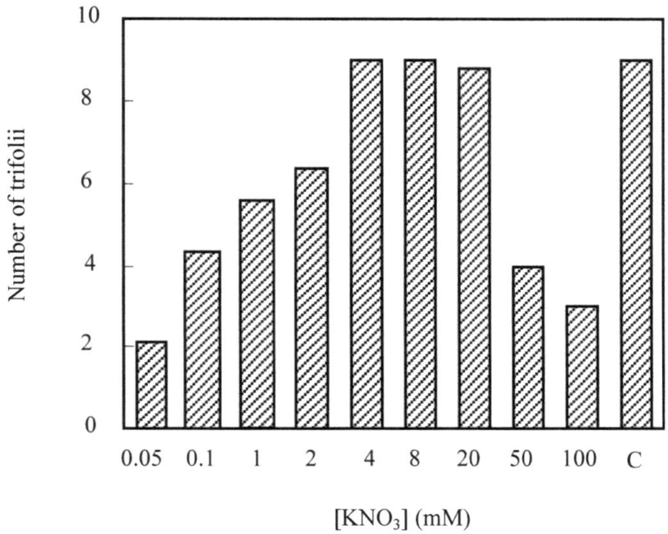

Figure 1. Effect of nitrate concentration in the culture medium on L japonicus *growth.* L japonicus *seeds were germinated in distilled water for 7 days. Plants were then transferred to pots and watered daily with Hornum base medium supplemented with* KNO_3 *at the indicated concentrations. The average number of trifolii per plant was determined after 35 days. Control plants (C) were grown in Hornum base medium supplemented with 95 mM KCl and 5 mM* KNO_3.

As shown in Figure 1, the number of trifolii increased with nitrate concentration in the culture medium, to reach the maximum growth at external nitrate concentration ranging between 4 and 8 mM. Higher nitrate concentrations induced a decrease in growth, most pronounced at nitrate concentrations higher than 50 mM. The negative effect of high nitrate concentrations on growth is not evident when the plants were grown in the presence of equimolar KCl concentrations instead of nitrate (Column C in Figure 1), thus indicating that growth inhibition is due to a specific effect of nitrate and not to an osmotic effect. However, a deleterious effect of the salt was observed on the seed germination rate. Increasing salt concentration in the germinating solution induced a parallel decrease in the germination rate to reach

66% and 44% for 100 mM KCl or 100 mM KNO_3 respectively (Table 1). The data suggest that seed germination is more sensitive to salt concentration than plant growth. To circumvent this problem in the experiments presented in this work, seeds were germinated in distilled water.

Culture conditions		**% Germination**
[KNO_3] (mM)	**[KCl] (mM)**	
5	0	72
20	0	88
50	0	88
100	0	44
200	0	1.6
400	0	0
0	5	88
5	15	89
5	45	89
5	95	66
5	195	1.2
5	395	0

Table 1. Effect of nitrate concentration on L japonicus *seed germination.* L japonicus *plants were germinated (and grown) in Magenta boxes at the indicated concentrations of KNO_3 and KCl. The% germination was determined 10 days after sowing.*

On the other hand, the effect of different nitrogen sources on growth was also studied. As shown in Figure 2, *L japonicus* plants irrigated with Hornum medium, containing 5 mM NH_4NO_3 and 3 mM KNO_3, showed better growth of the aerial parts of the plants than when cultivated with nitrate or ammonium as single nitrogen sources. Plants cultivated in the presence of different ammonium salts as single N source, showed in general lower shoot growth rates compared to plants cultivated with equimolar concentrations of KNO_3 (Figure 2A). No significant difference was obtained when using inorganic or organic sources of ammonium. This behaviour is different than that observed in tobacco plants that need succinate in order to grow in the presence of ammonium (Pelsy *et al.*, 1991). The shoot/root ratio was very similar for plants grown in the presence of ammonium and it was independent on the particular ammonium salt supplied to the plants (Figure 2C). A slight increase was observed when using higher ammonium concentrations due to a better growth of the aerial parts of the plants. However, the shoot/root ratio was substantially affected by the concentration of nitrate in the culture medium. Plants grown with 5 mM KNO_3, showed a 50% reduced root growth compared to plants cultivated with 0.1 mM KNO_3, while the opposite situation was observed for the aerial parts of the plant. This behaviour originates that the shoot/root ratio at 5 mM KNO_3 is 3-fold higher than at 0.1 mM KNO_3.

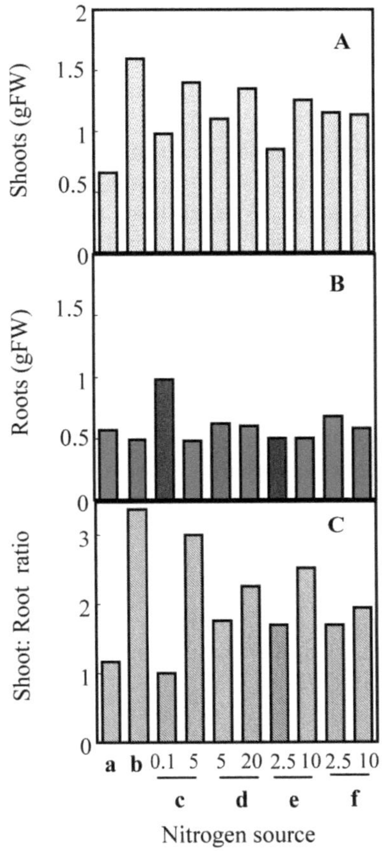

Figure 2. Effect of the nitrogen source on growth of shoots and roots of L japonicus *plants.* L japonicus *seeds were germinated in distilled water and the plants were then grown in pots with vermiculite (4 plants per pot) in the greenhouse, being daily irrigated with 50 ml of the following solutions: Hornum base medium (without nitrogen source) (a), Hornum medium (5 mM NH_4NO_3 and 3 mM KNO_3) (b) and Hornum base medium supplemented with KNO_3 (c), NH_4Cl (d), $(NH_4)_2SO_4$ (e) and $(NH_4)_2$ succinate (f) at the indicated concentrations (mM). The nitrification inhibitor dicyandiamide was included at 7.5 µg ml^{-1} in all the media containing NH_4^+ as single N source. After 37 days of growth, the fresh weight from shoots (A) and roots (B) and the shoot/root ratio (C) were determined collectively for the 4 plants grown on each pot.*

From the data presented, we can conclude that *L japonicus* plants grow better on KNO_3 than on NH_4^+. Optimum growth for *L japonicus* in the presence of KNO_3 takes place at concentrations ranging between 4 and 8 mM. Higher nitrate concentrations have a deleterious effect for the plants. The maximum growth was obtained when nitrate and ammonium are simultaneously supplied to the plants. Growth in the presence of ammonium as the sole nitrogen source is independent on succinate addition.

Effect of external nitrate concentration on nitrate assimilation

In higher plants, nitrate assimilation can take place in leaves as well as in roots, the contribution of each of these organs being dependent on the species and on the external nitrate concentration (Andrews, 1986). We have studied several aspects of nitrate assimilation in *L japonicus* plants grown in different conditions of nitrogen supply in order to determine the relative contribution of roots and leaves to the global process of nitrate assimilation. Among these aspects, we have characterized the accumulation of nitrate and the activity and protein levels of nitrate reductase (NR) and nitrite reductase (NiR) in leaves, roots and stems of *L japonicus* plants grown in Magenta boxes with increasing concentrations of KNO_3 and NH_4NO_3.

Plants cultivated at increasing concentrations of KNO_3 showed a parallel increase in nitrate content in leaves, stems and roots, reaching the maximum level of accumulation at an external nitrate concentration of 24 mM (Figure 3A). The relative proportion of internal nitrate accumulation in each organ did not change when the external supply of nitrate was increased in the growing solution. In all cases, the level of nitrate accumulation was maximal in stems, intermediate in roots and lower in leaves in the range of external nitrate concentrations studied (Figure 3A). A similar behaviour was observed when *L japonicus* plants were cultivated at increasing concentrations of NH_4NO_3 (Figure 3B). However, although the pattern of nitrate accumulation was similar than for plants grown with KNO_3 as the sole nitrogen source, the maximum level of accumulation was reached at a lower external nitrate concentration (8 mM NH_4NO_3).

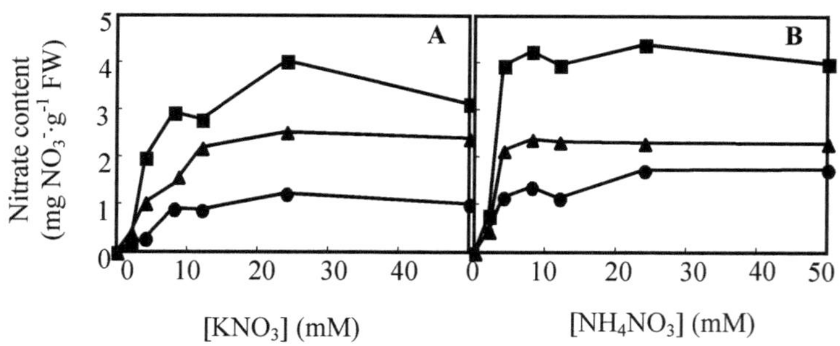

Figure 3. Nitrate content in leaves, stems and roots from L japonicus *plants grown at different concentrations of KNO_3 and NH_4NO_3. Plants were grown in Magenta boxes for 20 days in Hornum medium supplemented with KNO_3 (A) and NH_4NO_3 (B) at the indicated concentrations. Leaves (diamonds), stems (--squares) and roots (triangles) samples were harvested at the third light hour of the photoperiod. Nitrate content was determined as described in Experimental procedures.*

The levels of NR and NiR enzyme activities in crude extracts from *L japonicus* plant did also increase with the nitrate concentration present in the nutrient solution (Figure 4A and C). For all nitrate concentrations assayed, root NR and NiR activities per unit of fresh weight were much higher than in stems and leaves. The same pattern was observed for plants grown with NH_4NO_3, except that the level of enzyme activities were slightly higher in all tissues (Figure 4B and D). In our growth conditions, plants typically show their fresh weight biomass equally distributed among roots, stems and leaves (a third of the total plant fresh weight corresponds to each of these organs). On this basis, data from Figure 4 indicates that total root NR activity per plant is always higher (at least the double) than that in the shoots (leaves + stems). This means that increasing external nitrate concentrations up to 24 mM (much beyond the optimal concentration for growth), does not originate a shift in the partitioning of nitrate assimilation among roots and shoots.

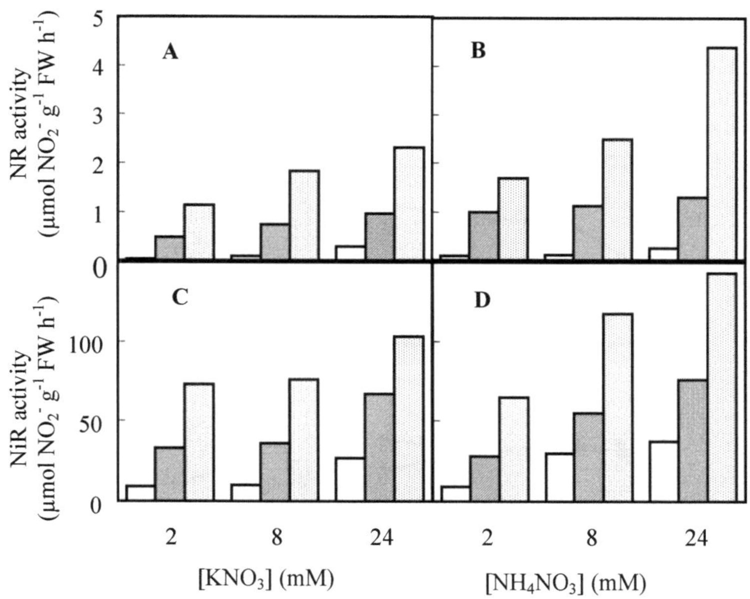

Figure 4. Levels of NR and NiR enzyme activities in leaves, stems and roots from L japonicus *plants grown in different concentrations of KNO_3 and NH_4NO_3. Plants were grown as described in Figure 3. NR (A y B) and NiR (C y D) activities were determined in leaves (white), stems (black), and roots (stippled) from* L japonicus *plants as described in experimental procedures.*

According to the data, *L japonicus* assimilate nitrate mainly in roots, as it has been described for temperate legumes (Andrews, 1986). However, differently to other temperate legumes, increasing the concentration of nitrate in the nutrient solution in *L japonicus* does not produce a global shift of nitrate assimilation to leaves. Roots appear to be the main tissue for nitrate assimilation both at low and high external nitrate concentrations.

Induction of nitrate assimilation

Nitrate is the primary signal for the induction of its own transport system, as well as for the enzymes implicated in nitrate assimilation (Crawford, 1995). In order to analyse the induction of nitrate assimilation by nitrate in *L japonicus*, plants grown with ammonium as the sole nitrogen source were transferred to growing media containing increasing concentrations of nitrate. After nitrate addition, the level of nitrate accumulation, NR and NiR activity and the level of NR and NiR proteins were analyzed in roots and leaves.

In the presence of ammonium as the sole nitrogen source, the levels of nitrate accumulation and NR activity were not detectable, while NiR activity showed basal levels both in roots and leaves (Figure 5). Nitrate accumulation in roots was detected 4 h after nitrate addition, and 8 h after nitrate addition in the case of leaves. At a low external nitrate concentration (0.1 mM), the maximum level of nitrate accumulation both in roots and in leaves was very low and reached 24 h after nitrate addition. However, at external nitrate concentrations ranging between 1 and 8 mM, the maximum level of nitrate accumulation was reached 48 h after nitrate addition (Figure 5A and B).

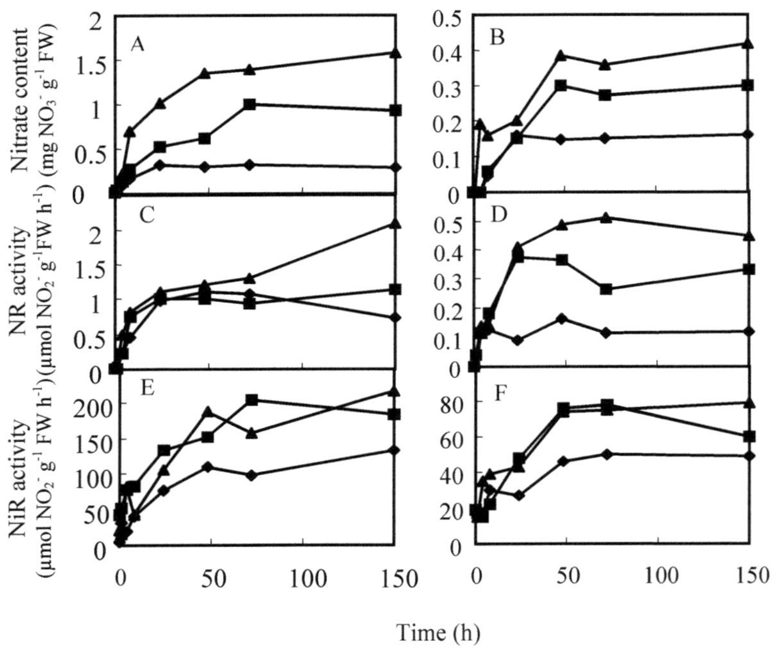

Figure 5. Induction of NR and NiR activities and nitrate accumulation by nitrate in leaves and roots from L japonicus *plants. Plants were grown in seed-trays in Hornum base medium supplemented with NH_4Cl (4 mM) as sole nitrogen source for 30 days. They were then transferred to hydroponic culture containing the same nutrient solution for 2 days. The assays were started by transfer of the plants to culture solutions containing 0.1 mM (diamonds), 1 mM (square) and 8 mM (triangles) KNO_3, respectively. NR and NiR activities and nitrate content were determined at the indicated times after transfer of the plants: Nitrate content in roots (A) and leaves (B); NR activity in roots (C) and leaves (D); NiR activity in roots (E) and leaves (F).*

Induction of NR activity showed a different pattern in roots in relation to leaves (Figure 5C and D). The levels of NR activity induction in roots were practically independent on the external nitrate concentration at the short-term, and only at longer times NR activity increased in relation to external nitrate concentration (Figure 5C). In leaves, the extent of the induction of NR activity was dependent on the external nitrate concentration both in the short and in the long-term (Figure 5D). The levels of NiR activity were similar at the three external nitrate concentrations assayed, although at 0.1 mM the level of activity was slightly lower (Figure 5 E and F). In contrast to NR activity, the kinetic of NiR activity induction was similar in roots and leaves (Figure 5E and F). In all cases, the maximum levels of nitrate accumulation and NR and NiR activity were higher in roots than in leaves (Figure 5).

NR and NiR protein levels were determined by immunoblot (Figure 6), in parallel to the previously studied induction of the enzyme activities. It is quite remarkable that ammonium-grown plants showed significant levels of both proteins, although NR activity was not detectable, and only basal levels of NiR activity could be found. After nitrate addition, the protein levels of both enzymes suffer a notable increase that was independent on external nitrate concentration (Figure 6).

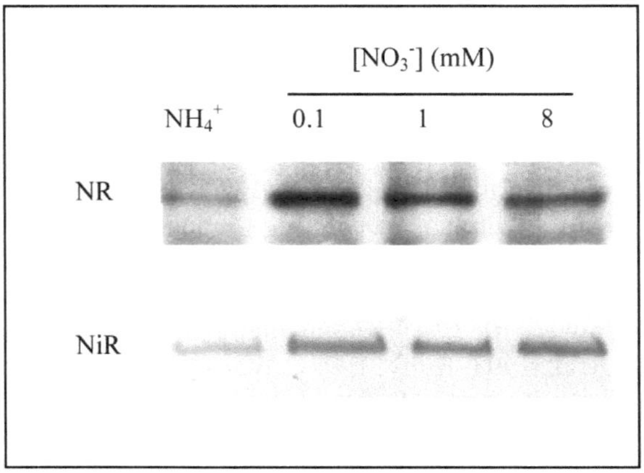

Figure 6. Induction of NR and NiR proteins by different nitrate concentrations in roots from L japonicus *plants. Plants were grown as described in Figure 5. Levels of NR and NiR proteins in root crude extracts from plants were analysed by western blot at zero time (NH_4^+ growing conditions) and 48 h after induction with KNO_3 (0.1, 1 and 8 mM). Other details as indicated in Experimental procedures.*

DISCUSSION

In this paper, we have characterized different aspects of the nitrate assimilatory process in *L japonicus* grown under diverse nitrogen supplies.

Firstly, we studied *L japonicus* growth in different nitrogen sources. In plants grown on nitrate as nitrogen source, the optimum growth was obtained at concentrations ranging between 4 and 8 mM. Concentrations higher than 20 mM were toxic and inhibited the germination of seeds. Nitrate toxicity is a well-known phenomenon in plants and resistance to high nitrate concentrations has been proposed as a putative method to isolate mutants affected in nitrate assimilation (Pelsy and Caboche, 1992). Debeys *et al.* (1990) reported the isolation of *Arabidopsis thaliana* mutants resistant to high nitrate concentrations.

L japonicus plants grown with ammonium nitrate showed a higher growth rate (for the aerial part of the plants) than plants cultivated with only nitrate or ammonium. Besides, the levels of NR and NiR activity as well as the level of nitrate accumulation were also higher. These results are in accordance with previously published results for other plants indicating that the presence of both nitrogen sources induce a better growth and higher levels of NR activity (Pilbeam and Kirby, 1992; Oaks, 1992). In barley, both the production of dry matter and the nitrogen content in plants cultivated with 1 mM ammonium nitrate were higher than in the presence of 2 mM nitrate or 2 mM ammonium (Lewis and Chadwick, 1983). The same happens in wheat, where the dry weight increase by 50% when the plants are cultivated with both nitrogen sources (Cox and Reisenauer, 1973). In maize, NR activity in roots and leaves is also higher when cultivated with ammonium nitrate than with nitrate as the sole nitrogen source (Oaks, 1992). On the other hand, and differently to other plants as tobacco (Pelsy *et al.*, 1991), *L japonicus* can grow with ammonium as the sole nitrogen source without being necessary the supply of organic acids like succinate, showing a similar growth rate independently on the ammonium salt used in the culture medium.

When we studied the ratio of shoot to root fresh weight in different nitrogen nutrition conditions, we observed that at a low concentration of KNO_3 (0.1 mM) the ratio was 3-fold lower than for plants cultivated with 5 mM KNO_3. This difference is due to the fact that low nitrate concentrations induce a higher root growth and a lower shoot growth. The inhibition of the aerial growth and the stimulation of root growth in the presence of limiting nitrate concentrations have also been described for other plants (Brower, 1962; Agren and Ingestad, 1987; Fichtner and Schulze, 1992; Wagner and Beck, 1993; Rufty, 1997). When the nitrogen supply is adequate, the growth of the aerial part of the plant is enhanced, providing a higher photosynthetic capacity to the plant. However, when nitrogen is limiting, root growth provide a better capacity to acquire nutrients from the soil (Bloom at el., 1985; Schulze and Chapin, 1987). The mechanisms controlling these responses are unknown. Brower (1962, 1963) proposed that root growth is favoured under these conditions since most of the absorbed nitrogen is retained and assimilated in the roots. A flow of amino acids takes then place between roots and leaves mediated by the xylem and the phloem (Cooper and Clarkson, 1989; Komor, 1994). This transport of amino acids is not affected at limiting concentrations of nitrogen (Agrell *et al.*, 1994; Ericcson, 1995), thus indicating that the relative growth of the roots is not controlled by the percentage of nitrogen assimilated in the roots. When the nitrogen supply is increased, a large number of metabolic and physiological changes take place in the plant (Rufty, 1997), making difficult to determine the signals controlling the relative growth of each part of the plant. Studies carried out with mutants and transgenic plants showing reduced levels of NR expression have shown a correlation between nitrate accumulated in leaves and the distribution of growth between the aerial part of the plant and the root (Scheible *et al.*, 1997). These plants accumulate more nitrate in leaves and show a higher shoot/root ratio than the wild type.

It has also been shown that the increase in the level of nitrate accumulation in leaves does not produce a decrease in the content of amino acids or proteins in the roots, but a decrease in starch content in leaves. This produces a decrease in the levels of sugar in the roots, existing a correlation between root growth and sugar content (Scheible *et a.*, 1997). Based in this data, it has been proposed that alterations in carbon partitioning could contribute to the changes in growth between shoot and root, as it has also been proposed by other authors (Jackson *et al.*, 1980; Huber, 1983; Rufty *et al.*, 1988; Farrar and Williams, 1991; Chu *et al.*, 1992). Significant emphasis has been recently made in the role of nitrate as a signal for root growth (Forde, 2002; Stitt, 1999). The data presented for *L japonicus* show that nitrate accumulation in leaves increase when external nitrate is also increased. On the other hand, at low nitrate concentrations root growth is favoured, being possible that in *L japonicus* the growth of the root could also be controlled, directly or indirectly, by the nitrate content of the leaves.

It is well established that an increase in nitrate availability induces an increase in nitrate transport and assimilation (Redinbaugh and Campbell, 1991; Hoff *et al.*, 1994; Crawford, 1995; Daniel-Vedele and Caboche, 1996). The partitioning of nitrate assimilation between roots and shoots changes as a result, with increased assimilation in the aerial part of the plant (Smirnoff and Stewart, 1985; Wallace, 1986; Andrews, 1986; Gojon *et al.*, 1994; Peuke *et al.*, 1996). This behaviour has been proposed for temperate legumes. Temperate legumes assimilate nitrate mainly in roots at low nitrate concentrations (1 mM), but when the external nitrate concentration increase (1-20 mM), nitrate assimilation capability in shoots become gradually more significant. A shift in the partitioning of nitrate assimilation can take place at high nitrate concentrations, under which leaves become the main tissue for nitrate assimilation (Andrews, 1986).

We have studied nitrate accumulation and the level of NR and NiR activity and proteins in *L japonicus* plants grown in the presence of increasing nitrate concentrations. According to the data presented above, *L japonicus* carries out nitrate assimilation mainly in roots in a manner similar to that been proposed for temperate legumes (Andrews, 1986; Woodall and Forde, 1996). However, *L japonicus* differs from other temperate legumes in that the global partitioning of nitrate assimilation among roots and shoots does not markedly shift under high nitrate conditions. Although nitrate content and NR and NiR enzyme activities increase in leaves, stems, and roots following increases in media nitrate concentration, assimilation occurs predominantly in roots. This situation has also been shown in trees (Gojon *et al.*, 1994), for which the low levels of NR activity detected in leaves at high nitrate concentrations are due to a low flux of nitrate from roots to leaves, in such a way that no nitrate accumulation can be detected. However, in *L japonicus* the levels of nitrate accumulation in illuminated leaves are quite high and comparable to the levels in roots, although NR activity is 10-times lower. This indicates that NR activity is not dependent on the level of nitrate content of the leaves. NR activity has been proposed to be dependent on the cytosolic concentration of nitrate and not on the vacuolar nitrate content (Clarkson, 1986). With our data we can not exclude that, although nitrate content seems to be very high in leaves, a low nitrate concentration could be present in the cytoplasm, thus explaining the low NR activity detected in this organ. Alternatively, other NR regulatory signals, in addition to nitrate, must be operating in *L japonicus* leaves. It is unlikely that the low levels of NR activity detected in this organ could be due the currently known mechanisms of NR post-translational regulation by phosphorylation and binding of 14-3-3 proteins, since maximal NR activity measurements (in the presence of EDTA and lack of exogenously added Mg^{2+}) were carried out (see Materials and Methods). In addition, NR activity following this regulatory mechanism gets maximal levels in illuminated leaves (Kaiser and Brendle-Behnisch, 1991; Campbell, 2002), the same type of tissue used for our studies.

We have also studied the induction by nitrate of nitrate accumulation, and NR and NiR enzyme activities, in roots and leaves. Plants grown with ammonium as a nitrogen source completely lack of detectable NR activity either in roots or in leaves of the plant, although NR protein is clearly visible by western blot (Figure 6). Although a nitrification inhibitor (dicyandiamide) was present in the NH_4^+-growing media, it is still possible that traces of nitrate may exist in this medium, far beyond our detection limits (<2 μM). A post-translational regulatory mechanism is proposed by which the absolute lack or trace amounts of nitrate would be sufficient to get a semi-constitutive or partially induced synthesis of NR protein, but higher nitrate concentrations are really necessary to get fully induced active NR (Figure 7).

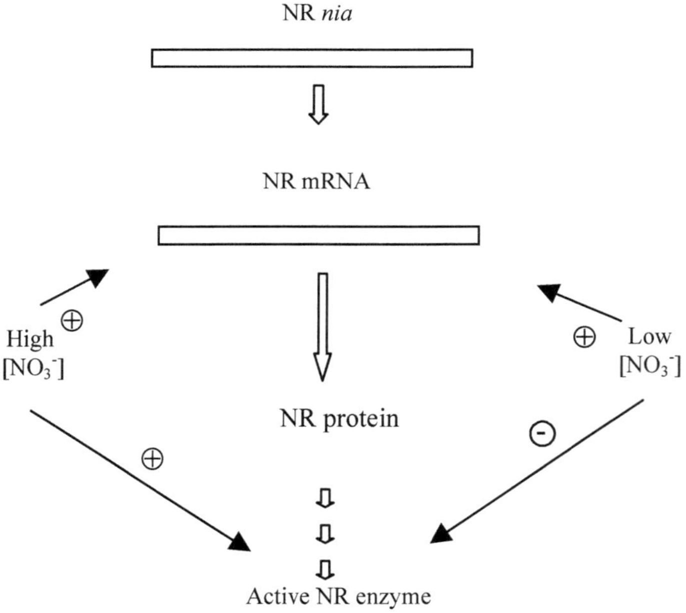

Figure 7. Differential regulation of NR protein synthesis and NR activity induction in L japonicus. *The presence of the NR polypeptide in* NH_4^+*-grown plants (Figure 6) suggests that the absence of nitrate or trace amounts of it (less than 2 μM) permit partially-induced NR protein synthesis, while higher concentrations of nitrate are essentially required to get detectable levels of NR enzyme activity.*

Contrarily to NR, low activity and protein levels of NiR can be detected in ammonium grown plants. The presence of basal but significant levels of NiR activity has also been observed in spinach (Back *et al.*, 1988), maize (Kramer *et al.*, 1989), and tobacco (Neininger *et al.*, 1992), and in lower-eukaryots (Pajuelo *et al.*, 1995). This low constitutive levels of NiR in the absence of exogenously added nitrate could be a protective mechanism addressed to metabolise any form of nitrite that could be produced, which is very toxic for the plant.

Nitrate induction of NR activity in roots was found to be independent on the external concentration of nitrate on the short term. Only 6 days after nitrate addition a dependence on external nitrate concentration could be observed. In contrast, a clear nitrate concentration dependence was observed in the induction of NR activity in leaves. Nitrate is transported by the cortical or epidermal cell of the roots and can be assimilated in the same root, accumulated in the root vacuoles or be transported via xylem to the leaves, where nitrate can be assimilated or accumulated in the vacuoles (Crawford, 1995; Daniel-Vedele and Caboche, 1996). The capacity of leaf nitrate assimilation is dependent on the amount of nitrate transported to the leaves (Shaner and Boyer, 1976; Beavers and Hageman, 1980; Gojon *et al.*, 1991). When induction of nitrate assimilation was performed using 0.1 mM of NO_3^-, very low nitrate accumulation was detected in leaves and roots, indicating that most transported nitrate is assimilated in the roots, and that the low NR activity in leaves could be the result of a very low leaf nitrate concentration. At 1 and 8 mM external nitrate concentrations a higher level of nitrate, as well as NR activity were detected in leaves, although they were never higher than in roots. In the case of NiR the induction kinetics were similar in roots and leaves, with similar levels of induction in both tissues, being slightly lower at 0.1 mM external nitrate. These results would indicate that NiR in *L japonicus* leaves responds with more sensitivity than NR to nitrate induction at low external nitrate concentrations.

Experimental procedures

Plant material and growth conditions

Seeds of *Lotus japonicus* cv. Gifu (GIFU B-129-S9) were obtained from Prof Jens Stougaard (University of Aarhus, Denmark) and then propagated by self-fertilization. Unless otherwise stated, plants were grown in a Sanyo SGR193.S26A growth chamber and at 20/18 °C light/dark temperatures and 70% humidity under a 16 h/8 h light/dark regime. Light intensity at the plant level was around 200 μmol m^{-2} s^{-1}, supplied by white fluorescent Philips colour 83 lamps combined with tungsten lamps to provide red supplement and dawn-dusk effects. Seeds were first scarified by a gentle treatment with Emery paper and then sterilized in 75% (v/v) ethanol for 5 min and 0.5% (w/v) hypochlorite for 20 min, followed by several washes with sterile distilled water for one day. Germination and growth were carried out in seed trays using vermiculite, as solid support, or in Magenta box using phytagel or vermiculite, as inert support. Five seeds were planted in each well of the seed tray or Magenta box. Hornum medium was used as the nutrient solution for the Magenta box and for daily irrigation of seed trays, as described by Handberg and Stougaard (1992). This medium without inorganic nitrogen (without 3 mM KNO_3

and without 5 mM NH_4NO_3) was named Hornum base medium and the nitrogen source or other supplements were added as indicated. Plant material was collected at the indicated times. Whole roots, leaves, and stems from plants were quickly frozen in liquid nitrogen, homogenised and then kept at –80°C until use.

Enzyme activity assays

Roots, leaves or stems material (1 g) was ground in a mortar at 4 °C and resuspended in extraction buffer (6 ml). Extraction buffer contained: 100 mM Tris-HCl pH 8, 10 µM FAD, 1 mM EDTA, 1 mM dithiothreitol (DTT), 1 mM phenylmethylsulphonylfluoride (PMSF), 3% (w/v) bovine serum albumin (BSA) and 0.1% (v/v) Triton X-100. The homogenates were centrifuged for 15 min at 15,000 x *g* and insoluble material was removed. The clear supernatants were immediately assayed for NR (Pajuelo *et al.*, 2002) or NiR (Pajuelo *et al.*, 1993) activities.

NR activity assay was carried out as follows: 500 µl final volume reaction mixtures were placed in 1.5 ml eppendorf tubes containing: 50 µl of 0.5 M potassium phosphate buffer, pH 7.5; 5 µl of 100 mM EDTA; 50 µl 100 mM KNO_3; 195 µl distilled water and 150 µl crude extract. The reaction was started by adding 50 µl of 0.7 mM NADH and carried out at 30°C for 30 min. Nitrite formation was measured as described elsewhere (de la Rosa *et al.* 1982), using sulphanilamide and N-(1-naphtyl)-ethylenediamine dihydrochloride reagents. Nitrite formation rate was verified to be linear with time during assay. Blank tubes were processed as samples except that the reaction was stopped at zero time.

NiR reaction mixtures (final volume 1 ml) contained the following: 0.5 ml of assay buffer (0.2 M Tris-HCl, pH 8,0 plus 6 mM KNO_2 plus 20 mM methyl viologen), the adequate amount of extract containing 0,4-2 nkat of enzyme, and 15 mg Na-dithionite dissolved in 0.1 ml of 0.3 M Na-bicarbonate. Reaction mixtures were incubated at 40ºC for 20 min, after which the reaction was stopped by vigorous stirring to oxidize dithionite. 30 µl of the reaction mixture was diluted to 3 ml with distilled water and used to analytically determine consumed nitrite. 1 ml each of the reagents sulphanilamide and ethylendiamine dihydrochloride were added, as for NR. Dithionite-free reactions were used as control.

Nitrate content

Colorimetric determinations of nitrate content in root, leaf, or stem crude extracts were performed as described by Cataldo *et al.* (1975), with slight modifications (Pajuelo *et al.*, 2002): 200 µl of 5% (w/v) salicylic acid dissolved in 96% (w/v) sulphuric acid was added to aliquots of 50 µl from the root crude extracts and left to react for 20 min. 4.75 ml of 2 M NaOH were added to the reaction mixtures and then the absorbance was read at 405 nm after cooling. A calibration curve of known amounts of nitrate dissolved in the standard extraction buffer was used for analytical determinations. Blanks were set up without salicylic acid.

Western blots

Immunodetection of NR and NiR proteins was carried out as described previously by Pajuelo *et al.* (2002) and Orea *et al.* (2001), respectively. Root or leaf NR and NiR proteins were analysed in crude extracts obtained in 2 ml g^{-1} FW of the corresponding extraction buffer described above (without BSA). 300-500 μg and 15- 20 μg of total protein (determined according to Bradford, 1976) were analysed by electrophoresis on 10% and 12% (w/v) SDS-polyacrylamide gels (Laemmli, 1970) for NR and NiR, respectively, and transferred onto polyvinylidene difluoride (PVDF) sheets (Immobilon P, Millipore) as described by Pajuelo *et al.* (1995). After transfer, filters were rinsed with TBS (tris-buffered saline, 20 mM Tris-HCl, pH 7.5; 0.5 mM NaCl) and then saturated with blocking solution [5% (w/v) skimmed dry milk powder in TBS] for 16 to 20 h at 4 °C. Anti-peptide-NR antiserum (1:250 dilution) (Pajuelo *et al.* 2002) or anti- spinach NiR antibodies (3.75 μg ml^{-1} final dilution) (kindly provided by Dr Shoji Ida, University of Kyoto, Japan) were added to the blocking solution, being adsorbed for 1.5 h at room temperature and then exhaustively washed in TTBS [0.1% (v/v) Tween 20 in TBS] (for NR protein) or TBS (for NiR protein). After a short treatment for 10 min in blocking solution, the filters were incubated for 1 h in a 1:1000 dilution of the goat anti-rabbit peroxidase conjugate (for NR immunodetection) or goat anti-rabbit alkaline phosphatase conjugate (for NiR immunodetection) (Sigma). The excess of secondary antibodies was eliminated by successive washes in TTBS or TBS. In the case of NR protein after being incubated in TBS for 10 min, the filters were immersed for 2 min in 10 ml of the detection reagents mixture from the ECL immunodetection system (ECL Western Blotting Protocols, Amersham), freshly prepared according to the manufacturer instructions. For NiR immunodetection, filters were pre-equilibrated in alkaline-phosphatase buffer (0.2 M Tris-HCl, pH 9.5, 10 mM $MgCl_2$) and the cross-reacting bands were visualised by reaction of filters in the presence of 17 mg nitrobluetetrazolium and 7 mg 5-bromo-4-chloro-3-indolyl-phosphate dissolved in 50 ml alkaline phosphatase buffer. The reaction was stopped by washing the filters with water.

ACKNOWLEDGEMENTS

Authors wish to thank funding from research projects BMC2001-3162 and BFU2004-02753 from MCYT/MEC-FEDER (Spain), and HPRN-CT2000-00086 and MRTN-CT-2003-505227 from European Union, as well as support given by Junta de Andalucía to group CVI-163. Technical assistance of MJ Cubas is also gratefully acknowledged.

REFERENCES

Andrews M. (1986b) **The partitioning of nitrate assimilation between root and shoot of higher plants.** *Plant, Cell and Environment* 9, 511-586.

Agrell D, Oscarson, P. and Larsson C-M (1994) **Translocation of N to and from barley roots: its dependence on local nitrate supply in split-root cultures.** *Physiologia Plantarum* 90, 467-474.

Agren GI and Ingestad T. (1987) **Root:shoot is a balance between nitrogen productivity and photosynthesis**. *Plant, Cell and Environment*, 10, 579-586.

Back E, Burkhardt W, Moyer M, Privalle L and Rothstein S. (1988) **Isolation of cDNA clones coding for nitrite reductase: complete sequence and nitrate induction**. *Molecular General Genetics* 212, 20-26.

Beavers L, and Hageman RH. (1980) **Nitrate and nitrite reduction**. In: *The Biochemistry of Plants* (B.J. Miflin, Ed). New York: Academic Press, vol. 5, pp. 115-168.

Bloom AJ, Chapin FS, and Mooney HA. (1985) **Resource limitation in plants-an economic analogy**. *Annual Review of Ecology and. Systematics* 16, 363-392.

Bradford MM. (1976) **A rapid and sensitive method for quantitation of microgram quantities of protein utilizing the principle of protein-dye binding**. *Analytical Biochemistry* 72, 248-254.

Brower R. (1962) **Nutrient influences on the distribution of the dry matter in the plant**. *Nhetherlands Joural of Agricultural Science* 10, 399-408.

Brower R. (1963) **Some aspect of the equilibrium between overground and underground plant parts**. *Jaarbook Instituut Biologisch Scheikundig Onderzoeg*, pp. 31-39.

Caba JM, Lluch C, and Ligero F. (1995) **Distribution of nitrate reductase activity in *Vicia faba*: Effect of nitrate supplies**. *Physiologia Plantarum* 93, 667-672.

Campbell WH. (2002) **Molecular control of nitrate reductase and other enzymes involved in nitrate assimilation**. In: *Photosynthetic nitrogen assimilation and associated carbon and respiratory metabolism* (Foyer CH, and Noctor G, Eds.), Kluwer Academic Publishers, Dordrecht, pp.35-48.

Cataldo DA, Haroon M, Schrader LE, and Youngs VL (1975) **Rapid colorimetric determination of nitrate in plant tissue by nitration of salicylic acid**. *Communications in. Soil Science and Plant Analysis* 6, 71-80.

Chu CC, Coleman JS, and Mooney HA. (1992) **Control of biomass partitioning between roots and shoots: atmospheric carbon dioxide enrichment and the acquisition and allocation of carbon and nitrogen in wild radish**. *Oecologia* 89, 580-587.

Clarkson DT. (1986) **Regulation of the absorption and release of nitrate by plant cells: A review of current ideas and methodology**. In: *Fundamental, Ecological and Agricultural Aspects of Nitrogen Metabolismin Higher Plants* (Lambers H, Neeteson JJ, Stulen I, Eds.). Dordrecht: Martinus Nijhoff Publisher, pp. 3-27.

Colebatch G, Trebaskis B, and Udvardi M. (2002) **Symbiotic nitrogen fixation in the postgenomics era**. *New Phytologist* 153, 37-42.

Cooper HD, and Clarkson DT. (1989) **Cycling of amino-nitrogen and other nutrients between shoots and roots in cereals: a possible mechanism integrating shoot and root in the regulation of nutrient uptake**. *Journal of Experimental Botany* 40, 753-762.

Cox WJ and Reisenauer HM. (1973) **Growth and ion uptake by wheat supplied nitrogen as nitrate, or ammonium, or both**. *Plant and Soil* 38, 363-380.

Crawford NM. (1995) **Nitrate: nutrient and signal for plant growth**. *Plant Cell* 7, 859-868.

Daniel-Vedele F and Caboche M. (1996) **Molecular analysis of nitrate assimilation in higher plants**. *Comptes Rendus de l'Académie des Sciences de Paris , Sciences de la vi* 319, 961-968.

Debeys D, Cammaerts D, and Jacobs M. (1990) **Isolation of mutants resistant to high nitrate concentrations**. In: *Abstracts International Symposium on Nitrate Assimilation, Molecular and Genetics Aspects,* Bombannes, INRA-Versalles, France, pp. 118.

De la Rosa MA, Márquez AJ, and Vega JM. (1982) **Dissociation of FAD from NADH-nitrate reductase complex from *Ankistrodesmus braunii* and role of flavin in catalysis**. *Zeitschrift fur Naturforschung C* 37, 24-30.

Ericsson T. (1995) **Growth and shoot:root allocation of seedling in relation to nutrient availability**. *Plant and Soi*, 168, 205-214.

Farrar JF, and Williams JHH. (1991) **Control of rate of respiration in roots: compartmentation, demand and supply of substrate**. In: *Compartmentation of Plant Metabolism in Non-photosynthetic Tissues* (Emes MJ, Ed.). New York: Cambridge University Press, pp. 167-188.

Fichtner K, and Schulze E-D. (1992) **The effect of nitrogen nutrition on annuals originating from habitats of different nitrogen availability**. *Oecologia*, 92, 236-241.

Forde BG. (2002) **Local and long-range signalling pathways regulating plant responses to nitrate**. *Annual Review of Plant Biology* 53, 203-224.

Forde BG. and Clarkson DT. (1999) **Nitrate and ammonium nutrition of plants: physiological and molecular perspectives**. In: *Advances in Botanical Research, Vol. 30*, Academic Press, New York, pp. 1-90.

Gojon A, Bussi C, Grignon C, and Salsac L. (1991) **Distribution of NO_3^- reduction between roots and shoots of peach tree seedlings as affected by NO_3^- uptake rate**. *Physiologia Plantarum* 82, 502-512.

Gojon A, Plassard C, and Bussi C. (1994) **Root/shoot distribution of NO_3^- assimilation in herbaceous and woody species**. In: *A Whole Plant Perpective on Carbon-Nitrogen Interaction* (Roy J, and Garnier E, Eds.). The Hague: SPB Academic Publishing, pp. 131-147.

Heldt HW. (1997) **Nitrate assimilation**. In: *Plant Biochemestry and molecular Biology* (Heldt, HW, Ed.). New York: Oxford University Press, pp. 247-276.

Hirel B., and Lea PJ. (2002) **The Biochemistry, Molecular Biology, and Genetic manipulation of primary ammonia assimilation**. In: *Photosynthetic nitrogen assimilation and associated carbon and respiratory metabolism.*(Foyer CH, and Noctor G, Eds.), Kluwer Academic Publishers, Dordrecht, pp. 71-92.

Hoff T, Troung H-N., and Caboche M. (1994) **The use of mutants and transgenic plants to study nitrate assimilation**. *Plant Cell and Evironment* 17, 489-506.

Huber, SC. (1983) **Relation between photosynthetic starch formation and dry-weight partitioning between the shoot and root**. *Canadian Journal of Botany* 61, 2709-2716.

Ireland RJ., and Lea PJ. (1999) **The enzymes of glutamine, glutamate, asparagine, and aspartate metabolism**. In: *Plant Amino Acids: Biochemistry and Biotechnology*. (Singh, BK, Ed.). New York: Marcel Dekker, Inc, pp. 49-109.

Jackson WA, Volk RJ, and Israel DW. (1980) **Energy supply and nitrate assimilation in root systems**. In: *Carbon-Nitrogen Interaction in Crop Production* (Tanaka, Ed.). Tokyo: Japan Society for Promotion of Science, pp. 25-40.

Kaiser WM, and Brendle-Behnisch E. (1991) **Rapid modulation of spinach leaf nitrate reductase activity by photosynthesis. I. Modulation *in vivo* by CO_2 availability**. *Plant Physiology* 96, 363-367.

Komor E. (1994) **Regulation of futile cycles: the transport of carbon and nitrogen in plants**. In: *Flux Control in Biological System* (Schulze, E-D, Ed). San Diego, CA: Academic Press, pp. 153-202.

Kramer V, Lahners K, Back E, Privalle LS, and Rothstein S. (1989) **Transient accumulation of nitrite reductase mRNA in maiz following the addition of nitrate**. *Plant Physiology* 90 1214-1220.

Laemmli UK. (1970) **Cleavage of structural proteins during the assembly of the head of bacteriophage T4.** *Nature* 251, 614-616.

Lewis OAM, and Chadwick S. (1983) **A ^{15}N investigation into nitrogen assimilation in hydroponically growth barley (*Hordeum vulgare* L cv. Clipper) in response to nitrate, ammonium and mixed nitrate and ammonium nutrition.** *New Phytologist* 95, 653-646.

Neininger A, Kronenberger J, and Mohr H. (1992) **Co-action of light, nitrate, and a plastidic factor in controlling nitrite-reductase gene expression in tobacco.** *Planta* 187, 381-387.

Oaks A. (1992) **A re-evaluation of nitrogen assimilation in roots.** *Bioscience* 42, 103-111.

Oostinder-Braaksma FJ, and Feenstra WJ. (1973) **Isolation and characterization of chlorate-resistant mutants of *Arabidopsis thaliana*.** *Mutation Research* 19, 175-185.

Orea A, Pajuelo P, Pajuelo E, Márquez AJ and Romero JM (2001) **Characterisation and expression studies of a root cDNA encoding for ferredoxin-nitrite reductase from *Lotus japonicus*.** *Physiologia Plantarum* 113,193-202.

Pajuelo E, Borrero, JA, and Márquez AJ (1993) **Immunological approach to subunit composition of ferredoxin-nitrite reductase from *Chamydomonas rein*hardtii.** *Plant Science* 95, 9-21.

Pajuelo E, Pajuelo P, Clemente MT and Márquez AJ (1995) **Regulation of expression of ferredoxin-nitrite reductase in synchronous culture of *Chlamydomonas reinhardtii*.** *Biochimica et Biophysica Acta*, 1249, 72-78.

Pajuelo P, Pajuelo E, Orea A, Romero JM, and Márquez AJ (2002) **Influence of plant age and growth conditions on nitrate assimilation in roots of *Lotus japonicus* plants.** *Functional. Plant Biology* 29, 485-494.

Pelsy F. and Caboche M. (1992) **Molecular genetics of nitrate reductase in higher plants.** *Advances in Genetics* 30, 1-40.

Pelsy F, Kronenberger J, and Caboche M. (1991) **M2 seed screening for nitrate reductase deficiency in *Nicotiana plumbaginifolia*.** *Plant Science* 76, 109-114.

Peuke AD, Glaab J, Kaiser, WM and Jeschlke WD. (1996) **The uptake and flow of C and n and ions between roots and shoots in *Ricinus communis* L IV. Flow and metabolism of inorganic nitrogen and malate depending on nitrogen nutrient and salt treatment.** *Journal of Experimental Botany* 47, 377-385.

Pilbeam DJ, and Kirby EA. (1992) **Some aspects of the utilization of nitrate and ammonium by plants.** In: *Nitrogen Metabolism of Plants* (Mengel K, and Pilbeam, DJ, Eds.). Oxford: Oxford Science Publications, pp. 55-70.

Radutoiu S, Madsen LH, Madsen EB, Felle HH, Umehara Y, Gronlund M, Sato S, Nakamuro Y, Tabata S, Sandal N, and Stougaard J. (2003) **Plant recognition of symbiotic bacteria requires two LysM receptor-like kinases.** *Nature* 425, 585-591.

Redinbaugh MG, and Campbell WM. (1991) **Higher plant responses to environmental nitrate.** *Physiologia Plantarum* 82, 640-650.

Rufty T.W. (1997) **Probing the carbon and nitrogen interaction: a whole plant perspective.** In: *A Molecular Approach to Prymary Metabolism in Higher Plants* (Foyer CH and Quick WP, Eds.). London: Taylor & Francis, pp. 255-273.

Rufty TW, Huber, SC, and Volk, RJ (1988) **Alterations in leaf carbohydrate metabolism in response of nitrogen stress.** *Plant Physiology* 88, 725-730.

Scheible W-R, Lauerer M, Schulze E-D, Caboche M, and Stitt M. (1997) **Accumulation of nitrate in the shoot acts as a signal to regulate shoo-root allocation in tobacco.** *The Plant Journal* 11, 671-691.

Schulze E-D, and Chapin FS. (1987) **Plant specialization to environments of different resource availability**. In: *Potential and Limitations of Ecosystem Analysis, Ecological Studies* (E.-D. Schulze and H. Zwölfer, Eds.). Berlin: Springer-Verlag, vol. 61, pp. 120-148.

Shaner DL, and Boyer J.S. (1976) **Nitrate reductase in maiz (*Zea mays* L) leaves**. I. Regulation by nitrate flux. *Plant Physiology* 58, 499-504.

Smirnoff N, and Stewart GR. (1985) **Nitration assimilation and translocation by higher plants: Comparative physiology and ecological consequences**. *Physiologia Plantarum* 64, 133-140.

Stitt M. (1999) **Nitrate regulation of metabolism and growth**. *Current Opinion in. Plant Biology* 2, 178-86.

Stougaard J, Szczyglowski, K, de Brujin FJ, and Parniske M. (1999) **Genetic nomenclature guidelines for the model legume *Lotus japonicus***. *Trends in Plant Science* 4, 300-301.

Szczyglowski, K, Hamburger D, Kapronov P, and Brujin FJ. (1997) **Construction of a *Lotus japonicus* late nodulin expressed sequence tag library and identification of novel nodule specific genes**. *Plant Physiology* 114, 1335-1346.

Wagner BM, and Beck E. (1993) **Cytokinins in the perennial herb *Urtica dioca* L as influenced by its nitrogen status**. *Planta* 190, 511-518.

Wallace W. (1986) **Distribution of nitrate assimilation between the root and shoot of legumes and a comparison with wheat**. *Physiologia Plantarum* 66, 630-636.

Woodall J. (1994) **Studies of glutamine synthetase in *Lotus japonicus* species and characterization of the occurrence of root GS2 in other legumes**. PhD thesis, University College, London.

Woodall J and Forde BG. (1996) **Glutamine synthetase polypeptide in the roots of 55 legume species in relation to their climatic origin and partitioning of nitrate assimilation**. *Plant Cell and Environ,* 19, 848-858.

Chapter 7.2

NITRATE AND AMMONIUM ASSIMILATORY ENZYMES

Antonio J Márquez[1†], Marco Betti[1], Margarita García-Calderón[1], Guillermo Estivill[1], Alfredo Credali[1], Purificación Pajuelo[1,3], Alicia Orea[1,2], M Teresa Clemente[1], Eloísa Pajuelo[1,4], and Francisco Galván[1]

[1]*Departamento de Bioquímica Vegetal y Biología Molecular, Facultad de Química, Universidad de Sevilla; Apartado 553, 41080-Sevilla SPAIN*; [2]*Instituto de Bioquímica Vegetal y Fotosíntesis, Centro de Investigaciones Científicas Isla de la Cartuja, Avda. Américo Vespucio, 49, 41092-Sevilla, SPAIN.* [3]*Present address: Escuela Universitaria de Ingeniería Técnica Agrícola; Carretera de Utrera, Km. 1; 41012-Sevilla, SPAIN.* [4]*Present address: Departamento de Microbiología y Parasitología, Facultad de Farmacia, C/ Profesor García González s/n, 41012-Sevilla, SPAIN.* [†]*Corresponding author.*

Email: cabeza@us.es

Phone: +34 954 557 145 Fax: +34 954 626 853

Keywords: Nitrate reductase, nitrite reductase, glutamine synthetase, glutamate synthase

Protocols for extraction and assay of the key enzymes of nitrate and ammonium assimilation in Lotus japonicus *are given in this chapter, together with the peculiarities observed for these assays in different organs and physiological situations of the plant.*

INTRODUCTION

The major source of inorganic nitrogen available for plants is a mixture of nitrate and ammonium, with nitrate being the predominant form in well-aerated soils because of nitrification. The nitrate present in the soils is then taken up by the roots of the plant and converted to ammonium by the sequential reductive action of the enzymes nitrate reductase (NR) and nitrite reductase (NiR). Some plants, most notably legumes, can also obtain nitrogen from atmospheric N_2. The symbiotic interaction of these plants with rhizobia, which are able in the nodules to reduce N_2 to NH_4^+ by the action of nitrogenase, makes it possible that ammonium is transferred from the microbe to the plant. Despite the ability of legumes to form nitrogen-fixing symbiosis with *Rhizobium* and *Bradyrhizobium* spp. there are many reasons why nodulation may not occur, reason why combined forms of inorganic nitrogen are

A.J. Márquez (ed.), Lotus japonicus Handbook, 315-328.

important nitrogen sources for legumes as they are for non-legumes. Plants also produce significant amounts of ammonium from photorespiration, phenylpropanoid biosynthesis and amino acid catabolism. Consequently, the reduced form of inorganic nitrogen ultimately available for the plants is ammonium, irrespective of the primary nitrogen source utilized. Assimilation of ammonium produced from external nitrogen sources is commonly described as primary ammonium assimilation, to be distinguished from other forms of ammonium released endogenously in the plants, which is known as secondary ammonium assimilation. Glutamine synthetase (GS) and glutamate synthases (GOGAT) are the key enzymes responsible for all forms of ammonium assimilation that take place in higher plants. A set of different isoforms of these enzymes, having specific patterns of expression in several tissues, is a distinctive peculiarity in higher plants. Glutamine and glutamate are the first organonitrogen compunds produced as a result of the GS-GOGAT biosynthetic pathway. Nitrogen is then redistributed from these two compounds to the rest of N-containing metabolites and macromolecules (Forde and Clarkson, 1999; Hirel and Lea, 2002).

The proportion of nitrate that can be assimilated in the shoot and in the root of different plant species varies considerably (Andrews, 1986). Three main groups of plants exist in this respect: a) those that assimilate nitrate significantly both in shoots and roots (temperate annual non-legume species) ; b) those that assimilate nitrate preferentially in leaves (tropical and subtropical legume and non-legume species); c) those that assimilate nitrate preferentially in roots (temperate perennial and annual legume species). Many temperate legumes, growing in non-agricultural soils with low nitrate concentrations (1 mM) use their roots to assimilate most of the nitrate they absorb, although shoot assimilation becomes more important at higher external concentrations. Other legume species, particularly amongst the tropical and subtropical legumes, have been found to assimilate most of their nitrate in the shoot, regardless of the external nitrate concentration. The model legume *Lotus japonicus*, as a temperate legume, is a good example of a plant that assimilates nitrate predominantly in the roots (Orea *et al.*, 2001; Pajuelo et al. 2002; Márquez *et al.*, 2004; and chapters 7.1 and 7.3 in this book). Measurements of enzyme activities following the protocols described in this chapter should reflect this behaviour. We describe here the routine *in vitro* assays commonly used in our laboratory for the main N-assimilatory enzymes involved in the assimilation of nitrate, and primary and secondary ammonium assimilation in *L japonicus*. Other methodological reviews describe more in deep the assays for these enzymes (Wray and Fido, 1990; Lea *et al.,* 1990).

Nitrate reductase

NR catalyses the first enzymatic, and possibly the rate-controlling step of the nitrate assimilation pathway, involving the two-electron reduction of nitrate to nitrite. The enzyme from plants is typically cytosolic and uses pyridine nucleotides as physiological electron donor according to the following reaction:

$$NO_3^- + NAD(P)H + H^+ \rightarrow NO_2^- + NAD(P)^+ + H_2O$$

Different types of NADH-specific (EC 1.6.6.1) and NAD(P)H-bispecific (EC 1.6.6.2) isoenzymes have been reported in the literature. In *L japonicus* leaf or root extracts the NR activity uses predominantly NADH as reductant, since no significant activity can be obtained using NADPH. The enzyme is encoded by a single *Nia* gene that has been sequenced (accession number X80670) and is predominantly expressed in roots (Prosser *et al.,* manuscript in preparation; Pajuelo *et al.*, 2002). Anti-peptide antibodies addressed against the deduced *L japonicus* NR sequence (Pajuelo *et al.,* 2002) were able to recognise on SDS-PAGE a band of around 100 kDa, the size expected for plant NR polypeptide chains.

Crude extract preparation and NR enzyme assay

1. Grind the plant material to a fine powder in a mortar in the presence of liquid nitrogen.
2. Transfer the frozen powder to cold Eppendorf tubes, thaw and homogenize with a pellet pestle homogenizer in the presence of the following extraction buffer (at a ratio of 6 ml per g FW) (I Prosser, IACR-Long Ashton, UK, personal communication): 100 mM Tris-HCl, pH 8,0; 10 µM FAD, 1 mM EDTA, 3% (w/v) BSA, 0.1 % (v/v) Triton X-100, 1 mM DTT, 1mM PMSF (from a 100 mM stock in isopropanol).
3. Centrifuge for 15 min at 15,000 rpm (27,000 x *g*) and 4°C to get the supernatant (crude extract).
4. Assay (final volume: 0.5 ml)

	Blank	Sample
Assay buffer: 100 mM K-P_i, pH 7.5; 2 mM EDTA; 20 µM FAD	250 µl	250 µl
100 mM KNO_3	50 µl	50µl
0.7 mM NADH (dissolved in 10 mM K-Pi, pH 7.5)	-	50 µl
H_2O	50 µl	-
Crude extract (+ extraction buffer)	150 µl	150 µl

5. Incubate 30 min at 30°C.
6. To stop the reaction, oxidize the excess NADH by adding 50 µl of 0.5 M zinc acetate supplemented with 0.15 mM phenazine methosulphate an incubate 20 min at 30°C (this step is not essentially required, if going straight forward on the colour developing reaction, according to the next step).
7. Add 0.5 ml sulphanilamide (1.5% v/v in 2.5 M HCl) and 0.5 ml N-(1-naphtyl)-ethylendiamine dihydrochloride (NNEDA, 0.02 % w/v in water) for nitrite colour development, and read the absorbance at 540 nm ($\varepsilon_{540} = 46$ mM^{-1} cm^{-1}) (spin-down previously if turbidity appears).

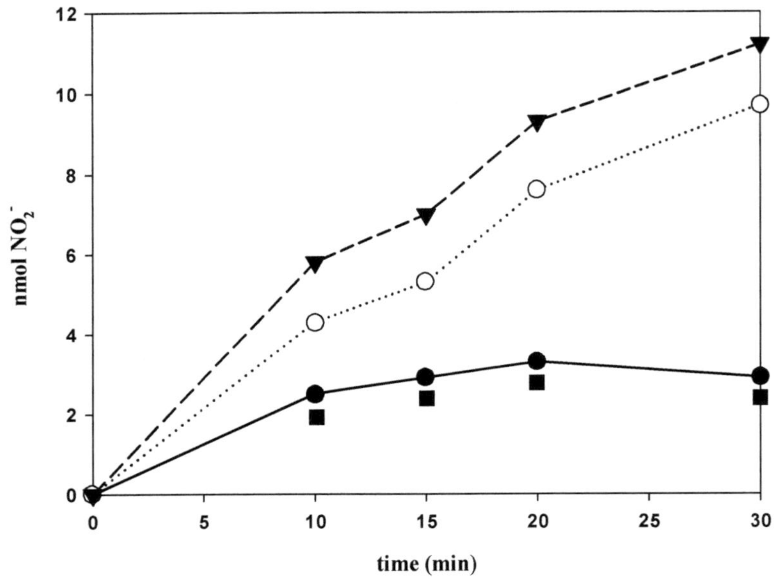

Figure 1. NR inactivating compounds detected in leaves from L japonicus *plants. The figure shows a kinetics of NR enzyme assay using 100 µl of a crude extract from roots (▲). Addition of just 5µl of a* L japonicus *leaf crude extract to the former root crude extract assay was able to produce 53 % and 77 % inhibition of the NR activity detectable in roots after 10 or 30 min of assay respectively (●). This inhibition was almost complete when the amount of leaf crude extract is increased up to 20-50 µl (not shown). The previous dialysis of the leaf crude extract prevented from this inhibitory effect (○), indicating that low molecular weight compounds are in charge of this inhibition. This result was also confirmed using the low molecular weight fraction obtained after filtration of the leaf crude extract through a Diaflo PM-10 AMICON membrane (10,000 M_r exclusion size), which completely retained the inhibitory properties of the leaf crude extract (■). TCA-deproteinized leaf extracts still produced the inhibitory effect, again confirming that low molecular weight compounds are responsible for the inhibition (not shown). For these experiments, leaf and root crude extracts were prepared using the following, commonly described, NR extraction buffer: 100 mM HEPES-KOH, pH 7.5; KNO_3 2 mM, 10 mM cysteine, 10 µM FAD, 10 µM leupeptine, 3% (w/v) polyvinylpolypyrrolidone (PVPP). The enzyme assay, in 500 µl final volume, contained: 25 µl of 0.5 M HEPES-KOH buffer (pH 7.5), 50 µl 100 mM EDTA, 50 µl 100 mM KNO_3, 25 µlH_2O, 200 µl of extraction buffer, and the amount of root and/or leaf crude extracts formerly indicated.*

Observations

- It is better check for each assay that the amount of NO_2^- produced is linear with time and with the amount of extract used.
- NR is post-translationally regulated. Maximal or actual NR enzyme activities, and extent of Mg^{2+} inhibition of the enzyme activity are commonly determined. Enzyme activity assays are then performed in duplicate in the presence and absence of Mg^{2+}. Crude extract preparations and assays for this purpose can be found in Pajuelo *et al.* (2002).
- Detectable root enzyme activity per unit FW is usually 10-20 fold higher in roots than in leaves (Table 1). Total root enzyme activity per plant is usually double than the total enzyme activity from the shoot (leaves + stems).
- The level of NR enzyme activity detected in roots can be greatly influenced by the age of the plant and amount of space provided for root growth (Pajuelo *et al.,* 2002).
- Addition of 30 mg ml^{-1} soluble PVPP (polyvinylpolypyrrolidone) is recommended when using highly concentrated extracts for NR assay.
- Low molecular weight NR inactivating compounds are commonly present in *L japonicus* leaf crude extracts (Figure 1). These compounds are also equally able to inactivate leaf NR from spinach or barley, but do not affect NiR, GS, or GOGAT enzyme activities (not shown). Care should be taken with the type of leaf extraction and assay buffers used. Buffers described above prevent from this inactivation.

Nitrite reductase

Ferredoxin-nitrite reductase (EC 1.7.7.1) is a plastid enzyme that catalyses de reduction of nitrite to ammonium using reduced ferredoxin as electron donor, according to the following reaction:

$$NO_2^- + 6\ Fd_{red} + 8\ H^+ \rightarrow NH_4^+ + 6\ Fd_{ox} + 2\ H_2O$$

In vivo, reduced ferredoxin can be generated in the green tissue by photosynthesis, or, in non-green tissue, from NADPH via oxidative pentose phosphate cycle and ferredoxin-$NADP^+$ oxidoreductase. *In vitro*, ferredoxin can be substituted by dithionite-reduced methyl viologen (MV) as the electron donor for the enzyme reaction. A full-length NiR cDNA has been isolated from *L japonicus* (accession number AJ 293240) encoding a precursor protein of 582 amino acid residues with a predicted molecular mass of 64,825 Da and a transit peptide consisting of 25 amino acid residues. This cDNA corresponds to a mature protein of 62,112 Da, well in accordance with the size of 63 kDa observed by SDS-PAGE and immunoblot (Orea *et al.,* 2001). The NiR gene (*Nii*) is present as a single copy in this plant and is expressed in roots and leaves, although the level of expression is much higher in roots, in accordance with NR expression levels and the fact that *L japonicus* assimilates nitrate mainly in roots (Orea *et al.,* 2001).

Crude extract preparation and NiR enzyme assay

1. Grind and extract the leaf or root material as described above for NR (for NiR studies FAD can be omitted from the buffer).
2. Assay (final volume: 1 ml) (Pajuelo *et al.,* 1993; Pajuelo *et al.,* 1995; Orea *et al.,* 2001).

	Reference	**Sample**
Assay buffer (200 mM Tris-HCl, pH 8.0; 6 mM KNO_2; 20 mM MV)	500 µl	500 µl
Crude extract	-	Up to 100 µl
H_2O	400 µl	Bring to 900 µl
150 mg ml^{-1} $Na_2S_2O_4$ (freshly prepared in 0.3 M $NaHCO_3$)	100 µl	100 µl

3. Incubate at 40°C for 20 min
4. Shake vigorously to oxidize excess dithionite (blue color disappears).
5. Measurement of nitrite consumed along the reaction:
   5.1. Dilute an aliquot of 30 µl from the assay reaction up to 3 ml with H_2O.
   5.2. Add 1 ml sulfanilamide and 1 ml NNEDA reagents prepared as described above for NR (5 ml final volume colorimetric reaction).
   5.3. Read absorbance at 540 nm. Calculations are made subtracting the sample absorbance from that of the reference tube ($\varepsilon_{540} = 53\ mM^{-1}\ cm^{-1}$).

Observations:

1. Root NiR enzyme activity is also affected by the age of the plant and limitation of space for root growth, as described for NR, although to a lesser extent. Typical values of root and leaf NiR enzyme activity are shown in Table 1.

Glutamine synthetase

GS (EC 6.3.1.2) catalyses the biosynthesis of L-glutamine from L-glutamate and NH_4^+, at the expenses of ATP and requirement for metal divalent cations through the following reaction:

$$\text{L-glutamate} + NH_4^+ + ATP \xrightarrow{Me^{2+}} \text{L-glutamine} + ADP + P_i$$

This reaction is also used to assay for GS enzyme activity *in vitro* in what is termed "biosynthetic activity."

In addition to this reaction, GS is commonly assayed through the "transferase activity" of the enzyme according to the following reaction:

$$\text{L-glutamine} + NH_2OH \xrightarrow[\text{ADP, arseniate}]{Me^{2+}} \gamma\text{-glutamylhydroxamate} + NH_4^+$$

In *L japonicus,* as in other plant species, different cytosolic and plastidic GS isoforms are present. Chromatography on DEAE-Sephacel can be used to separate the cytosolic (GS_1) and plastidic (GS_2) isoenzymes in leaves of this plant showing the peculiarity that the plastidic GS isoform eluted at lower ionic strength than the cytosolic isoform, contrarily to what happens in most C_3 plants (Orea *et al.,* 2002). Other authors were not able to separate both type of isoenzymes in wild-type *L japonicus* using the same technique (Ortega *et al.*, 2004). Isolation of *L japonicus* photorespiratory mutants specifically deficient in GS_2 demonstrates the essential role of this particular isoenzyme in the re-assimilation of ammonium released by photorespiration. Genetic evidence indicates that a single gene, which is expressed predominantly in leaves, but also in roots, encodes this plastidic GS isoform (Orea *et al.*, 2002). Comparison of WT and GS mutants enabled to say that the plastidic GS isoform may account for 70-80 % of the total GS activity in leaves, but it is almost negligible in roots. cDNA clones for GS_2 have been recently obtained by RT-PCR and sequenced (accession number: AY187004) (Betti *et al.,* manuscript in preparation), encoding for a precursor protein of 430 amino acid residues containing an N-terminal transit peptide of 43 residues, which after removal releases a mature protein of 42.7 kDa, well in agreement with the 45 kDa protein band detected in western blots, and quite different from the cytosolic polypeptide band corresponding to 39 kDa (Orea *et al.*, 2002). Cytosolic GS EST sequences available can be classified into three groups (TC3934, TC4466 and TC7702 in TIGRGI data base), possibly corresponding to three different genes responsible for the differential expression in different organs and/or tissues of three different cytosolic polypeptides in this plant, one of them being a characteristic nodulin (Forde and Cullimore, 1989, Hirel and Lea, 2002). One of these cytosolic genes has been further characterized at the molecular level in *Lotus japonicus* (Thykjaer *et al.*, 1997; Ruiz *et al.*, 1999).

Crude extract preparation

Grind and extract the leaf or root material as described above for NR and NiR enzymes but using the following extraction buffer (5 ml g^{-1} FW): 10 mM MOPS, pH 7,0; 10 mM $MgCl_2$; 10 % (v/v) glycerol; 1 mM DTE; 0.1 % (v/v) β-mercaptoethanol; 0.05 % (v/v) Triton X-100; 1 mM PMSF (from a 100 mM stock in isopropanol).

GS enzyme assays

Assays as described by Clemente and Márquez (1999a-b) , Betti *et al.* (2002).

GS biosynthetic enzyme assay *(final volume 100 μl)*

	Blank	**Sample**
1 M Tris-HCl, pH 7.5	10 μl	10 μl
1 M L-glutamate	-	10 μl
1M $MgCl_2$	5 μl	5 μl
2M NH_4Cl	2.5 μl	2.5 μl
100 mM ATP (dissolved in 100 mM Tris-HCl, pH 7.5, and neutralized with NaOH)	4 μl	4 μl
H_2O	74.5 μl	64.5 μl
Crude Extract (1:10)	4 μl	4 μl

1. Incubate 15 min at 37°C
2. Add 800 μl malachite mix* to each tube and wait 1 min.
3. Add 100 μl 34% (w/v) trisodium citrate ($\cdot 2H_2O$)
4. Wait 10-15 min for colour development and read the absorbance at 660 nm (ε_{660} = 50 mM^{-1} cm^{-1})

*Malachite mix preparation (for determining P_i produced)

1. 3 volumes of 337 mg l^{-1} malachite green (dissolve in H_2O, filter and keep at 4°C protected from light)
2. 1 volume $Mo_7 O_{24} (NH_4)_6$ 4 H_2O at 42 g l^{-1} (dissolve in 4 M HCl and keep at 4°C).
3. Mix both solutions at the indicated proportions and filter this mix.
4. Add 2 g l^{-1} CHAPS and keep at 4°C in dark (1-2 weeks as maximum)

Observations

For standard crude extracts use on each assay 4 μl from a 10-fold dilution of the extract in extraction buffer. If different amounts of extract are used for the assay, the amount of H_2O should be varied accordingly to get the same final volume of reaction. The use of higher amounts of extract will result in a very high background. Sephadex G-25 desalting of the extract does not result in reduction of this background indicating that it is mainly due to other ATPases present in the extract. High backgrounds can also be due to the use of a partially hydrolysed ATP substrate stock. Normal GS biosynthetic activity ranges in the values shown in Table 1.

GS transferase enzyme assay *(750 μl final volume)*

	Blank	**Sample**
0.3 M Glutamine (dissolved in 0.4 M MOPS, pH 7.0)	225 μl	225 μl
24 mM $MnCl_2$	75 μl	75 μl
1.5 mM ADP (in 10 mM Tris-HCl, pH 7.0)	15 μl	15 μl
1.2 M Hydroxylammonium ($NH_3OH{\cdot}HCl$)	75 μl	75 μl
1.2 M NaOH	37.5 μl	37.5 μl
Extract	4 μl	4 μl
H_2O	318.5 μl	243.5 μl
0.5 M $Na_2HAsO_4^{2-}$	-	75 μl

1. Incubate 15 min at 37°C.
2. Add 250 μl of γ-glutamyl hydroxamate developing mix*
3. After 1 min centrifuge 5 min at 15,000 rpm (27,000 x *g*)
4. Read absorbance at 500 nm (ε_{500} = 1.08 mM^{-1} cm^{-1})

*γ-glutamyl hydroxamate developing mix preparation

1. 106.64 g $FeCl_3 \cdot 6\ H_2O$
2. 24.72 ml 35 % (w/v) HCl
3. 38.4 g trichloroacetic acid
4. Complete to 1 l with H_2O

Observations

If different amounts of extract are used for the assay, the amount of H_2O should be varied accordingly to get the same final volume of reaction. Normal GS transferase activity values are shown in Table 1.

NaOH is used in the assay to neutralize $NH_3OH{\cdot}HCl$.

In our hands, the biosynthetic assay gives always much more reproducible results than the transferase assay, and follows better the pattern of GS protein in the extracts (as can be determined by western blots or ELISA techniques). Other enzymes have been reported to interfere with the transferase assay of GS in crude extracts (Stewart *et al.*, 1980; Miflin *et al.*, 1981; Moorhead *et al.*, 2003).

Glutamate synthase

This enzyme catalyses the transfer of the amide group of glutamine to 2-oxoglutarate, yielding two molecules of glutamate, a reaction requiring additionally two electrons, according to the following:

$$\text{L-glutamine} + \text{2-oxoglutarate} + 2\,e + 2\,H^+ \rightarrow 2\ \text{L-glutamate} + 2\,H_2O$$

Two different types of enzyme have been reported in higher plants and must be present also in *L japonicus.* One uses ferredoxin (Fd) and the other pyridine-nucleotides (basically NADH) as reductants. Ferredoxin-dependent glutamate synthase (Fd-GOGAT) (EC 1.4.7.1) is a monomeric protein of molecular weight around 160 kDa, which contains iron-sulfur and flavin as prosthetic groups (Galván *et al.*, 1984; Márquez *et al.,* 1986; Márquez *et al.,* 1988; van den Heuvel *et al.*, 2004). Up to date the largest sequence data available for this protein in *Lotus japonicus* corresponds to accession number AP004474. Depending on the plant species, one or two genes may encode for Fd-GOGAT, a protein that is expressed both in leaves and in roots and requires respectively photosynthetically or non-photosynthetically reduced Fd as described previously for Fd-NiR. *In vitro,* methyl viologen (MV) can be used as a substitute for ferredoxin for this assay, although care must be taken since it has been reported that dithionite-reduced MV is also adequate electron donor for NADH-GOGAT (Márquez *et al.*, 1984). NADH-GOGAT (EC 1.4.1.14) is also a monomeric iron-sulphur flavoprotein, although larger than Fd-GOGAT, having a molecular mass around 220 kDa. Plant NADH-GOGAT is encoded by a single gene (EST accession number for *L japonicus:* AW720305), which is mostly expressed in non-green tissues, and in young tissues until photosynthesis and Fd-GOGAT become fully developed (van den Heuvel *et al.,* 2004; Hirel and Lea, 2002).

Crude extract preparation

Grind and extract the leaf or root material as described above for NR and NiR enzymes but using the following extraction buffer (5 ml g^{-1} FW) (Márquez *et al.,* 1988, Pajuelo *et al.,* 1997; Pajuelo *et al.,* 2000): 50 mM K-phosphate buffer, pH 7.5; 100 mM KCl; 5 mM Na_2-EDTA; 0.1% (v/v) β-mercaptoethanol; 2 mM 2-oxoglutarate; 1 mM PMSF (from a 100 mM stock in isopropanol); 20% (v/v) ethylene glycol.

GOGAT enzyme assay (2 ml final volume reaction)

	Blank	**MV-GOGAT**	**NADH-GOGAT**
Assay buffer (0.22 M K-P_i, pH 7.5, 11 mM glutamine; 11 mM Na_2-Oxoglutarate)	0.9 ml	0.9 ml	0.9 ml
50 mM aminooxyacetate	0.2 ml	0.2 ml	0.2 ml
Crude extract + H_2O	0.5 ml	0.5 ml	0.5 ml
H_2O	0.4 ml	-	0.2 ml
100 mM MV	-	0.2 ml	-
20 mg ml^{-1} $Na_2S_2O_4$ (freshly prepared in 0.3 M $NaHCO_3$)	-	0.2 ml	-
15 mM NADH	-	-	0.2 ml

1. Incubate 20 min at 30°C
2. Boil 1 min to stop the reactions and shake vigorously to oxidize dithionite (blue colour disappears).
3. Measurement of glutamate produced:
   a. Load each assay on a 2 ml Dowex AG1-X8 column equilibrated in water.
   b. Drain and wash each column with 10 ml water (glutamine elution).
   c. Wash each column with 8 ml of 2 N acetic acid, collecting the glutamate eluate in a clean tube.
   d. On separate tubes, take 1 ml of the glutamate eluate from each column, add 2 ml of Cd-ninhydrin reagent*, heat at 80 °C for 10 min and then place the tubes on ice.

      *Cd-ninhydrin reagentDissolve altogether 4 g ninhydrin and 1 g Cd-acetate in a solvent made of 80 ml ethanol plus 10 ml glacial acetic acid plus 20 ml distilled water.
   e. Read the absorbance at 506 nm ($\varepsilon_{506} = 6\ mM^{-1}\ cm^{-1}$).

Observations

Fd- or NADH-GOGAT can be greatly influenced by light, nitrogen source and developmental stage of the plant (Pajuelo *et al.,* 1997). Average activities measured using this protocol are shown in Table 1.

Measurements of MV-GOGAT usually reproduce faithfully the activity of Fd-GOGAT that can be obtained with purified Fd as reductant. However care should be taken since it has been reported that proteolysis of Fd-GOGAT may result in a decrease of the activity with Fd while MV-dependent activity is not affected (Pajuelo *et al.*, 2000).

	ENZYME ACTIVITY			
	μmol g^{-1} FW h^{-1}		mU mg^{-1} protein	
	roots	*leaves*	*roots*	*leaves*
NADH-NR	2-4	0.1-0.2	5-15	0-0.1
MV-NiR	80-130	60-80	250-400	50-70
GS biosynthetic	70-100	800-1000	200-300	700-900
GS transferase	2000-3000	4000-4500	6000-9000	3500-4000
MV-GOGAT	30-40	300-500	80-100	200-450
NADH-GOGAT	15-25	0-10	40-60	0-10

Table 1. Typical range of values obtained for different enzyme assays in crude extracts from L japonicus leaves or roots. Plant growth and harvesting conditions are the standard conditions described by Orea et al., (2001, 2002) and Pajuelo et al., (2002) as well as Orea et al., in other chapters of this book.

ACKNOWLEDGEMENTS

Authors wish to thank funding from research projects BMC2001-3162 and BFU2004-02753 from MCYT/MEC-FEDER (Spain), and BIO2-CT93-0400, HPRN-CT2000-00086 and MRTN-CT-2003-505227 from European Union, as well as support given by Junta de Andalucía to group CVI-163. Technical assistance of MJ Cubas is also gratefully acknowledged.

REFERENCES

Andrews M. (1986) **The partitioning of nitrate assimilation between root and shoot of higher plants.** *Plant Cell Environment* 9, 511-586.

Betti M, Márquez AJ, Yanes C, and Maestre A. (2002) **ATP binding to purified homopolymeric plant glutamine synthetase studied by isothermal titration calorimetry.** *Termochimica Acta* 394, 63-71.

Clemente MT, and Márquez AJ.(1999 a) **Functional importance of Asp-56 from the α-polypeptide of *Phaseolus vulgaris* glutamine synthetase: an essential residue for transferase but not for biosynthetic enzyme activity.** *European Journal of Biochemistry* 264, 453-460.

Clemente MT, and Márquez, AJ. (1999 b) **Site-directed mutagenesis of Glu-297 from the α-polypeptide of *Phaseolus vulgaris* glutamine synthetase alters kinetic and structural**

properties and confers resistance to L-methionine sulfoximine. *Plant Molecular. Biology* 40, 835-845.

Forde BG, and Cullimore JV. (1989) **The molecular biology of glutamine synthetase in higher plants.** *Oxford Surveys of Plant Molecular and Cellular Biology* 6, 247-296.

Forde BG and Clarkson DT. (1999) **Nitrate and ammonium nutrition of plants: physiological and molecular perspectives.** *Advances in Botanical Research* 30, 1-90.

Galván F, Márquez AJ, and Vega JM. (1984) **Purification and molecular properties of ferredoxin-glutamate synthase from *Chlamydomonas reinhradtii*.** *Planta* 162, 180-187.

Hirel BG and Lea PJ. (2002) **The biochemistry, molecular biology, and genetic manipulation of primary ammonia assimilation.** In: *Photosynthetic nitrogen assimilation and associated carbon and respiratory metabolism* (Foyer CH, and Noctor G, Eds.), Kluwer Academic Publishers, pp. 71-92.

Lea PJ, Blackwell RD, Chen F-L and Hecht U (1990) **Enzymes of ammonia assimilation.** In: *Methods in Plant Biochemistry* (Lea PJ, Ed.), Academic Press, pp. 257-277.

Márquez AJ, Galván F, Vega JM. (1984) **Purification and characterization of the NADH-glutamate synthase from *Chlamydomonas reinhradii*.** *Plant Science Letters* 34, 305-314.

Márquez AJ, Gotor C, Romero LC, Galván F, and Vega JM. (1986) **Ferredoxin-glutamate synthase from *Chlamydomonas reinhardii*. Prosthetic groups and preliminary studies of mechanism.** *International Journal of Biochemistry* 18, 531-535.

Márquez AJ, Avila C, Forde BG, and Wallsgrove, RM. (1988) **Ferredoxin-glutamate synthase from barley leaves: rapid purification and partial characterization.** *Plant Physiology and Biochemistry* 26, 645-651.

Márquez AJ, Orea A, Pajuelo P, Pajuelo E, Romero JM, Arcondéguy T, Betti M, García-Calderón M, Estivill G, and Pal'ove-Balang P. (2004) **Nitrate assimilation in roots of the model legume *Lotus japonicus*.** *Biologia Bratislava* 59/Suppl. 13, 69-76.

Miflin BJ, Wallsgrove RM, and Lea PJ. (1981) **Glutamate metabolism in higher plants.** In: *Topics in cellular regulation, Vol. 20, Academic Press,* pp. 1-43.

Moorhead GBG, Meek SEM, Douglas P, Bridges D, Smith CS, Morrice N, MacKintosh C. (2003) **Purification of a plant nucleotide pyrophosphatase as a protein that interferes with nitrate reductase and glutamine synthetase assays.** *European Journal of Biochemistry* 270, 1356-1362.

Orea A, Pajuelo P, Pajuelo E, Márquez AJ, and Romero JM. (2001) **Characterisation and expression studies of a root cDNA encoding for ferredoxin-nitrite reductase from *Lotus japonicus*.** *Physiologia Plantarum* 113, 193-202.

Orea A, Pajuelo P, Pajuelo E, Quidiello C, Romero JM, and Márquez AJ. (2002) **Isolation of photorespiratory mutants from *Lotus japonicus* deficient in glutamine synthetase.** *Physiologia Plantarum* 115, 352-361.

Ortega JL, Temple SJ, Bagga S, Ghoshroy S, and Sengupta-Gopalan C. (2004) **Biochemical and molecular characterization of transgeic *Lotus japonicus* plants constitutively over-expressing a cytosolic glutamine synthetase gene.** *Planta* 219, 807-818.

Pajuelo E, Borrero JA, and Márquez AJ. (1993) **Immunological approach to subunit composition of ferredoxin-nitrite reductase from *Chlamydomonas reinhardtii*.** *Plant Science* 95, 9-21.

Pajuelo E, Pajuelo P, Clemente MT, and Márquez AJ. (1995) **Regulation of the expresión of ferredoxin-nitrite reductase in synchronous cultures of *Chlamydomonas reinhardtii*.** *Biochimica et. Biophysica Acta* 1249, 72-78.

Pajuelo P, Pajuelo E, Forde BG, and Márquez, AJ. (1997) **Regulation of the expression of ferredoxin-glutamte synthase in barley.** *Planta* 203, 517-525.

Pajuelo P, Pajuelo E, and Márquez AJ. (2000) **Proteolysis of barley (*Hordeum vulgare* L) ferredoxin-glutamate synthase affects ferredoxin- and methyl viologen-dependent enzyme activities differently.** *Journal of Plant Physiology* 157, 575-579.

Pajuelo P, Pajuelo E, Orea A, Romero JM, and Márquez AJ. (2002) **Influence of plant age and growth conditions on nitrate assimilation in roots of *Lotus japonicus* plants.** *Functional Plant Biology* 29, 485-494.

Ruiz MT, Prosser I, Hirel B, and Clarkson DT. (1999) **A cDNA Sequence (Accession No. X94299) encoding a cytosolic subunit of glutamine synthetase in the model legume *Lotus japonicus.*** *Plant Physiology* 119, 1148.

Thykjaer T, Danielsen D, She Q, and Stougaard J. (1997) **Organization and expression of genes in the genomic region surrounding the glutamine synthetase gene *Gln1* from *Lotus japonicus*.** *Molecular. General. Genetics* 255, 628-636.

Stewart GR, Mann AF, and Fentem PA. (1980) **Enzymes of glutamate formation: glutamate dehydrogenase, glutamine synthetase, and glutamate synthase.** In: *The Biochemistry of Plants, Vol. 4 (*Miflin, B.J, ed.*),* Academic Press, pp. 271-327.

van den Heuvel, RHH, Curti B, Vanoni MA, and Mattevi A. (2004) **Glutamate synthase: a fascinating pathway from L-glutamine to L-glutamate**. *Cell and Molecular Life Sciences* 61, 669-681.

Wray JL, and Fido J. (1990) **Nitrate reductase and nitrite reductase.** In: *Methods in Plant Biochemistry, Vol. 3* (Lea, P.J, Ed.), Academic Press, pp. 241-256.

Chapter 7.3

TRANSGENIC PLANTS AFFECTED IN NITRATE ASSIMILATION

Alicia Orea[1,2], Ian Prosser[3,4], José M Romero[2], and Antonio J Márquez[1*]

[1]*Departamento de Bioquímica Vegetal y Biología Molecular, Facultad de Química, Universidad de Sevilla, Apartado 553, 41080-Sevilla,* SPAIN; [2]*Instituto de Bioquímica Vegetal y Fotosíntesis, Centro de Investigaciones Científicas Isla de la Cartuja, C/ Américo Vespucio 49, E-41092-Sevilla,* SPAIN; [3]*IACR-Long-Ashton Research Station, Department of Agricultural Sciences, University of Bristol, Long Ashton, Bristol, BS18 9AF,* UNITED KINGDOM. [4]*Present address: Rothamsted Research, Harpenden, Herts, AL5 2JQ,* UNITED KINGDOM; **Corresponding author.*

Email: cabeza@us.es

Phone: +34 954 557 145 Fax: +34 954 626 853

Keywords: root and shoot nitrate assimilation, nitrate reductase, transgenics, leaf nitrate reductase inhibitory compounds

Plant nitrate assimilation is a process that takes place both in roots and in shoots. The partitioning of nitrate assimilation between both organs depends on the particular specie studied and environmental factors. Lotus japonicus *is a temperate legume that assimilates nitrate mainly in roots. We have analysed the levels of nitrate reductase (NR) activity in roots and leaves from* L japonicus *transgenic plants with altered NR expression produced by means of constitutive, leaf-specific or root-specific promoters. Although NR expression was effectively modified in some of the transgenic lines, none of them showed higher levels of NR activity in leaves than in roots. Some type of mechanism may be operating in* L japonicus *to maintain low levels of NR enzyme activity in leaves of this plant.*

INTRODUCTION

Nitrate assimilation is a fundamental process of higher plants (see for review, Hoff *et al.*, 1994; Crawford, 1995; Daniel-Vedele and Caboche, 1996; Lea and Ireland, 1999). The plant kingdom obtains nearly 100-fold more nitrogen through nitrate assimilation than from the biological fixation of molecular nitrogen (Guerrero *et al.*, 1981). Legumes in particular have the ability to form nitrogen-fixing symbiosis with *Rhizobium* and *Bradyrhizobium* spp. However there are many reasons why nodulation may not occur, so nitrate is an important nitrogen source for legumes as it

A.J. Márquez (ed.), Lotus japonicus Handbook, 329-340.

is for non-legumes. Nitrate is actively transported into root cells where it stored in the vacuole, reduced or transported to the leaves where it could be also stored in the vacuole or reduced. Nitrate is reduced to ammonium in two consecutive enzymatic steps. First, nitrate is reduced to nitrite in the cytoplasm (reaction catalysed by nitrate reductase) (NR, EC 1.6.6.1.) and then nitrite is reduced to ammonium by nitrite reductase (NiR, EC 1.7.7.1.). Subsequently, ammonium is incorporated into amino acids by glutamine synthetase (GS, EC 6.3.1.2.) and glutamate synthase (GOGAT, EC 1.4.7.1.) (Lea and Forde, 1994; Ireland and Lea, 1999).

Both roots and shoots from higher plants have the capacity of assimilating nitrate (Smirnoff and Stewart, 1985). The partitioning of nitrate assimilation between roots and shoots in higher plants could be defined by the relative proportion of the total plant NR activity in these two plant organs or/and by the relative concentration of nitrate and reduced nitrogen in the xylem (Andrews, 1986a). This is because in plants grown on nitrate both nitrite and ammonium are rapidly assimilated and only nitrate is stored or transported (Crawford, 1995). Most herbaceous plants assimilate nitrate in leaves. However, woody plants and temperate legume species assimilate nitrate mainly in roots (Andrews, 1986b; Oaks, 1992; Gojon *et al.*, 1994;Woodall and Forde, 1996). Besides, the partitioning of nitrate assimilation between leaves and roots can change in a plant along growth and in relation to environmental factors, such as light, temperature or external nitrate concentration (Smirnoff and Stewart, 1985; Wallace, 1986; Andrews, 1986b; Gojon *et al.*, 1994; Peuke *et al.*, 1996).

It is well known that nitrate fertilisation leads to an increase in the levels of nitrate uptake and nitrate assimilation (Redinbaugh and Campbell, 1991; Hoff *et al.*, 1994; Crawford, 1995; Daniel-Vedele and Caboche, 1996) and changes of the partitioning of nitrate assimilation (Smirnoff and Stewart, 1985; Wallace, 1986; Andrews, 1986b; Gojon *et al.*, 1994; Peuke *et al.*, 1996). Temperate perennial legumes carry out most of their nitrate assimilation in roots when they are grown on low external nitrate concentration (about 1mM). As external nitrate concentrations increase (in fertilised soil, 1-20 mM), shoot nitrate assimilation becomes more important, getting even higher than in roots (Andrews, 1986b). *Lotus japonicus* is a temperate perennial legume that assimilates nitrate mainly in roots (Woodall and Forde, 1996; Pajuelo *et al.*, 2002; Orea *et al.*, 2001 and Chapter 7.1, this book). In this plant, NR activity is lower in shoots than in roots, being particularly low in leaves, even when plants are grown in high external nitrate concentrations. This happens in spite that the levels of nitrate accumulation in leaves were high and reached similar levels than in roots (see Chapter 7.1).

Theoretical considerations suggest that leaf assimilation is energetically more efficient than root assimilation when photosynthesis is light –saturated, and competition between carbon dioxide and nitrate for photochemical energy is minimised (Smirnoff and Stewart, 1985).

The use of mutants and transgenic plants has been a powerful tool to study and modify plant nitrogen metabolism (Hoff *et al.*, 1994; Lea and Forde, 1994). In this work, we have studied NR activity and protein expression in roots and shoots of a series of transgenic plants from *L japonicus* carrying NR constructs in either sense or antisense orientations. Possible changes in the partitioning of the nitrate assimilation in these transgenic plants were examined.

RESULTS AND DISCUSSION

We have studied the levels of NR activity and NR protein in transgenic plants carrying different types of *Nia* constructs in either sense or antisense orientations under the control of different types of promoters. Table 1 summarises the different types of transgenic plants studied. Three types of promoters were used: 1) a constitutive CaMV 35 S promoter that should be operative both in leaves and in roots; 2) Promoters that are specific or preferentially expressed in roots, like the *RolD* promoter from *Agrobacterium rhizogenes* or *Glnβ* promoter from *Phaseolus vulgaris*; 3) a promoter specific of green tissues like the *Cab* promoter, that is induced by light.

Transgenic lines	Promoter	Gene (orientation)	Expression
CaMV8.u4.a1	CaMV 35 S	*L japonicus nia* (antisense)	Constitutive
CaMV8.u4.a2			
CaMV8.u4.f1			
CaMV6.NR.a	CaMV 35 S	Spinach *nia* (sense)	
CaMV6.NR.6c			
Rold.3b.1c	*RolD*	*L japonicus nia* (antisense)	Roots
Rold.1c.2a			
Glnβ.u4.g1	*Glnβ*	*L japonicus nia* (antisense)	
Glnβ.u4.f1			
Cab.a1	*Cab*	Spinach *nia* (sense)	Green tissues

Table 1. Constructions used to transform L japonicus *plants.*

Table 2 shows the results obtained from maximal NR enzyme activity determination (in the presence of EDTA and absence of Mg^{2+}) in the different transgenic lines studied. Several of these lines (noticed in bold) showed an increase or decrease in NR enzyme activity in the way expected from the constructs respectively introduced. One of the transgenic lines transformed with CaMV 35S promoter and a *L japonicus* antisense *Nia* cDNA (CAMV8.u4.f1) showed NR activity level lower than wild type and control plants (transformed with pGTV-kan vector) both in leaves and in roots. Another transgenic plant expressing the *Nia* spinach gene in sense orientation under CaMV 35S promoter showed higher NR activity level than wild type and control plants in leaves without significant changes in roots (CaMV6NR.a). However, root-specific or preferentially expressed promoters produced different results. None of the plants transformed with *L japonicus Nia* antisense gene under the *RolD* promoter showed altered levels of NR activity but one of the lines carrying *L japonicus* antisense *Nia* gene under the control of Glnβ promoter produced effectively low levels of NR activity in roots (Glnb.u4.F1).

Constructions	Plants		NR activity ($\mu mol\ NO_2^-\ g^{-1}\ FW\ h^{-1}$)	
			Leaves	Roots
-	WT		0.2-0.35	2-4.5
pGTV-kan vector	pGPTV.control		0.49	3.4
CaMV 35S promoter + *L japonicus Nia* (antisense)	CaMV8.u4.a1		0.42	4.4
	CaMV8.u4.a2		0.21	3.6
	CaMV8.u4.f1		**0.15**	**1.7**
CaMV 35S promoter + spinach *Nia* (sense)	**CaMV6.NR.a**		**0.87**	3.2
	CaMV6.NR.6c		0.23	3.3
RolD promoter + *L japonicus Nia* (antisense)	Rold.3b.1c		0.24	3.9
	Rold.1c.2a		0.46	2.6
Glnß promoter + *L japonicus Nia* (antisense)	Glnß .u4.g1		0.24	4.5
	Glnß.u4.f1		0.19	**1.4**
Cab promoter + spinach *Nia* (sense)	*Cab.a1*	L2	**0.7**	3.8
		D8	0.5	3.1

Table 2. Level of NR activity in leaves and roots of L japonicus *transgenic plants. Samples of leaves and roots were harvested at the third light hour of photoperiod. In the case of Cab.a1 plant the samples from leaves and roots were harvested both in the second light hour (L2) and in the last dark hour (D8) of photoperiod. Results from plants showing expected alterations in the level of NR activity are written in bold.*

Cab.a1 transformants (carrying the *Nia* spinach gene under the control of *Cab* promoter) showed higher NR activity levels in leaves than wild type and control plants during the light period of the day, while NR activity was similar to that of control plants during the night. NR activity levels detected in roots from Cab.a1 transformants were the same as those in wild type and control plants, either in light or dark conditions.

To follow up the activity data, the amount of NR protein from the transgenic plants that showed changes in the level of NR activity, was examined qualitatively with respect to the wild type and control plants (Table 3). CaMV8.u4.fl transformants, carrying *L japonicus Nia* gene in antisense orientation, showed also lower levels of NR protein in roots and in shoots, in agreement with the lower NR activity found in this plant. Similarly, Glnβ.u4.F1 transformants (under control of the *Glnβ* promoter, preferentially expressed in roots) showed a decrease in NR protein levels in roots, while CaMV8.u4.fl (under control of CaMV 35S constitutive promoter) showed a decreased amount of NR protein both in leaves and in roots. Moreover, transformants carrying the spinach *Nia* gene in sense orientation showed higher NR protein levels. CaMV6.NR.a (carrying spinach *Nia* gene under control of CaMV 35S promoter) presented a significant increase in the amount of NR protein in leaves and a slight increase in roots. In the case of Cab.a1 (under control of *Cab* promoter) the plants showed a 9-fold increase of NR protein specifically in illuminated leaves, although this increase is significantly reduced in the dark.

Constructions	**Plants**		**% NR protein**	
			Leaves	**Roots**
-	wild type		100 (0.75)	100 (22.2)
pGPTV-kan vector	pGPTV.control		96 (0.72)	101 (22.4)
CaMV35S promoter + *L japonicus nia* (antisense)	CaMV8.u4.fl		51 (0.37)	46 (10.2)
CaMV 35S promoter + spinach *nia* (sense)	CaMV6.NR.a		300 (2.25)	120 (26.6)
Glnß promoter + *L japonicus nia* (antisense)	Glnß.u4.f 1		93 (0.69)	65 (14.4)
Cab promoter + spinach *nia* (sense)	Cab.a1	L2	900 (6.75)	90 (19.9)
		D8	600 (4.50)	98 (21.7)

Table 3. NR protein levels in leaves and roots of L japonicus *transgenic plants. Plants were grown and harvested as is described in Table 3. NR protein in leaf and root crude extracts (500 µg and 300 µg of total protein from leaves and roots, respectively) was monitored by western blot using antibodies polyclonal against* L japonicus *NR T-19-S peptide (Pajuelo* et al. *2002). Immunoreactive bands were quantified using a "Bio Image Intelligent Quantifier" system densitometer (figures in brackets) and the relative NR protein content was expressed as percentage of the value detected in leaves and roots of wild type.*

The results shown above are clearly indicative that several of the transformants analysed effectively changed the level of NR activity and protein in accordance with the genetic alterations introduced in them. The absence of alterations in other transgenic lines could be due to gene silencing or other types of mechanisms that impaired the expression of the inserted gene (Matzke and Matzke, 1995; Meyer, 1995).

It is remarkable that, although some of the lines studied showed altered NR expression, the changes produced were not apparently sufficient to shift the partitioning of nitrate assimilation among roots and shoots of this plant. So, although for example Glnβ.u4.fl showed a decrease in NR protein and activity in roots, this did not bring about an increase in NR activity in leaves. In addition, the increase of NR activity and protein in leaves (CaMV6.NR.a and Cab.a1 transformants) did not produce a concomitant decrease of NR activity in roots. In all cases, the NR activity was always lower in leaves than in roots. In particular, the Cab.a1 transformants, where the level of NR protein in light was 9-fold higher in leaves than wild type and control plants, yet NR activity was less than 2-fold increased and was still lower than that in the roots.

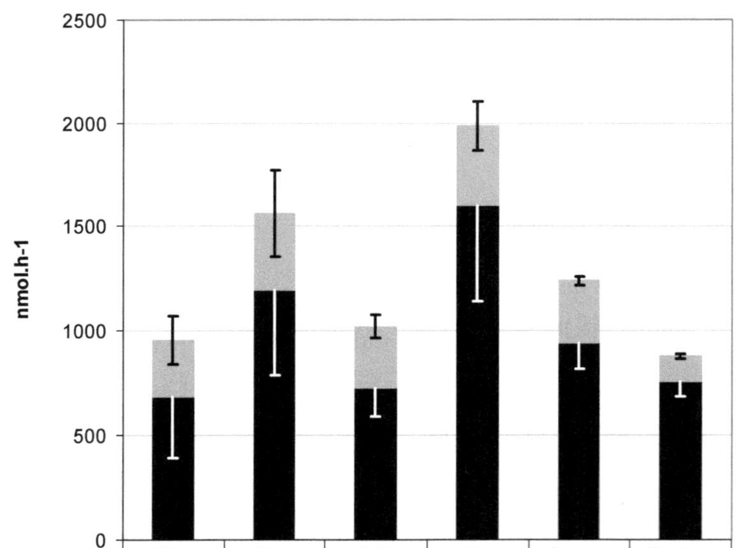

	WT	WT	WT	WT	Cab.a1	Control
Shoot-Mg	267.3	371	293.2	390.1	303.1	117.4
Root -Mg	684.2	1191.6	723.2	1600.6	936.7	753.9

Figure 1. Integrated NRA (nitrate reductase activity) of WT, transformed control and Cab.a1 *plants grown using aeroponic culture in complete nutrients and under greenhouse mid-summer natural light conditions. Plant material was harvested at 10 a.m. Integrated NRA activity with the actual quantity of tissue for the whole roots and shoots systems per plant was measured. The results for the WT plants represent four separate experiments. Each experiment is the mean from seven plants assayed in triplicate.*

In Figure 1, the relative, potential contribution of the whole root and shoot systems to nitrate reduction shows clearly that even under high light, sufficient nitrate nutrition, there is no concerted shift to shoot reduction. Under these conditions, the shoot NR is higher than when grown in the lower light conditions of a CE facility (up to a maximum of 40%). However, in the Cab.a1 line, the shoot contribution was 33% and therefore not significantly different from that of the WT or control plants despite the raised NR level due to expression of the spinach NR in the shoot.

All these results suggest the existence of a global mechanism of nitrate assimilation in *L japonicus* that prevents a shift of nitrate assimilation to leaves. In the case of the Cab.a1 transformant, the lack of correspondence between the levels of NR protein and NR activity could be indicative of post-translational regulation in leaves of *L japonicus*. Since NRA (nitrate reductase activity) was determined in the presence of EDTA and therefore is a measure of NRA_{max} it is unlikely that the low activity is due to the activation state of the enzyme in the leaves (Man *et al.* 1999). These results would be in agreement with results shown on a previous chapter (Orea *et al.,* chapter 7.1) were it was shown that an increase on external nitrate was neither able to shift the partitioning of nitrate assimilation from roots to leaves of this plant.

EXPERIMENTAL PROCEDURES

Plant material and growth conditions

Seeds of *Lotus japonicus* cv. Gifu (GIFU B-129-S9) were obtained from Prof. Jens Stougaard (University of Aarhus, Denmark) and propagated by self-fertilization. Unless otherwise stated, cuttings from transgenic plants and wild type controls were grown in hydroponics cultures to produce roots. For this purpose cuttings of 15 cm of length (without the bottom leaves) were placed in plastic containers (22 x 16 x 20 cm) submerging the base tip in the nutrient solution. The nutrient solution was half-strength Hornum medium (Handberg and Stougaard, 1992), continuously aerated with two inlets from an air pump. Once significant roots were produced, the plants were transferred to pots using vermiculite as solid support and were irrigated daily for 30 days with Hornum medium. Growth was carried out in a Sanyo SGR193.S26A growth chamber and at 20/18°C light/dark temperatures and 70% humidity under a 16 h/8 h light/dark regime. Light intensity at the plant level was around 200 µmol m^{-2} s^{-1}, supplied by white fluorescent Philips colour 83 lamps combined with tungsten lamps to provide red supplement and dawn-dusk effects. Whole roots and leaves from plants were quickly frozen in liquid nitrogen, homogenised and then kept at -80°C until use.

For growing plants in aeroponic culture, the rooted cuttings were transferred to a mist tank in a greenhouse (Waterhouse *et al.* 1996) and were provided with 1x Letcombe nutrient solution for 14 days (replaced weekly). These experiments were conducted during the summer so the plants had long days and a higher light intensity than in the CE rooms. Plants were removed, the roots washed briefly in 5l deionised water and gently blotted on tissue paper and the root and shoot material harvested separately and stored at -80°C.

Isolation of the *L japonicus* NR cDNA

A nearly full length (short by 405 bp to ATG) cDNA for nitrate reductase was isolated from a *L japonicus* root cDNA library prepared in λZAPII (Stratagene) and termed pBS:LjNR3. The clone was excised from the λ ZAP and the cDNA was sequenced to confirm its identity with the genomic sequence. To prepare a full-length version, the 5'end of the genomic sequence was amplified using the primer combination TCCTTTCATTT**GG**T**A**CCTCTTATTCACA***ATG*** introducing a *Kpn*I site at the 5' end and TCCATACTAAACCGGGTCGGTGTTTTCACC beyond the *Bst*EII site. The amplified fragment was digested with *Kpn*I and *Bst*EII and ligated with similarly digested pBS:LjNR3 and *Kpn*I digested pBluescript to form pBS:LjNR.

Construction of Glnβ promoter: LjNR antisense in pGPTV-kan

The *P. vulgaris* glutamine synthetase gene β subunit promoter was amplified from plasmid pR1H-Bgl (1.7 kbp *Hin*dIII-*Bgl*II gene fragment in pUC8) (Turton *et al.*1988) with the M13 forward primer (20bp) and a primer GAATCCTG**G**GAT**C**CAACTCG. The underlined bases indicate differences to introduce a *Bam*HI site in the product. The product was digested with *Hin*dIII and *Bam*HI and cloned into similarly cut pUC18 and the sequence of the cloned PCR product was verified. The fragment was subcloned into *Hin*dIII- *Sac*I digested pGPTV-kan (Gleave, 1992). The binary vector was digested with *Kpn*I prior to ligation with *Kpn*I digested pBS:LjNR cDNA. The antisense orientation was selected by restriction mapping.

Construction of CaMV 35S promoter:LjNR antisense in pGPTV-kan

The 35S promoter from pBI120.7 was cloned into pGPTV-kan as a *Hin*dIII/ *Sac*I fragment to make pLARS120. The binary vector was digested with *Kpn*I prior to ligation with *Kpn*I digested pBS:LjNR cDNA. The antisense orientation was selected by restriction mapping.

Construction of Cab promoter: spinach NR in pGPTV-kan

This construct was prepared as described in Prosser (1994) except that pGPTV-kan was used in place of pBI120. The construct contains an *Arabidopsis* Cab promoter fused to the spinach NR cDNA (Prosser and Lazarus 1990).

Construction of CaMV 35S promoter: spinach NR in pGPTV-kan

The spinach NR cDNA was inserted into pLARS120 (see above) as a *Kpn*I/ *Sac*I fragment derived from pIMP6 (Prosser 1994).

Construction of rolD promoter: LjNR in pGPTV-kan

A rolD-GPTV-kan cassette was prepared as described in Limami *et al.*(1999). The *LjNR* was cloned into the cassette as a *Kpn*I fragment as for generating the CaMV:LjNR construct. The antisense orientation was selected by restriction mapping.

Plant transformation

The transformation protocol was essentially that of Handberg and Stougaard (1992) using the various constructs and selection with G418.

NR activity measurement

Roots or leaves material (1 g) was ground in a mortar at 4°C and resuspended in extraction buffer (6 ml). Extraction buffer contained: 100 mM Tris-HCl pH 8, 10 μM FAD, 1 mM EDTA, 1 mM dithiothreitol (DTT), 1 mM phenylmethylsulphonylfluoride (PMSF), 3% (w/v) bovine serum albumi n(BSA) and 0.1% (v/v) Triton X-100. The homogenates were centrifuged for 15 min at 15,000 x *g* and insoluble material was removed. The clear supernatants were immediately assayed for NR (Pajuelo *et al.*, 2002). NR activity assay was carried out as follows: 500 μl final volume reaction mixtures were placed in 1.5 ml eppendorf tubes containing: 50 μl of 0.5 M potassium phosphate buffer, pH 7.5; 5 μl of 100 mM EDTA; 50 μl 100 mM KNO_3; 195 μl distilled water and 150 μl root or leaf crude extract. The reaction was started by adding 50 μl of 0.7 mM NADH and carried out at 30°C for 30 min. Nitrite formation was measured as described elsewhere (de la Rosa *et al.* 1982), using sulphanilamide and N-(1-naphtyl)-ethylenediamine dihydrochloride reagents. Nitrite formation rate was verified to be linear with time during assay. Blank tubes were processed as samples except that the reaction was stopped at zero time.

Western blots

Immunodetection of NR was carried out as described previously by Pajuelo *et al.* (2002). Root and leaves NR were analysed in crude extracts obtained in 2 ml g^{-1} FW of the corresponding extraction buffer described above (without BSA). 300-500 μg of total protein (determined according to Bradford, 1976) were analysed by electrophoresis on 10% (w/v) SDS-polyacrylamide gels (Laemmli, 1970) and transferred onto polyvinylidene difluoride (PVDF) sheets (Immobilon P, Millipore) as described by Pajuelo *et al.* (1995). After transfer, filters were rinsed with TBS (tris-buffered saline, 20 mM Tris-HCl, pH 7.5; 0.5 mM NaCl) and then saturated with blocking solution [5% (w/v) skimmed dry milk powder in TBS] for 16 to 20 h at 4°C. Anti-NR-peptide antiserum (Pajuelo *et al.* 2002) was added at (1:250 dilution) in the blocking solution, being adsorbed for 1.5 h at room temperature and then exhaustively washed in TTBS [0.1% (v/v) Tween 20 in TBS]. After a short treatment for 10 min in blocking solution, the filters were incubated for 1 h in a

1:1000 dilution of the goat anti-rabbit peroxidase conjugate (for NR immunodetection) (Sigma). The excess of secondary antibodies was eliminated by successive washes in TTBS or TBS. After incubation in TBS for 10 min, the filters were immersed for 2 min in 10 ml of the detection reagents mixture from the ECL immunodetection system (ECL Western Blotting Protocols, Amersham), freshly prepared according to the manufacturer instructions. Cross-reacting bands were detected after exposure of blots to autoradiography for 20-50 s. In order to quantify the level of NR protein in crude extracts, the band visualized by autoradiography were analysed using a Millipore Bioimage system densitometer. Protein content was estimated from the integrated values of the areas corresponding to the different bands.

ACKNOWLEDGEMENTS

Authors wish to thank funding from research projects PB97-0714, BMC2001-3162 and BFU2004-02753/BMC from the Spanish programs of MEC/MCYT-FEDER , and BIO2-CT93-0400, HPRN-CT2000-00086 and MRTN-CT-2003-505227 from the Europen Union, as well as support given by Junta de Andalucía to group CVI-163. The technical assistance of MJ Cubas is also gratefully acknowledged.

REFERENCES

Andrews M. (1986a) **Nitrate and reduced-N concentrations in the xylem sap of *Stellaria media, Xanthium strumarium* and six legume species**. *Plant, Cell and Environment* 9, 605-608.

Andrews M. (1986b) **The partitioning of nitrate assimilation between root and shoot of higher plants**. *Plant, Cell and Environment* 9, 511-586.

Bradford MM. (1976) **A rapid and sensitive method for quantitation of microgram quantities of protein utilizing the principle of protein-dye binding**. *Analytical Biochemistry* 72, 248-254.

Crawford NM. (1995) **Nitrate:nutrient and signal for plant growth.** *Plant Cell* 7, 859-868.

Daniel-Vedele F, and Caboche M. (1996) **Molecular analysis of nitrate assimilation in higher plants**. *Comptes Rendus de l'Academy des Sciences, Sciences de la Vi* 319, 961-968.

De la Rosa MA, Márquez AJ, and Vega JM. (1982) **Dissociation of FAD from NADH-nitrate reductase complex from Ankistrodesmus braunii and role of flavin in catalysis**. *Zeitschrift fur Naturforschung C* 37, 24-30.

Gleave AP. (1992) **A versatile binary vector system with a T-DNA organisational structure conductive to efficient integration of cloned DNA into the plant genome**. *Plant Molecular Biology* 20, 1204.

Gojon R, Marion-Poll A, and Bussi C. (1994) **Root/shoot distribution of NO_3^- assimilation in herbaceous and woody species**. In: *A Whole Plant Perpective on Carbon-Nitrogen Interaction* (Roy J, and Garnier E, Eds.). SPB Academic Publishing, The Hague, pp. 131-147.

Guerrero, MG, Vega, JM, and Losada, M. (1981) **The assimilatory nitrate-reducing system and its regulation**. *Annual Review in Plant Physiology* 32, 169-204.

Handberg K, and Stougaard, J. (1992) ***Lotus japonicus* an autogamous, diploid legume species for classical and molecular genetics**. *Plant Journal* 2, 487-496.

Hoff T, Troung H-N, and Caboche M. (1994) **The use of mutants and transgenic plants to study nitrate assimilation**. *Plant, Cell and Environment* 17, 489-506.

Ireland R. and Lea PJ. (1999) **The enzymes of glutamine, glutamate, asparagine, and aspartate metabolism.** In: *Plant Amino Acids: Biochemistry and Biotechnology.* (Singh BK, Ed.). New York: Marcel Dekker, Inc, pp. 49-109.

Laemmli UK. (1970) **Cleavage of structural proteins during the assembly of the head of bacteriophage T4.** *Nature*, 251, 614-616.

Lea PJ, and Forde BG. (1994) **The use of mutant and transgenic plants to study amino acids metabolism.** *Plant, Cell and Environment* 17, 541-556.

Lea PJ, and Ireland RJ. (1999) **Nitrogen metabolism in higher plants.** In: *Plant Amino Acids: Biochemistry and Biotechnology* (Singh, BK, Ed.). New York: Marcel Dekker, Inc, pp. 1-47.

Limami A, Phillipson B, Ameziane R, Pernollet N, Jiang QJ, Poy R, Deleens E, Chaumont-Bonnet M, Gresshoff PM, and Hirel, B. (1999) **Does root glutamine synthetase control plant biomass production in *Lotus japonicus* L?** *Planta* 209, 495-502.

Man HM, Abd-El BGK, Stegmann P, Weiner H, and Kaiser WM. (1999) **The activation state of nitrate reductase is not always correlated with total nitrate reductase activity in leaves.** *Planta* 209, 462-468.

Oaks A. (1992) **A re-evaluation of nitrogen assimilation in roots.** *Bioscience* 42, 103-111.

Orea A, Pajuelo P, Pajuelo E, Márquez AJ, and Romero JM. (2001) **Characterisation and expression studies of a root cDNA encoding for ferredoxin-nitrite reductase from *Lotus japonicus*.** *Physiologia Plantarum* 113, 193-202.

Pajuelo E, Pajuelo P, Clemente MT, and Márquez AJ. (1995) **Regulation of expression of ferredoxin-nitrite reductase in synchronous culture of *Chlamydomonas reinhardtii*.** *Biochimica et Biophysica Acta* 1249, 72-78.

Pajuelo P, Pajuelo E, Orea A, Romero JM, and Márquez AJ. (2002) **Influence of plant age and growth conditions on nitrate assimilatiom in roots of Lotus japonicus plants.** *Functional. Plant Biology* 29, 485-494.

Peuke AD, Glaab J, Kaiser WM, and Jeschlke WD. (1996) **The uptake and flow of C and N and ions between roots and shoots in *Ricinus communis* L** IV. Flow and metabolism of inorganic nitrogen and malate depending on nitrogen nutrient and salt treatment. *Journal of Experimental Botany* 47, 377-385.

Prosser IM. (1994) ***Cloning, sequence and expression of spinach nitrate reductase.*** *PhD thesis.* University of Bristol.

Prosser IM, and Lazarus CM. (1990) **Nucleotide sequence of a spinach nitrate reductase cDNA.** *Plant Molecular Biology* 15, 187-190.

Redinbaugh MG, and Campbell WH. (1991) **Higher plant responses to environmental nitrate.** *Physiologia Plantarum* 82, 640-650.

Smirnoff N, and Stewart GR. (1985) **Nitration assimilation and translocation by higher plants: Comparative physiology and ecological consequences.** *Physiologia Plantarum* 64, 133-140.

Turton, JF, Hopley AP, and Forde BG. (1988) **5'-flanking sequence of a glutamine synthetase gen specifying the subunit of the cytosolic enzyme from *Phaseolus vulgaris* L** *Nucl. Acids Research* 16, 11367.

Wallace W. (1986) **Distribution of nitrate assimilation between the root and shoot of legumes and a comparison with wheat.** *Physiologia Plantarum* 66, 630-636.

Waterhouse RN, Smyth AJ, Massonneau A, Prosser IM, and Clarkson DT. (1996) **Molecular cloning and characterisation of asparagine synthetase from *Lotus japonicus*: Dynamics of asparagine synthesis in N-sufficient conditions.** *Plant Molecular Biology* 30, 883-897.

Woodall J, and Forde B. (1996) **Glutamine synthetase polypeptide in the roots of 55 legume species in relation to their climatic origin and partitioning of nitrate assimilation**. *Plant, Cell and Environment* 19, 848-858.

Chapter 7.4

SECONDARY METABOLITE PROFILING

Nicolas Rispail[*], Robert Nash[Φ], and K Judith Webb[**]

*[*]Institute of Grassland and Environmental Research, Plas Gogerddan, Aberystwyth, Ceredigion, SY23 3EB, Wales, UK; [Φ]Molecular Nature Ltd, Plas Gogerddan, Aberystwyth, Ceredigion, SY23 3EB, Wales, UK; *Corresponding author.*

Email: judith.webb@bbsrc.ac.uk

Phone: +44 1970 823 124   Fax: +44 1970 828 357

Keywords: phenolic compounds, terpenoids, alkaloids, secondary metabolites, GC-MS, HPLC-PDA/MS

Secondary metabolites include a very wide variety of compounds with different structures and chemical properties. In order to obtain an overview of the secondary metabolite content of a plant species such as Lotus japonicus, *a profiling technique coupling sequential extraction with different chromatographic methods (GC-MS, HPLA-DAD/MS) was established. This method allows the qualitative analysis of ionic (charged) compounds such as alkaloids and non-ionic (neutral) compounds such as terpenoids and phenolic compounds. This technique can also be used for the isolation of new compounds*

INTRODUCTION

Secondary metabolites are a wide range of compounds from different metabolite families that can be highly inducible in response to stresses. These compounds are not essential for cell structure and maintenance of life but are often involved in plant protection against biotic or abiotic stresses (Weisshaar and Jenkins, 1998; Hattenschwiler and Vitousek, 2000). Some secondary metabolite families such as carotenoids and flavonoids are also involved in cell pigmentation in flower and seed, which attract pollinators and seed dispersers. Therefore, they are also involved in plant reproduction (Winkel-Shirley, 2001).

Moreover, plant secondary metabolites present chemical and pharmaceutical properties interesting for human health (Raskin et al., 2002; Reddy et al., 2003). Compounds belonging to the terpenoids, alkaloids and flavonoids are currently used as drugs or as dietary supplements to cure or prevent various diseases (Raskin et al., 2002) and in particular some of these compounds seems to be efficient in preventing and inhibiting various types of cancer (Watson et al., 2001; Reddy et al., 2003).

A.J. Márquez (ed.), Lotus japonicus Handbook, 341-348.

Each secondary metabolite family has some specific chemical characteristics, which implies that specific extraction and analysis methods should be developed to study each family in detail. On the other hand, using their polarity properties, which range from polar for polyhydroxylated alkaloids or flavonoid glycosides to non-polar for terpenoids or lipids, and improved database systems, it is possible to profile an increasing number of compounds by using different analytical systems such as GC-MS, LC-MS and HPLC-DAD coupled to a sequential extraction.

The technique described below is aimed at profiling a wide range of compounds from different classes of secondary metabolites. These compounds can be easily extracted from freeze-dry materials using aqueous ethanol, methanol and dichloromethane allowing the extraction of a large range of compounds from water-soluble to non-polar compounds. Then, it is possible to separate them by using different chromatographic methods (reversed phase and ion exchange) and to analyse the fractions obtained by GC-MS and HPLC-DAD/MS (Figure 1). This secondary metabolite profiling is established after removing the major flavonoid glucosides, which are the predominant secondary metabolites along with insoluble tannins and lignin in Lotus plants, as they would mask all the others secondary metabolites present. These flavonoids are the subject of more detailed study and are therefore analysed separately using another extraction technique (for details see: Phenolic compounds: Extraction and Analysis Method).

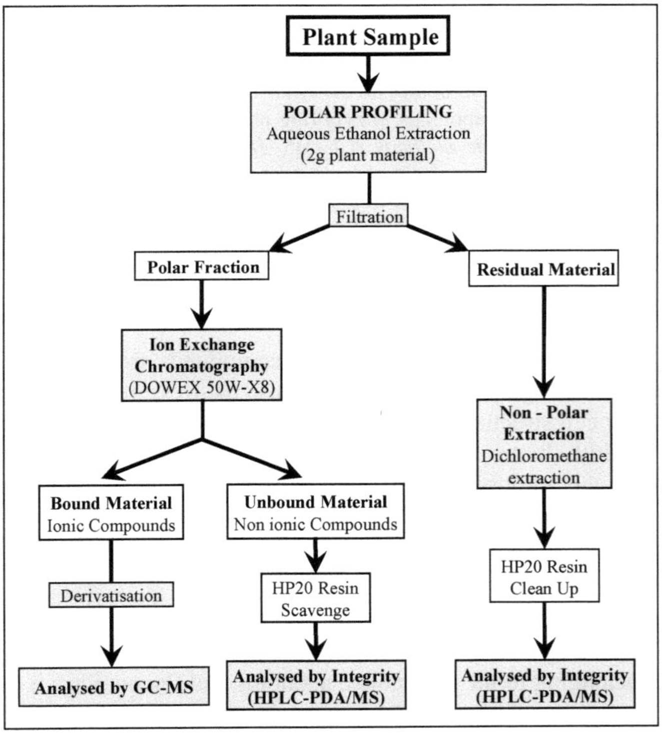

Figure 1. Schematic diagram of secondary metabolite profiling method.

The profiling method described here was originally designed by Molecular Nature Ltd. (International Patent Application Numbers, PCT/GB2003/000905, PCT/GB2003/000880, PCT/GB2003/000892, PCT/GB2003/000906) in order to identify and isolate new compounds which can present a pharmaceutical interest (Watson et al., 2001; Reddy et al., 2003). Since it allows the detection of a large number of compounds from different families of secondary metabolites, it is the method of choice to establish a general profile of secondary metabolites to obtain an overview of the secondary metabolite content of species such as *Lotus japonicus*.

SECONDARY METABOLITE PROFILING PROCEDURE

Materials

The procedure described below is set up for the use of a Perkin Elmer's GC-MS system (model Q-Mass 910; Perkin Elmer, UK) and a Waters' Integrity HPLC-PDA/MS system (Waters, UK) with Millenium 32® software for the data analysis. All glassware, consumables, and solutions used are described in detail in the protocols.

Secondary metabolite extraction

Secondary metabolite profiling entails a sequential extraction using 50% ethanol (v/v) to collect the polar compounds then a dichloromethane extraction for the non-polar compounds.

Polar extraction

Two grams of freeze-dried material is ground and extracted overnight with 50% ethanol (v/v) at room temperature in a 250 ml conical flask. Then, the extracted mixture is filtered on a Buchner funnel and the flask is washed once with 10 ml of 50% ethanol (v/v). The washing solution is filtered and added to the extraction solution. The residual material on the filter is then store at -20°C and freeze-dried before the dichloromethane extraction.

The extracted solution is fractionated by ion-exchange chromatography. First, the sample is applied to a 10x1 cm glass column containing 3 cm of DOWEX 50W-X8 (HCl form, Sigma, UK) resin previously regenerated by adding to it excess 2N HCl, soaking for 15 min, then washing it few times with deionised water until it reaches pH 7 and finally equilibrated with 25 ml of 50% ethanol (v/v). The unbound fraction of the sample, containing the non-ionic compounds, is collected on a 500 ml round bottom flask (RBF). The column is washed once with 75 ml of 50% ethanol (v/v) and with 75 ml of deionised water. The effluents are collected, added to the unbound fraction, which is stored at -20°C before subsequent fractionation and analysis.

After the washing steps, the column is eluted with 100 ml of 2M ammonium hydroxide (NH_4OH) and the effluent containing the ionic compounds is collected in a new 500 ml RBF. This fraction is then carefully evaporated at 35°C using a rotary

– evaporator (Büchi, Switzerland). The evaporation step is stopped when around 5 ml are left in the RBF. At this stage, the extraction liquid is transferred into 10 ml glass vial and, if a deposit is observed on the RBF, the flask is washed with few drops of 70% methanol (v/v) using a Pasteur pipette and this washing solution is added to the 10 ml vial. Then the vial is stored at -20°C and freeze-dried before GC-MS analysis.

Non–ionic fractionation

The unbound fraction from the ion-exchange chromatography previously collected is further scavenged by reversed-phase chromatography using an HP 20 column on a BioFlash chromatography system (Biotage).

Before the separation, the HP20 column is washed with 200 ml of 100% acetone and equilibrated with 25% methanol (v/v). Then the unbound fraction is applied onto the column and the HP20 column is washed with 25% methanol (v/v). The washed solution along with the unbound fraction containing mainly sugars and most of the hydrophilic flavonoid glycosides, is discarded.

The column is then eluted with 400 ml of 25% acetone in methanol (1:4, v/v). The effluent, collected in a 1 l RBF, is then carefully evaporated at 35°C with a rotary-evaporator (Büchi, Switzerland). When around 5 ml are left in the RBF, the evaporation step is stopped and the solution is transferred into a 10 ml glass vial. The RBF is then rinsed with few drops of 25% acetone in methanol (1:4, v/v) that are added to the 10 ml vial. Then the vial is stored at -20°C and freeze-dried before HPLC-DAD/MS analysis.

Non-polar extraction

The residual material from the polar extraction, previously freeze-dried, is placed in a filter paper thimble inside a glass Soxhlet. In parallel, a 500 ml RBF containing 200 ml of dichloromethane and two or three glass beads is placed on a heating mantle. The Soxhlet is attached on top of the RBF and 150 ml of dichloromethane is slowly added to the sample. Then, a refrigeration column is fixed to the Soxhlet apparatus and the cooling water is turned on. The heating mantle is then switched on ensuring a steady refluxing rate and left overnight to extract.

At the end of the extraction, the heating mantle is switched off and the Soxhlet apparatus is allowed to cool to room temperature before being dismantled. The dichloromethane extract is added to 100 ml of HP20 resin in a 1 l RBF and the resin is completely dried with a rotary-evaporator (Büchi, Switzerland) at 35°C. Then it is transferred in a 250 ml conical flask where it is eluted three times with 100 ml of 10% acetone in methanol (1:10, v/v). The solution is then filtered through tissue paper into a 500 ml RBF and evaporated up to 5 ml at 35°C. The solution left is then transferred in a 10 ml glass vial and the RBF is rinsed with few drops of 10% acetone in methanol (1:10, v/v) which are added to the 10 ml vial. Then the vial is stored at -20°C and freeze-dried before HPLC-DAD/MS analysis.

ANALYSIS

GC-MS analysis

Gas-liquid chromatography is designed to separate volatile compounds from a complex mixture. This technique uses the temperature of vaporisation specific to each compound to separate them from a solution by passing the sample through a heated column where it is partitioned between an inert gas under pressure and a thin layer of non-volatile liquid coated on an inert support inside the column.

Many compounds are difficult to vaporise, like the polyhydroxylated alkaloids. Their capacities to vaporise can be improved by replacing the hydroxyl groups by other chemical groups like trimethylsilyl groups before the injection onto the GC-MS. Therefore, before the injection onto the GC-MS, the ionic fraction obtained after the polar extraction is derivatised by silylation in order to ensure a good vaporisation of the sample. This step is realised by adding 100 µl of Sigma Sil A, a mixture of trimethylchlorosilane, hexamethyldisilazane and pyridine (1:3:9, v/v/v; Sigma Chemical Company, UK), to each mg of freeze-dried sample and heating for 15 min at 60°C.

This fraction is then injected onto the GC-MS system (model Q-Mass 910, Perkin Elmer) and separated by a temperature gradient (180°C for 5 min, then from 180°C to 300°C at 10°C/min and a further 5 min at 300 °C) on a capillary column coated with a non-polar liquid phase (25 m x 0.22 mm id x 0.25 µm BPX5 stationary phase, SGE Ltd.). Then, the effluent, containing the separated and vaporised compounds, pass through the mass spectrometer (TMD detector) where it is ionised and fragmented before been picked up by the analyser, which collect the mass spectral data between 71 and 600 m/z.

HPLC-PDA/MS

High-pressure liquid chromatography is designed to separate compounds from a mixture of compounds in solution. This technique uses the solubility and size properties of each compound to separate them with a solvent mixture that is pumped under pressure through the column. Compounds are detected by a photodiode array detector (Water 996 PDA detector, Waters Ass), which establish the light absorbance spectra from visible and UV wavelengths of each detected compound, and a spectrometer (TMD detector), collecting their corresponding mass spectra.

This technique is used to analyse the non-ionic fraction obtain after the polar extraction and the non-polar fraction from the dichloromethane extraction. Before injection, the samples previously freeze-dried are dissolved in 100% methanol and adjusted to the concentration of 10 $mg.ml^{-1}$. Then, three µl is injected onto an HPLC-PDA/MS (Waters, UK) and the separation is realised onto a C_8 HPLC column (50 mm x 2.1 mm id x 3.5 µm, Waters) with a linear gradient of two solvent, 100% deionised water and 100% acetonitrile (with 0.01% trifluoroacetic acid), at a flow rate of 0.35 $ml.min^{-1}$, starting at 90% deionised water, 10% acetonitrile (with 0.01% trifluoroacetic acid), rising to 100% acetonitrile over 6 min, which was held for a further 6.5 min. Then, data of the eluted peak are collected from 200 to 600 nm and between 71 and 600 m/z by the photo-diode array detector and the mass spectrometer respectively. The data analysis is processed using an adapted processing method of Millennium 32® software and the spectra of the eluted peaks are matched with those contained in our own libraries.

COMMENTS

In this article, we describe a method for secondary metabolite profiling. Other profiling methods are also available for profiling of secondary metabolites and will give different profiles. Thus, the method used must be carefully chosen and adapted according to the facilities available and the aim of the research undertaken.

The method, described here, was originally established for the isolation of unknown secondary metabolites. Therefore, it may not be the most suitable for secondary metabolite profiling as it necessitates a large quantity of dried material (at least two g). However, this technique was used successfully to obtain an overview of the secondary metabolite content of *Lotus japonicus* and *Lotus corniculatus*, a related species with a better-documented secondary metabolite content.

Although these two species are very similar at the metabolic level, we observed some differences in their profile, allowing us to discriminate between these two species (Figure 2).

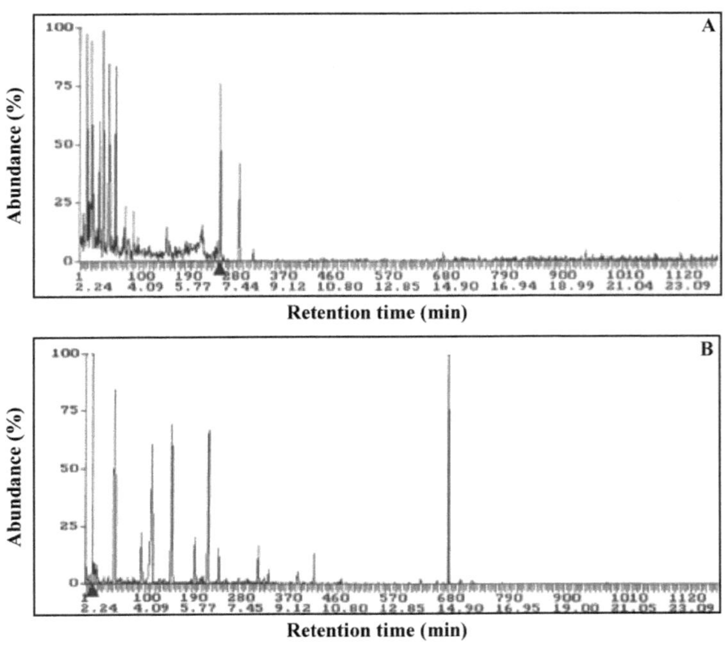

Figure 2. Comparison of GC-MS chromatograms of (A) L. japonicus *and (B)* L. corniculatus *roots. Chromatograms were obtained by separation on a capillary column coated with a non-polar liquid phase using a temperature gradient of 180 °C for 5 min, then 180 °C to 300°C at 10°C/min and further 5 min at 300 °C.*

Moreover, profiling allowed us to detect a novel alkaloid in stems of both species and at a smaller quantity in roots and leaves of *L. japonicus*. This metabolite seems to be a new polyhydroxylated pyrrolidine alkaloid related to DMDP (2R, 5R-dihydroxymethyl- 3R, 4R- dihydroxypyrrolidine) and HomoDMDP (Figure 3) and is likely to be a glycosidase enzyme inhibitor, which may play a role in the plant defence reaction against insect or nematode attacks. Thus, further studies of this compound are currently in process in collaboration with Dr Nash and Molecular Nature Ltd.

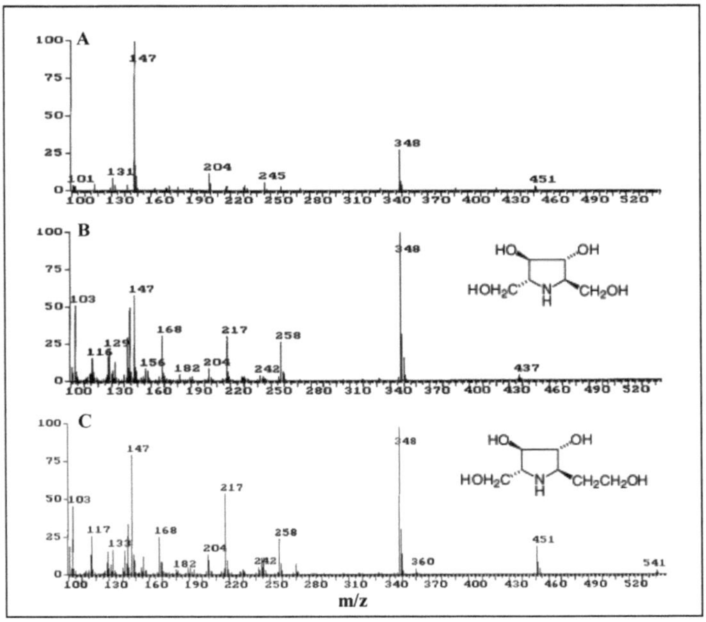

Figure 3. Mass spectra of the novel compound (A) and comparison with the mass spectra of DMDP (B) and HomoDMDP (C). Each spectra was collected with a mass spectrometer (TMD detector) between 71 and 600 m/z after GC-MS separation on a capillary column coated with a non-polar liquid phase using a temperature gradient of 180 °C for 5 min, then 180 °C to 300 °C at 10 °C/min and further 5 min at 300 °C.

References

Hattenschwiler S, and Vitousek P. (2000) **The role of polyphenols in terrestrial ecosystem nutrient cycling**. *Trends in Ecology and Evolution* 15, 238-243.

Raskin I, Ribnicky DM, Komarnytsky S, Ilic N, Poulev A, Borisjuk N, Brinker A, Moreno DA, Ripoll C, and Yakoby N. (2002) **Plants and human health in the twenty-first century**. *Trends in Biotechnology* 20, 522-531.

Reddy L, Odhav B, and Bhoola KD. (2003) **Natural products for cancer prevention: a global perspective**. *Pharmacology & Therapeutics* 99, 1-13.

Watson AA, Fleet GWJ, Asano N, Molyneux RJ, and Nash RJ. (2001) **Polyhydroxylated alkaloids -- natural occurrence and therapeutic applications**. *Phytochemistry* 56, 265-295.

Weisshaar B, and Jenkins G. (1998) **Phenylpropanoid biosynthesis and its regulation**. *Current Opinion in Plant Biology* 1, 251-257.

Winkel-Shirley B. (2001) **Flavonoid Biosynthesis. A Colorful Model for Genetics, Biochemistry, Cell Biology, and Biotechnology**. *Plant Physiology* 126, 485-493.

Chapter 7.5

PHENOLIC COMPOUNDS: EXTRACTION AND ANALYSIS

Nicolas Rispail, Phillip Morris, and K Judith Webb

Institute of Grassland and Environmental Research, Plas Gogerddan, Aberystwyth, Ceredigion, SY23 3EB, Wales, UK; Corresponding author.

Email: judith.webb@bbsrc.ac.uk

Phone: +44 1970 823 124 Fax: +44 1970 828 357

Keywords: Soluble phenolic compounds, flavonoids, phenolic acids, phenylpropanoids, *Lotus japonicus*, HPLC-PDA, secondary metabolites

Phenolic compounds make up one of the major families of secondary metabolites in plants, and they represent a diverse group of compounds. Phenolics can be broadly divided into non-soluble compounds such as condensed tannins, lignins, and cell-wall bound hydroxycinammic acids, and soluble phenolics such as phenolic acids, phenylpropanoids, flavonoids and quinones. In this article, a method for extracting and analysing the soluble fraction of these phenolics, using HPLC coupled to photodiode array (PDA) detector, is described along with its uses to profile the flavonoid content of Lotus japonicus.

INTRODUCTION

Phenolic compounds are a large and diverse group of molecules, which includes many different families of aromatic secondary metabolites in plants. These phenolics are the most abundant secondary metabolites in plants and can be classified into non-soluble compounds such as condensed tannins, lignins, cell-wall bound hydroxycinammic acids, and soluble compounds such as phenolic acids, phenylpropanoids, flavonoids and quinones. All these groups are involved in many processes in plants and animals. One family, the flavonoids, is of particular interest because of its multiple roles in plants and its impact on human health (Harborne and Williams, 2000).

In plants, flavonoids play a role in flower and seed pigmentation, in plant fertility and reproduction, and in various defence reactions to protect against abiotic stresses like UV light or biotic stresses such as predator and pathogen attacks (Weisshaar and Jenkins, 1998; Winkel-Shirley, 2001; Forkmann and Martens, 2001). Evidence also

A.J. Márquez (ed.), Lotus japonicus Handbook, 349-354.

suggests their involvement in plant growth and development since they have been shown to regulate polar auxin transport (Murphy et al., 2000; Brown et al., 2001).

In legumes, they play a significant role in the nitrogen-fixing symbiosis at two different levels. First, they regulate the synthesis of Nod factors in *Rhizobium*, which is responsible for the inception of the nodule (Peters et al., 1986; Zaat et al., 1988; Recourt et al., 1992). Second, they act as molecule signals in the transduction pathways of the Nod factor leading to the inhibition of auxin transport within the root cortical cells and therefore to the formation of nodule primordia (Mathesius et al., 1998). Moreover, there is some evidence indicating that flavonoids may also be involved in arbuscular mycorrhiza symbiosis (Harrison and Dixon, 1993; Harrison and Dixon, 1994; Akiyama et al., 2002) although in a non essential manner (Becard et al., 1995).

The flavonoid family is subdivided into different sub-families such as flavones, isoflavones, flavanones, flavonols, anthocyanins, chalcones and condensed tannins. All of these types of compounds comprise three aromatic rings harbouring different substitutions such as methylation and hydroxylation (Figure 1; Aoki et al., 2000). Most of the flavonoid aglycones are found in a glycosylated form in plant cells, this is assumed to protect them from degradation, to reduce their toxic effects and to aid their transport across membranes by increasing their water solubility (Jones and Vogt, 2001). These compounds, in their aglycone or glycosylated forms, are located in the vacuoles within plant cells and are in the polar soluble fraction. Therefore, flavonoids can be easily extracted with polar solvents such as methanol, which is not the case for insoluble lignins and tannins that bind to proteins on cell disruption during the extraction. Then, flavonoids and other phenolics present in the polar soluble fraction can be separated by reverse phase HPLC equipped with a photo diode array detector (HPLC-PDA) which establishes the light absorbance spectrum from visible and UV wavelengths of each detected compound, as phenolic compounds are aromatic and show intense absorption in the UV region of the spectrum.

In this article, we describe a method that was developed to profile soluble phenolics, which are mainly flavonoid glycosides, and its application in profiling the flavonoid content of our model species, *Lotus japonicus*.

PROCEDURE

Materials

The procedure described below is set up for the analysis of the major phenolic compounds on a Water Ass HPLC system. This system is equipped with μ-Nova-Pak reverse phase C_{18} 8x10 RCM column (Water Ass), a photodiode array detector (Water 996 PDA detector, Water Ass) with Millenium 32® software, for the separation, detection and data analysis respectively. All glassware, consumables, and solutions used are described in details in the protocol.

Prior to the extraction of the major phenolic compounds, the plant samples to be analysed are freeze-dried and ground to a fine powder.

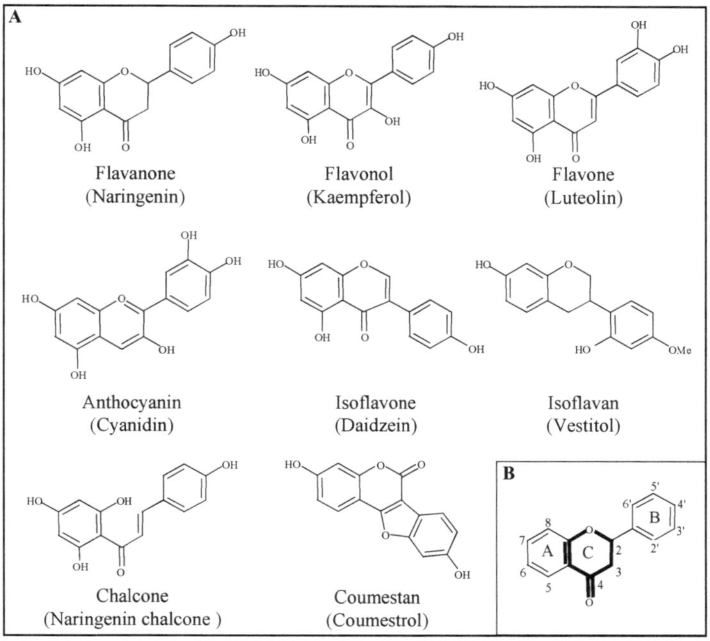

Figure 1. Flavonoid structures (Adapted from Aoki et al.*, 2000).* ***A:*** *Chemical structure of selected subclasses and biologically active flavonoids.* ***B:*** *General structure of flavonoid compounds. The different flavonoid classes are defined according to the structure of the C-ring and the functions carried by C_3 and C_4 (in bold).*

Extraction of Phenolic Compounds

Flavonoid compounds are extracted from 50 or 100 mg of freeze-dried sample in 5 ml of 70% methanol HPLC grade (Fisher Chemical, UK), and shaken for 15 min at room temperature. After centrifugation (3,000 rpm, 5 min), the supernatant is filtered on Fisherbrand QL125 90 mm filter paper (Fisher Science Corporation, UK). The pellet is then re-extracted with 5 ml of 70% methanol and finally rinsed with 5 ml of 100% methanol. All three supernatants are pooled together before removal of the methanol under vacuum with a rota-evaporator at 70°C (Büchi, Switzerland). When the methanol has evaporated, the extract is concentrated and purified using an activated Sep Pack C_{18} column (Sep-Pak®RC Cartridge, Water) and eluted with 4 ml of 100% HPLC grade methanol before being stored at 4°C prior to analysis.

Analysis of Phenolic compounds

Between 0.4 to 4 ml of extracted sample is dried using a SC-3 sample concentrator (Techne) and resuspended in 30 μl of 70% methanol.

Samples are analysed by high performance liquid chromatography on a Waters Millennium 32® system. Flavonoids are separated by gradient HPLC on a μ-Nova-Pak reverse phase C_{18} (8x10) RCM column (Water Ass) with a linear methanol/acetic acid (5%) gradient from 0 to 100% methanol in 50 min at a flow rate of 2 ml/min. Eluting peaks are monitored with a photodiode array detector (Water 996 PDA detector, Water Ass) at 280, 340, and 380 nm and spectra are recorded between 240 to 400 nm. Detection is realised using these three wavelengths in order to observe a wide range of phenolic classes: phenolic acids and isoflavonoids at 280nm, flavonoids and coumestans at 340nm and chalcones at 380nm. Data analysis is processed using an adapted processing method using Millennium 32® software and the UV/visible spectra of the eluted peaks were matched with those contained in our own libraries containing previously analysed standards and unknown compounds.

COMMENTS

The method described here allows separation and analysis of the soluble fraction of the major phenolic compounds present in plant tissue. It also allows the quantification of the different detected compounds. Therefore, it is the method of choice to profile the phenolic content of plant species like *Lotus japonicus* but also for observing changes in the phenolic profile after challenging the plant with symbionts, pathogens or other stresses.

Although the extraction method used here is one of the easiest and quickest, other methods are available based on differential solvent extraction with increasing polarity, which may be more useful for profiling minor phenolic components. Therefore, the method used must be carefully chosen according to the research aim.

This technique provides us with the UV spectrum of the different compounds detected. In the absence of authentic standards, these UV spectra only allow determination of the compound class. Therefore, this method must be coupled to HPLC-MS and/or NMR analyses in order to identify completely the compounds detected and get some insight into their structure. However, this method was successfully used to obtain the phenolic profile of different tissue of *Lotus japonicus*. For instance, the leaf profile revealed mainly flavonol glycosides derived from kaempferol and quercetin, deduced by comparison with their respective UV spectra (Figure 2).

We also used this technique to evaluate the effect of the arbuscular mycorrhizal fungus colonisation on the phenolic content of *Lotus japonicus* and found some significant qualitative differences (Figure 3).

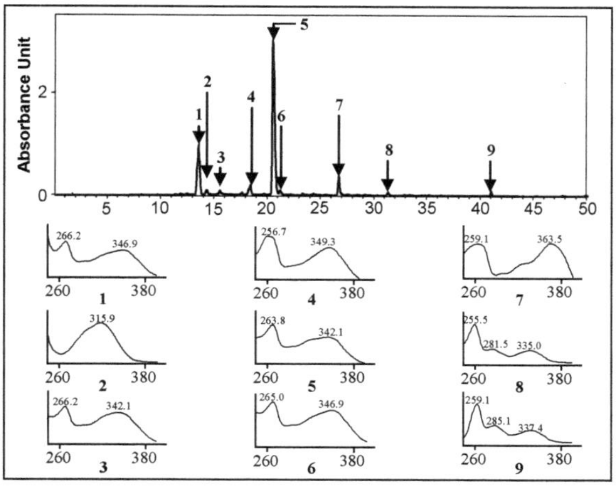

Figure 2. Phenolic compound profile of Lotus japonicus *leaves. Chromatograms were obtained* via *a gradient of 0 to 100 % methanol in 50 minutes using 5 % acetic acid and 100 % methanol after injection onto a μ-Nova-Pak reverse phase C18 column. The UV spectrum of each peak, established by the photo diode array detector, was shown underneath the chromatogram. Compounds 1,3,4,5 and 6: Flavonol Glycoside. 2: Hydroxycinnamic Acid Ester. 7: Kaempferol. Compounds 8 and 9: Unknown.*

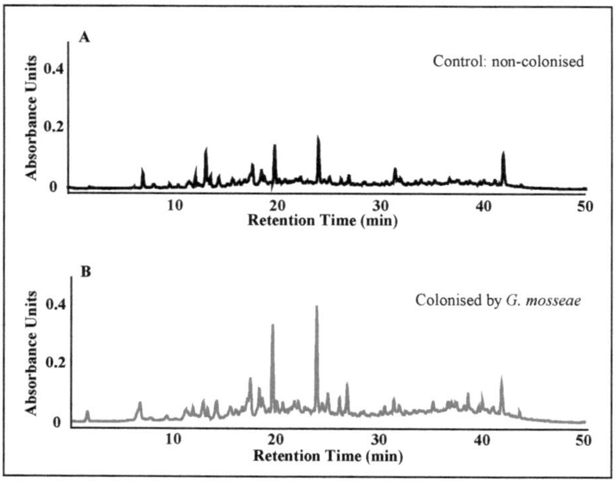

Figure 3. Influence of inoculation with Glomus mosseae *(AM fungus) on flavonoid content of* Lotus japonicus *roots. Comparison of HPLC-PDA chromatograms of* Lotus japonicus *var Gifu non-colonised (A) and colonised by* Glomus mosseae *(B). Chromatograms were obtained* via *a gradient of 0 to 100 % methanol in 50 minutes using 5 % acetic acid and 100 % methanol after injection onto a μ-Nova-Pak reverse phase C_{18} column.*

REFERENCES

Akiyama K, Matsuoka H, and Hayashi H. (2002) **Isolation and identification of a phosphate deficiency-induced C-glycosylflavonoid that stimulates arbuscular mycorrhiza formation in melon roots.** *Molecular Plant-Microbe Interactions* 15, 334-340.

Aoki T, Akash, T, and Ayabe S. (2000) **Flavonoids of leguminous plants: Structure, biological activity, and biosynthesis.** *Journal of Plant Research* 113, 475-488.

Becard G, Taylor LP, Douds DD, Pfeffer PE, and Doner LW. (1995) **Flavonoids Are Not Necessary Plant Signal Compounds in Arbuscular Mycorrhizal Symbioses.** *Molecular Plant-Microbe Interactions* 8, 252-258.

Brown DE, Rashotte AM, Murphy AS, Normanly J, Tague BW, Peer WA, Taiz L, and Muday GK. (2001) **Flavonoids act as negative regulators of auxin transport *in vivo* in Arabidopsis.** *Plant Physiology* 126, 524-535.

Forkmann G and Martens S. (2001) **Metabolic engineering and applications of flavonoids.** *Current Opinion in Biotechnology* 12, 155-160.

Harborne JB and Williams CA. (2000) **Advances in flavonoid research since 1992.** *Phytochemistry* 55, 481-504.

Harrison MJ, and Dixon RA. (1993) **Isoflavonoid Accumulation and Expression of Defence Gene Transcripts During the Establishment of Vesicular-Arbuscular Mycorrhizal Associations in Roots of Medicago-Truncatula.** *Molecular Plant-Microbe Interactions* 6, 643-654.

Harrison MJ, and Dixon, RA. (1994) **Spatial Patterns of Expression of Flavonoid/Isoflavonoid Pathway Genes During Interactions Between Roots of Medicago- Truncatula and the Mycorrhizal Fungus Glomus Versiforme.** *The Plant Journal* 6, 9-20.

Jones P, and Vogt, T. (2001) **Glycosyltransferases in secondary plant metabolism: Tranquilizers and stimulant controllers.** *Planta* 213, 164-174.

Mathesius U, Schlaman HRM, Spaink HP, Sautter C, Rolfe BG, and Djordjevic, MA. (1998) **Auxin transport inhibition precedes root nodule formation in white clover roots and is regulated by flavonoids and derivatives of chitin oligosaccharides.** *The Plant Journal* 14, 23-34.

Murphy A, Peer WA, and Taiz L. (2000) **Regulation of auxin transport by aminopeptidases and endogenous flavonoids.** *Planta* 211, 315-324.

Peters NK, Frost JW, and Long SR. (1986) **A plant flavone, luteolin, induces expression of *Rhizobium meliloti* nodulation genes.** *Science* 233, 977-980.

Recourt K, Verkerke M, Schripsema J, van Brussel AA, Lugtenberg BJ, and Kijne JW. (1992) **Major flavonoids in uninoculated and inoculated roots of Vicia sativa subsp. nigra are four conjugates of the nodulation gene-inhibitor kaempferol.** *Plant Molecular Biology* 18, 505-513.

Weisshaar B, and Jenkins G. (1998) **Phenylpropanoid biosynthesis and its regulation.** *Current. Opinion Plant Biology* 1, 251-257.

Winkel-Shirley B. (2001) **Flavonoid Biosynthesis. A Colorful Model for Genetics, Biochemistry, Cell Biology, and Biotechnology.** *Plant Physiology* 126, 485-493.

Zaat SAJ, Wijffelman,CA, Mulders IHM, van Brussel AAN, and Lugtenberg BJJ. (1988) **Root exudates of various host plant of *Rhizobium leguminosarum* contain different sets of inducers of *Rhizobium* nodulation genes.** *Plant Physiology* 86, 1298-1303.

Chapter 7.6

ELICITATION OF ISOFLAVAN PHYTOALEXINS

Alexandra Lanot and Phillip Morris

Institute of Grassland and Environmental Research, Plas Gogerddan, Aberystwyth, Ceredigion, SY23 3EB, Wales, UK

Email: phillip.morris@bbsrc.ac.uk

Phone: +44 1970 823 113 Fax: +44 1970 823 242

Keywords: elicitation, phytoalexins, vestitol, screening, flavonoid analysis, *Lotus japonicus*, isoflavans

Lotus japonicus *makes a good model legume species for the study of induced isoflavonoid phytoalexin biosynthesis. As part of its defence response against pathogens, Lotus leaves have been reported to produce a number of isoflavans, a specific class of isoflavonoids. Model systems for Lotus elicitation have been set up using the thiol reagent reduced glutathione as elicitor and the accumulation of phytoalexins has been monitored by analysis of vestitol, which is excreted into the media. In this article, we present two protocols for the elicitation of Lotus leaves and three methods for the analysis of the induced isoflavan vestitol.*

INTRODUCTION

In plants, the flavonoid pathway has been well-characterised and involves numerous branches leading to the accumulation of a wide range of end products with a spectrum of functions: colourful anthocyanins, antimicrobial isoflavonoids, UV protecting flavonols, deterrent condensed tannins (Koes *et al.*, 1994, Aoki *et al.*, 2000; Iwashina, 2000). They have also been the object of significant interest as many flavonoid compounds exhibit commercially relevant properties (Di Carlo *et al.*, 1999; Humphreys and Chapple, 2000).

In legumes, flavonoids also play an important role in the interactions between plants and micro-organisms (Shirley, 1996; Harborne and Williams, 2000). Flavones and chalcones, for example, are important inducers in the mechanisms of nodulation by symbiotic bacteria, a process almost specifically confined to legumes (Stafford, 1997). Isoflavonoids, which represent a different branch of the flavonoid pathway, are characterised by a migration of the phenyl ring and occur principally in legumes. They are involved in the defence response of plants against pathogens (phytoalexins) (Dixon *et al.*, 1995; Kuc, 1995). Recently, numerous articles have been published

A.J. Márquez (ed.), Lotus japonicus Handbook, 355-361.

describing the growing health-promoting potentials of these molecules (Dixon and Steele, 1999; Dixon *et al.*, 2002).

Phytoalexins from the Leguminosae have already been described and characterised from a range of legume plants (Ingham, 1982). Model experiment systems have also been developed where the defence response is generated by elicitors of biotic or abiotic origin such as cell wall preparations of pathogens, heavy metals or thiol reagents (Wingate *et al.*, 1988; Edwards *et al.*, 1991; Robbins *et al.*, 1995).

In Lotus species, the main phytoalexin biosynthesised is vestitol, which belong to the class of isoflavans (Figure 1). Vestitol has been described in Lotus seedlings after elicitation with a thiol reagent, reduced glutathione (Shimada *et al.*, 2000) while vestitol, and its methylated/demethylated versions sativan and demethylvestitol, were observed in *L corniculatus* leaves and hairy root cultures (Bonde *et al.*, 1973; Robbins *et al.*, 1991). They have also been reported to be induced in *L corniculatus, L uligonus, L hipidus, L edulis* and *L angustissimus* leaves following treatment with fungal pathogens (Ingham, 1977; Ingham and Dewick, 1979; Ingham and Dewick, 1980).

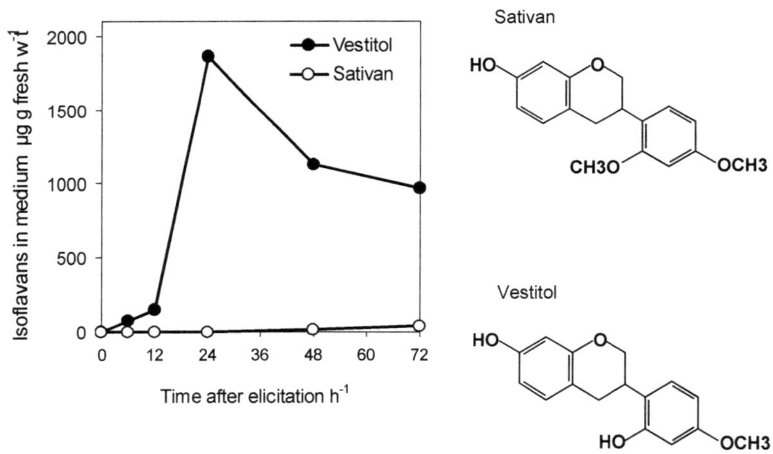

Figure 1. Kinetics of glutathione induced vestitol and sativan accumulation in the incubation medium following elicitation of Lotus japonicus *leaves.*

Glutathione elicitation o*f* Lotus is a good model system to study isoflavonoid phytoalexin biosynthesis and its regulation because of the existence of a collection of mutants (Perry *et al.*, 2003), an EST database (www.kazusa.or.jp/en/plant/lotus/EST/, 21/09/03) and the amenability of Lotus to transformation (Handberg and Stougaard, 1992; Lombari *et al.*, 2003). It is also a good plant model for genetic manipulation of legume flavonoid genes (Weisshaar and Jenkins, 1998; Forkmann and Martens, 2001).

This article describes two protocols for elicitation of Lotus leaves and three methods of analysis of vestitol and sativan: High Pressure Liquid Chromatography (HPLC), Thin Layer Chromatography (TLC) and chemometric analysis of FT-IR spectra. The first protocol describes the elicitation of Lotus leaves with reduced glutathione and the quantification of vestitol and sativan by HPLC with an example of the kinetics of phytoalexin accumulation. The second protocol describes a high throughput method allowing the qualitative analysis of phytoalexin biosynthesis in a large number of plants used as part of a screening program for phytoalexin mutants.

PROCEDURES

Material

Plants are mature, non-flowering greenhouse grown Lotus *cv* GIFU, although similar results have also been obtained with seedling leaves. However, seedling roots and stems have so far failed to elicit isoflavans following glutathione elicitation. Reduced glutathione (Sigma G-4251) is prepared as a 100mM stock solution, filter sterilised and then used immediately. SDS is used as a 10% stock at a final concentration of 0.05% and is necessary to act as a wetting agent.

Kinetics of elicitation of leaves

Elicitation

Two grams of trifoliate leaves are excised from each plant and collected in a glass dish containing sterile water. Three 0.5g leaf aliquots are randomly sampled and placed into three 250 ml sterile Erlenmeyer flasks containing 50 ml of 10 mM glutathione and 0.05% SDS. The flasks are incubated at 25°C in room light and shaken at ca 150 rpm for up to 72 h.

Extraction

Following incubation, leaf tissue is removed and briefly rinsed and the medium samples passed through a glass microfibre filter. The filtered medium is loaded onto activated Sep-Pak C_{18} cartridges (Waters Ass) and eluted drop-wise. The column is then washed with 2 ml distilled water and the flavonoids eluted with 4 ml 100% methanol into evaporation-proof vials, which are stored at 4°C until analysis. For HPLC or TLC, 400 µl of each extract is taken to dryness in a centrifugal vacuum evaporator (Jouan RC 10.22).

High through put screen for vestitol induction in leaves

Elicitation

For high throughput screening, 100 to 200 milligrams of trifoliate leaves are placed in 1.8 ml of distilled water in wells of a 5x5 Repli-dishes (Sterilin 103). 200 µl of glutathione stock solution and 10 µl of stock SDS solution are added to each well. The plates are sealed with parafilm and incubated with shaking at ca 80 rpm in the dark at 25°C for 24 h.

Extraction

The leaves are then removed and allowed to drain and 1 ml of a 20 mg/ml suspension of Waters C_{18} preparative resin is added (Waters WAT020594), this resin had been pre-activated with 100% methanol and diluted to 5% methanol. After 30 min incubation, the medium is separated from the resin using a 96 well vacuum filtration unit. Alternatively, the medium is separated from the resin using a small diameter suction pipette leaving the resin in the plate wells. Isoflavans are eluted by adding 200 μl of 100% methanol to each well and gently mixing. These extracts are then transferred to a 96 well plate and allowed to dry. A second elution is performed with 200 μl 100% methanol and pooled with the first extract.

Analysis methods

High Pressure Liquid Chromatography (HPLC)

Aliquots of the extracts are dried and resuspended in 25 μl 100% methanol. They are then separated on a Waters reverse phase μ-Nova-Pak C_{18} (8x10) column in a Radial Compression Module using a Waters Millennium 32® system.

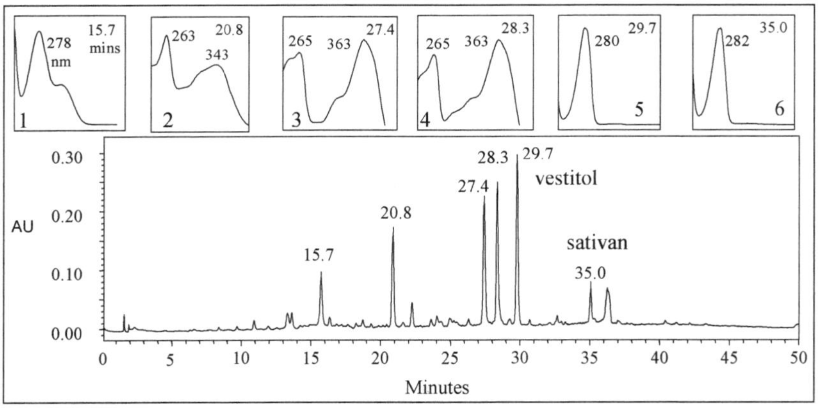

Figure 2. HPLC of medium from glutathione-elicited Lotus japonicus *leaves.*

Elution from the column is carried out with a linear gradient from 0 to 100% methanol in 50 min. The second solvent is 5% acetic acid and the flow rate 2 ml/min. The elution is monitored by a programmable 996 PDA detector (Waters) at 280 nm (isoflavans have a typical λmax at 280 nm, Figure 2) Data is collected from 240 to 400 nm and analysed with Millennium software (Waters, Milford, MA). Vestitol retention time is *ca* 29 min while sativan is eluted after *ca* 35 min (Figure 2).

Chemometric analysis with FT-IR

Extracts are resuspended in 20 μl 100% methanol and loaded onto 96 well aluminium plates that are dried at 60°C for 30 mins. Plates are scanned in reflection mode from 500 to 4000 cm^{-1} on an Equinox 55 FT-IR spectrometer (Brucker Optics Ltd) using a HTS XT microplate attachment and data analysed using OPUS software. Vestitol is quantified against a standard curve obtained with different quantities of vestitol standard spiked into negative control samples following integration of the C-H band (Figure 3).

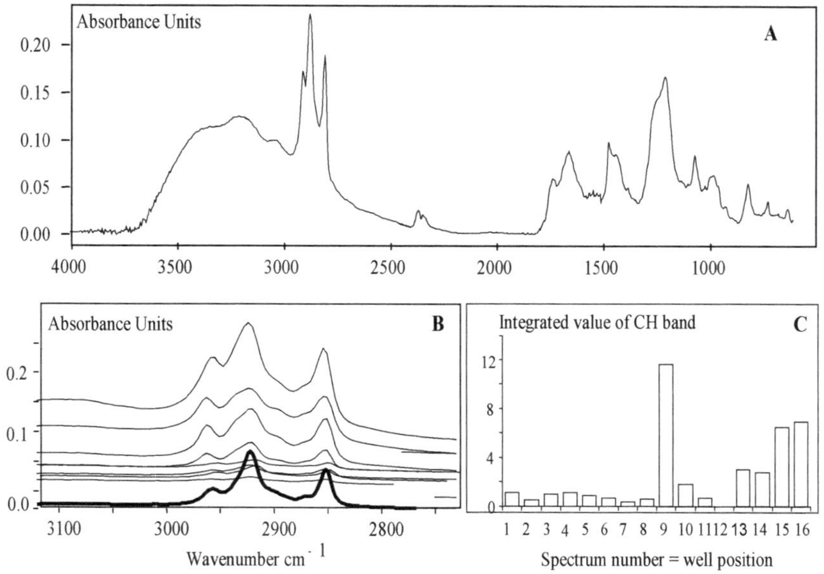

Figure 3. High throughput screening of medium from elicited Lotus leaves for vestitol using FT-IR spectroscopy. (A) FT-IR spectrum of vestitol. (B) C-H Stretching range, reference spectra vestitol (bold). (C) Integrated value of C-H band, wells 1-11 20 µl dried medium samples, wells 13-16 = 2 or 5 µg vestitol as a standard.

Thin Layer Chromatography (TLC)

TLC is carried out on 0.25mm Silica gel G plates (Macherey-Nagel) developed in $CHCl_3$:MeOH (15:1). Isoflavans are visualised by spraying with Gibbs reagent (0.4% dichloroquinone 4, chloroimide in methanol (SIGMA D-6511) followed by 20% Na_2CO_3) or with 1% diazotised p-nitroanaline. Vestitol spots immediately turn blue at rf 0.2-0.3 with Gibbs reagent. If only vestitol is of interest, a higher ratio of methanol ($CHCl_3$: MeOH (50:6)) is utilised to increase its rf value. Other isoflavans develop slowly with a blue or purple colour. Sativan is not visible with Gibbs reagent but appears yellow with diazotised p-nitroanaline at an rf value of 0.6-0.7.

DISCUSSION

Leaves of Lotus were elicited in order to identify and quantify the isoflavan phytoalexins biosynthesised and to investigate the kinetic characteristics of the response in this tissue (Figure 1). Analysis of the elicitation medium was sufficient as the isoflavans excreted into the medium accounted for more than 90% of the total isoflavan content (data not shown). The main isoflavan identified was vestitol, with maximum accumulation after 24 h. Other isoflavans were also identified: sativan was detected after 48 h incubation and its presence coincides with a decrease in vestitol concentration. It is probable that sativan, (methylated vestitol - Figure 1), was synthesised from vestitol. Smaller amounts of other isoflavans were also detected after 72 h, which have yet to be identified. Other classes of flavonoids were also observed on the HPLC chromatogram. Lotus leaves contain constitutively high concentration of kaempferol and quercetin glycosides, which are also lost from the leaves on treatment with glutathione and SDS (see spectra 2 to 4 in Figure 2).

This model system could be utilised for the identification (and/or cloning) of the flavonoid genes involved in defence response as reported by Shimada (Shimada, et al., 2000). It would also allow the investigation of the control mechanisms that direct flux toward the synthesis of specific classes of flavonoids. In particular, several genes from this pathway are present in plants as gene families: phenylalanine ammonia-lyase, chalcone synthase, chalcone isomerase. It has been proposed that individual members of these gene families are specifically involved in the biosynthesis of some flavonoids. By identifying which gene family members are specific to the defence response; it might be possible to genetically engineer new plants where the concentration of flavonoids of interest would be modulated. The second protocol describes a method to rapidly investigate the phytoalexin induction of a high number of plants and would be ideal for the screening of mutants or transgenics with modified phytoalexin responses (Figure 3).

ACKNOWLEDGEMENTS

The Institute of Grassland and Environmental Research is grant aided by the BBSRC and Alexandra Lanot is funded by a BBSRC Industrial CASE PhD studentship.

REFERENCES

Aoki T, Akashi T, and Ayabe S. (2000) **Flavonoids of leguminous plants: Structure, biological activity, and biosynthesis**. *Journal of Plant Research* 113, 475-488.

Bonde MR, Millar RL, and Ingham JL (1973) **Induction and identification of sativan and vestitol as two phytoalexins from *Lotus corniculatus***. *Phytochemistry* 12, 2957-2959.

Di Carlo G, Mascolo N, Izzo AA, and Capasso F. (1999) **Flavonoids: Old and new aspects of a class of natural therapeutic drugs**. *Life Sciences* 65, 337-353.

Dixon RA, Harrison MJ, and Paiva NL (1995) **The isoflavonoid phytoalexin pathway - from enzymes to genes to transcription factors**. *Physiologia Plantarum* 93, 385-392.

Dixon RA and Steele CL (1999) **Flavonoids and isoflavonoids - a gold mine for metabolic engineering**. *Trends in Plant Science* 4, 394-400.

Dixon RA, Achnine L, Kota P, Liu CJ, Reddy MSS, and Wang LJ. (2002) **The phenylpropanoid pathway and plant defence - a genomics perspective**. *Molecular Plant Pathology* 3, 371-390.

Edwards R, Blount JW, and Dixon RA. (1991) **Glutathione and elicitation of the phytoalexin response in legume cell-cultures**. *Planta* 184, 403-409.

Forkmann G, and Martens S. (2001) **Metabolic engineering and applications of flavonoids**. *Current Opinion in Biotechnology* 12,155-160.

Handberg K, and Stougaard J. (1992) ***Lotus japonicus*, an autogamous, diploid legume species for classical and molecular-genetics**. *The Plant Journal* 2, 487-496.

Harborne JB, and Williams CA. (2000) **Advances in flavonoid research since 1992**. *Phytochemistry* 55, 481-504.

Humphreys JM, and Chapple C. (2000) **Molecular 'pharming' with plant P450s**. *Trends in Plant Science* 5, 271-272.

Ingham JL (1977) **Isoflavan phytoalexins from *Anthyllis, Lotus* and *Tetragonolobus***. *Phytochemistry* 16, 1279-1282.

Ingham JL, and Dewick PM. (1979) **A new isoflavan phytoalexin from leaflets of *Lotus hipidus***. *Phytochemistry* 18, 1711-1714.

Ingham JL, and Dewick PM. (1980) **Isolation of a new isoflavan phytoalexin from 2 *Lotus* species**. *Phytochemistry* 19, 2799-2800.

Ingham JL (1982) **Phytoalexins of the Leguminosae** In: *Phytoalexins,* (Bailey JA, and Mansfield, JW Ed.). Glasgow: Blackie. pp. 21-80.

Iwashina T. (2000) **The structure and distribution of the flavonoids in plants**. *Journal of Plant Research* 113, 287-299.

Koes RE, Quattrocchio F, and Mol JNM. (1994) **The flavonoid biosynthetic-pathway in plants - Function and evolution**. *Bioessays* 16, 123-132.

Kuc J. (1995) **Phytoalexins, stress metabolism, and disease resistance in plants**. *Annual Review of Phytopathology* 33, 275-297.

Lombari P, Ercolano E, El Alaoui H, and Chiurazzi M. (2003) **A new transformation-regeneration procedure in the model legume *Lotus japonicus*: root explants as a source of large numbers of cells susceptible to *Agrobacterium*-mediated transformation**. *Plant Cell Reports* 21, 771-777.

Perry JA, Wang TL, Welham TJ, Gardner S, Pike JM, Yoshida S, and Parniske M. (2003) **A TILLING reverse genetics tool and a web-accessible collection of mutants of the legume *Lotus japonicus***. *Plant Physiology* 131, 866-871.

Robbins MP, Hartnoll J, and Morris P. (1991) **Phenylpropanoid defense responses in transgenic *Lotus corniculatus* .1. Glutathione elicitation of isoflavan phytoalexins in transformed root cultures**. *Plant Cell Reports* 10, 59-62.

Robbins MP, Thomas B, and Morris P. (1995) **Phenylpropanoid defense responses in transgenic *Lotus corniculatus*. 2. Modelling plant defence responses in transgenic root cultures using thiol and carbohydrate elicitors**. *Journal of Experimental Botany* 46, 513-524.

Shimada N, Akashi T, Aoki T, and Ayabe S. (2000) **Induction of isoflavonoid pathway in the model legume *Lotus japonicus*: molecular characterization of enzymes involved in phytoalexin biosynthesis**. *Plant Science* 160, 37-47.

Shirley BW. (1996) **Flavonoid biosynthesis: 'New' functions for an 'old' pathway**. *Trends Plant Science* 1, 377-382.

Stafford HA. (1997) **Roles of flavonoids in symbiotic and defence functions in legume roots**. *Botanical Reviews* 63, 27-39.

Weisshaar B and Jenkins GI. (1998) **Phenylpropanoid biosynthesis and its regulation**. *Current Opinion in Plant Biology* 1, 251-257.

Wingate VPM, Lawton MA, and Lamb CJ. (1988) **Glutathione causes a massive and selective induction of plant defence genes**. *Plant Physiology* 87, 206-210.

SECTION 8

LOTUS JAPONICUS LITERATURE

Chapter 8.1

An update of work published on *Lotus japonicus*

Marco Betti and Antonio J Márquez*

*Departamento de Bioquímica Vegetal y Biología Molecular, Facultad de Química, Universidad de Sevilla, Apartado 553, 41080-Sevilla, Spain. *Corresponding author.*

Email: cabeza@us.es

Phone: *+34 954 557 145* Fax: *+34 954 626 853*

The following selection of Lotus japonicus *literature is arranged in chronological order to provide historical context.*

References

Márquez AJ, Orea A, Pajuelo P, Pajuelo E, Romero JM, Arcondéguy T, Betti M, García-Calderón M, Estivill G, and Pal'ove-Balang P. (2004) **Nitrogen assimilation in roots of the model legume *Lotus japonicus*.** *Biologia, Bratislava,* 59/Suppl. 13, 69-76.

Imaizumi-Anraku H, Takeda N, Charpentier M, Perry J, Miwa H, Umehara Y, Kouchi H, Murakami Y, Mulder L, Vickers K, Pike J, Allan Downie J, Wang T, Sato S, Asamizu E, Tabata S, Yoshikawa M, Murooka Y, Wu GJ, Kawaguchi M, Kawasaki S, Parniske M, and Hayashi M. (2004) **Plastid proteins crucial for symbiotic fungal and bacterial entry into plant roots.** *Nature* (Epub 2004 Dec 22).

Hakozaki H, Endo M, Masuko H, Park JI, Ito H, Uchida M, Kamada M, Takahashi H, Higashitani A, and Watanabe M. (2004) **Cloning and expression pattern of a novel microspore-specific gene encoding hypersensitive-induced response protein (LjHIR1) from the model legume, Lotus japonicus.** *Genes & Genetic Systems* 79, 307-310.

Flemetakis E, Efrose RC, Desbrosses G, Dimou M, Delis C, Aivalakis G, Udvardi MK, and Katinakis P. (2004) **Induction and Spatial Organization of Polyamine Biosynthesis During Nodule Development in *Lotus japonicus*.** *Molecular Plant-Microbe Interactions* 17, 1283-1293.

Choi HK, Mun JH, Kim DJ, Zhu HY, Baek JM, Mudge J, Roe B, Ellis N, Doyle J, Kiss GB, Young ND, and Cook DR. (2004) **Estimating genome conservation between crop and model legume species.** *Proceeding of the National Academy of Sciences of the United States of America* 101, 15289-15294.

Suganuma N, Yamamoto A, Itou A, Hakoyama T, Banba M, Hata S, Kawaguchi M, and Kouchi H. (2004) **cDNA macroarray analysis of gene expression in ineffective nodules**

A.J. Márquez (ed.), Lotus japonicus Handbook, 365-381.

induced on the *Lotus japonicus* sen1 mutant. *Molecular Plant-Microbe Interactions* 17, 1223-1233.

Kouchi H, Shimomura K, Hata S, Hirota A, Wu GJ, Kumagai H, Tajima S, Suganuma N, Suzuki A, Aoki T, Hayashi M, Yokoyama T, Ohyama T, Asamizu E, Kuwata C, Shibata D, and Tabata S. (2004) **Large-scale analysis of gene expression profiles during early stages of root nodule formation in a model legume, *Lotus japonicus*.** *DNA Research* 4, 263-274.

Kalsi G, Roberts NJ, Day RB, Stacey G, and Etzler ME. (2004) **Expression of Dolichos biflorus LNP in *Lotus japonicus* broadens its host-strain specificity.** *Glycobiology* 14, 1130-1130.

Ortega JL, Temple SJ, Bagga S, Ghoshroy S, and Sengupta-Gopalan C. (2004) **Biochemical and molecular characterization of transgenic *Lotus japonicus* plants constitutively over-expressing a cytosolic glutamine synthetase gene.** *Planta* 219, 807-818.

Ferraioli S, Tate R, Rogato A, Chiurazzi M, and Patriarca EJ. (2004) **Development of ectopic roots from abortive nodule primordia.** *Molecular Plant-Microbe Interactions* 17, 1043-1050.

Kinoshita N, Ooki Y, Deguchi Y, Chechetka SA, Kouchi H, Umehara Y, Izui K, and Hata S. (2004) **Cloning and expression analysis of a MAPKKK gene and a novel nodulin gene of *Lotus japonicus*.** *Bioscience Biotechnology and Biochemistry* 68, 1805-1807.

Arimura G, Ozawa R, Kugimiya S, Takabayashi J, and Bohlmann J. (2004) **Herbivore-induced defense response in a model legume. Two-spotted spider mites induce emission of (E)-beta-ocimene and transcript accumulation of (E)-beta-ocimene synthase in *Lotus japonicus*.** *Plant Physiology* 135, 1976-1983.

Valot B, Gianinazzi S, and Eliane DG. (2004) **Sub-cellular proteomic analysis of a *Medicago truncatula* root microsomal fraction.** *Phytochemistry* 65, 1721-1732.

Suzuki A, Akune M, Kogiso M, Imagama Y, Osuki K, Uchiumi T, Higashi S, Han SY, Yoshida S, Asami T, and Abe M. (2004) **Control of nodule number by the phytohormone abscisic acid in the roots of two leguminous species.** *Plant and Cell Physiology* 45, 914-922.

Gaude N, Tippmann H, Flemetakis E, Katinakis P, Udvardi M, and Dormann P. (2004) **The galactolipid digalactosyldiacylglycerol accumulates in the peribacteroid membrane of nitrogen-fixing nodules of soybean and Lotus.** *Journal of Biological Chemistry* 279, 34624-34630.

Asamizu E, Nakamura Y, Sato S, and Tabata S. (2004) **Characteristics of the *Lotus japonicus* gene repertoire deduced from large-scale expressed sequence tag (EST) analysis.** *Plant Molecular Biology* 54, 405-414.

Colebatch G, Desbrosses G, Ott T, Krusell L, Montanari O, Kloska S, Kopka J, and Udvardi MK. (2004) **Global changes in transcription orchestrate metabolic differentiation during symbiotic nitrogen fixation in *Lotus japonicus*.** *Plant Journal* 39, 487-512.

Parniske M. (2004) **Molecular genetics of the arbuscular mycorrhizal symbiosis.** *Current Opinion in Plant Biology* 7, 414-421.

Graham MA, Silverstein KAT, Cannon SB, and VandenBosch KA. (2004) **Computational identification and characterization of novel genes from legumes.** *Plant Physiology* 135, 1179-1197.

Demchenko K, Winzer T, Stougaard J, Parniske M, and Pawlowski K. (2004) **Distinct roles of *Lotus japonicus* SYMRK and SYM15 in root colonization and arbuscule formation.** *New Phytologist* 163, 381-392.

Desbrosses G, Kopka C, Ott T, and Udvardi MK. (2004) ***Lotus japonicus*** **LjKUP is induced late during nodule development and encodes a potassium transporter of the plasma membrane.** *Molecular Plant-Microbe Interactions* 17, 789-797.

Tansengco ML, Imaizumi-Anraku H, Yoshikawa M, Takagi S, Kawaguchi M, Hayashi M, and Murooka Y. (2004) **Pollen development and tube growth are affected in the symbiotic mutant of *Lotus japonicus* crinkle.** *Plant and Cell Physiology* 45, 511-520.

Spaink HP. (2004) **Specific recognition of bacteria by plant LysM domain receptor kinases.** *Trend in Microbiology* 12, 201-204.

Forslund K, Moran M, Jorgensen B, Olsen CE, Asamizu E, Sato S, Tabata S, and Bak S. (2004) **Biosynthesis of the nitrile glucosides rhodiocyanoside A and D and the cyanogenic glucosides lotaustralin and linamarin in *Lotus japonicus*.** *Plant Physiology* 135, 71-84.

Nukui N, Ezura H, and Minamisawa K. (2004) **Transgenic *Lotus japonicus* with an ethylene receptor gene Cm-ERS1/H70A enhances formation of infection threads and nodule primordia.** *Plant and Cell Physiology* 45, 427-435.

D'Apuzzo E, Rogato A, Simon-Rosin U, El Alaoui H, Barbulova A, Betti M, Dimou M, Katinakis P, Marquez A, Marini AM, Udvardi MK, and Chiurazzi M. (2004) **Characterization of three functional high-affinity ammonium transporters in *Lotus japonicus* with differential transcriptional regulation spatial expression.** *Plant Physiology* 134, 1763-1774.

Aoki T, Sawai S, and Ayabe S. (2004) **Structures and functions of oxidosqualene cyclase gene in families in Lotus japonicus.** *Plant and Cell Physiology* 45, S5-S5.

Tabata S. (2004) **Resources for genomics and genetics in a legume, *Lotus japonicus*.** *Plant and Cell Physiology* 45, S15-S15.

Akashi R. (2004) **The National Bio Resource Project for the Legume, *Lotus japonicus* and *Glycine max*.** *Plant and Cell Physiology* 45, S15-S15.

Nakamura Y, Kaneko T, Asamizu E, Kato T, Sato S, and Tabata S. (2004) **Current status of a large scale genome analysis of a model legume, *Lotus japonicus*.** *Plant and Cell Physiology 45*, S37-S37.

Imaizumi-Anraku H, Takeda N, Umehara Y, Murakami Y, Yoshikawa M, Sato S, Asamizu E, Tabata S, Charpentier M, Mulder L, Perry J, Parniske M, Murooka Y, Kouchi H, Kawaguchi M, Hayashi M, and Kawasaki S. (2004) **TRINITY controls nodulation and mycorrhization concomitantly in *Lotus japonicus*.** *Plant and Cell Physiology* 45, S137-S137.

Yano K, Vickers K, Perry J, Sato S, Asamizu E, Tabata S, Kawaguchi M, Murooka Y, Parniske M, and Hayashi M. (2004) **SYM82 of *Lotus japonicus* is required for symbiosis of rhizobia and arbuscular mycorrhizal fungi.** *Plant and Cell Physiology* 45, S138-S138.

Hayashi M, Tansengco M, Yano K, Imaizumi-Anraku H, Kawasaki S, Sato S, Tabata S, Kawaguchi M, and Murooka Y. (2004) **Genetic analysis of infection thread mutants in *Lotus japonicus*.** *Plant and Cell Physiology* 45, S138-S138.

Ishikawa N, Sakai T, Inada S, and Tsukaya H. (2004) **Analysis of the stipule-less mutants in *Lotus japonicus*.** *Plant and Cell Physiology* 45, S163-S163.

Nishida H, Sakurai N, Suzuki H, and Shibata D. (2004) **cDNA macroarray analysis in *Lotus japonicus* suspension-cultured cells.** *Plant and Cell Physiology* 45, S166-S166.

Takikawa Y, Kakutani K, Moriura N, Yokoyama H, Nonomura T, Matsuda Y, and Toyoda H. (2004) **Expression analysis of transporter genes in *Lotus japonicus*.** *Plant and Cell Physiology* 45, S168-S168.

Mai TH, Nomura M, Hirashima Y, Isomoto M, Okamoto E, Asamizu E, Sato S, Kato T, Tabata S, Takegawa K, and Tajima S. (2004) **Functional analysis of Sed5-like gene in *Lotus japonicus*.** *Plant and Cell Physiology* 45, S169-S169.

Nomura M, Fujii M, Kenmotsu T, Hata S, and Tajima S. (2004) **Functional analysis of PEPC in *Lotus japonicus*.** *Plant and Cell Physiology* 45, S170-S170.

Shimoda Y, Suzuki A, Abe M, Higashi S, Sato S, Tabata S, and Uchiumi T. (2004) **Functions of nonsymbiotic globin genes in *Lotus japonicus*.** *Plant and Cell Physiology* 45, S171-S171.

Kumagai H, Umehara Y, Sato S, Kaneko T, Tabata S, and Kouchi H. (2004) **Phenotypic and genetic analysis of the *Lotus japonicus* symbiotic mutant 56M.** *Plant and Cell Physiology* 45, S187-S187.

Umehara Y, Chen WL, Hossain MS, Maekawa T, Hayashi M, Kojima T, Ohtomo R, Hayashi M, Harada K, and Kouchi H. (2004) **Symbiotic mutants of *Lotus japonicus* derived from cultured cells.** *Plant and Cell Physiology* 45, S188-S188.

Maekawa T, Hayashi M, Asamizu E, Tabata S, Kouchi H, Murooka Y. (2004) **Transcriptional analysis of nitrate response in root hairs of *Lotus japonicus*, jasmonic acid responsive genes are controlled by nitrate.** *Plant and Cell Physiology* 45, S194-S194.

Asamizu E, Nakamura Y, Sato S, and Tabata S. (2004) **Large-scale gene expression analysis of the *Lotus japonicus* nodulation process by SAGE, and analysis of the promoter region responsible for the induction.** *Plant and Cell Physiology* 45, S202-S202.

Yano K, Sakurai N, Nishida H, Namai K, Sakai Y, Suzuki H, Asamizu E, Tabata S, Saito K, and Shibata D. (2004) **Establishment and large-scale analysis of full-length cDNA libraries in *Lotus japonicus*.** *Plant and Cell Physiology* 45, S204-S204.

Nakano M, Yamamoto R, Fujisawa N, Inada S, Okada K, and Sakai T. (2004) **Comprehensive preparation of mutants in the model plant of legume, *Lotus japonicus*.** *Plant and Cell Physiology* 45, S206-S206.

Yoshikawa M, Kouchi H, Takagi S, Kawaguchi M, Hayashi M, and Murooka Y. (2004) **Root hair deformation and actin cytoskeleton behavior in early symbiotic response deficient mutants of *Lotus japonicus*.** *Plant and Cell Physiology* 45, S218-S218.

Lohar DP, Schaff JE, Laskey JG, Kieber JJ, Bilyeu KD, and Bird DM. (2004) **Cytokinins play opposite roles in lateral root formation, and nematode and Rhizobial symbioses.** *Plant Journal* 38, 203-214.

Uchiumi T, Ohwada T, Itakura M, Mitsui H, Nukui N, Dawadi P, Kaneko T, Tabata S, Yokoyama T, Tejima K, Saeki K, Omori H, Hayashi M, Maekawa T, Sriprang R, Murooka Y, Tajima S, Simomura K, Nomura M, Suzuki A, Shimoda Y, Sioya K, Abe M, and Minamisawa K. (2004) **Expression islands clustered on the symbiosis island of the *Mesorhizobium loti* genome.** *Journal of Bacteriology* 186, 2439-2448.

Edwards ME, Choo TS, Dickson CA, Scott C, Gidley MJ, and Reid JSG. (2004) **The seeds *of Lotus japonicus* lines transformed with sense, antisense, and sense/antisense galactomannan galactosyltransferase constructs have structurally altered galactomannans in their endosperm cell walls.** *Plant Physiology* 134, 1153-1162.

Kato T, Sato S, Nakamura Y, Kaneko T, Asamizu E, and Tabata S. (2003) **Structural analysis of a *Lotus japonicus* genome. V. Sequence features and mapping of sixty-four TAC clones which cover the 6.4 Mb regions of the genome.** *DNA Research* 10, 277-285.

Bartsev AV, Deakin WJ, Boukli NM, McAlvin CB, Stacey G, Malnoe P, Broughton WJ, and Staehelin C. (2004) **NopL, an effector protein of Rhizobium sp NGR234, thwarts activation of plant defense reactions.** *Plant Physiology* 134, 871-879.

Stracke S, Sato S, Sandal N, Koyama M, Kaneko T, Tabata S, and Parniske M. (2004) **Exploitation of colinear relationships between the genomes of *Lotus japonicus*, *Pisum sativum* and *Arabidopsis thaliana*, for positional cloning of a legume symbiosis gene.** *Theoretical and Applied Genetics* S108, 442-449.

Wandrey M, Trevaskis B, Brewin N, and Udvardi MK. (2004) **Molecular and cell biology of a family of voltage-dependent anion channel porins in *Lotus japonicus*.** *Plant Physiology* 134, 182-193.

Harrison J, Hirel B, and Limami AM. (2004) **Variation in nitrate uptake and assimilation between two ecotypes of *Lotus japonicus* and their recombinant inbred lines.** *Physiologia Plantarum* 120, 124-131.

Miyake K, Ito T, Senda M, Ishikawa R, Harada T, Niizeki M, and Akada S. (2003) **Isolation of a subfamily of genes for R2R3-MYB transcription factors showing up-regulated expression under nitrogen nutrient-limited conditions**. *Plant Molecular Biology* 53, 237-245.

Cannon SB, McCombie WR, Sato S, Tabata S, Denny R, Palmer L, Katari M, Young ND, and Stacey G. (2003) **Evolution and microsynteny of the apyrase gene family in three legume genomes.** *Molecular Genetics and Genomics* 270, 347-361.

Lohar DP, and Bird DM. (2003) ***Lotus japonicus*, A new model to study root-parasitic nematodes.** *Plant and Cell Physiology* 44, 1176-1184.

Doll J, Hause B, Demchenko K, Pawlowski K, and Krajinski F. (2003) **A member of the germin-like protein family is a highly conserved mycorrhiza-specific induced gene.** *Plant and Cell Physiology* 44, 1208-1214.

Bartsev AV, Boukli NM, Deakin WJ, Staehelin C, and Broughton WJ. (2003) **Purification and phosphorylation of the effector protein NopL from *Rhizobium* sp NGR234.** *FEBS Letters* 554, 271-274.

Pacios-Bras C, Schlaman HRM, Boot K, Admiraal P, Langerak JM, Stougaard J, Spaink HP. (2003) **Auxin distribution in *Lotus japonicus* during root nodule development**. *Plant Molecular Biology* 52, 1169-1180.

Andersen SU, Cvitanich C, Hougaard BK, Roussis A, Gronlund M, Jensen DB, Frokjaer LA, and Jensen EO. (2003) **The glucocorticoid-inducible GVG system causes severe growth defects in both root and shoot of the model legume *Lotus japonicus*.** *Molecular Plant-Microbe Interactions* 16, 1069-1076.

Matamoros MA, Clemente MR, Sato S, Asamizu E, Tabata S, Ramos J, Moran JF, Stiller J, Gresshoff PM, and Becana M. (2003) **Molecular analysis of the pathway for the synthesis of thiol tripeptides in the model legume *Lotus japonicus*.** *Molecular Plant-Microbe Interactions* 16, 1039-1046.

Sawbridge T, Ong EK, Binnion C, Emmerling M, Meath K, Nunan K, O'Neill M, O'Toole F, Simmonds J, Wearne K, Winkworth A, and Spangenberg G. (2003) **Generation and analysis of expressed sequence tags in white clover (Trifolium repens L.).** *Plant Science* 165, 1077-1087.

Bertioli DJ, Leal-Bertioli SCM, Lion MB, Santos VL, Pappas G, Cannon SB, and Guimaraes PM. (2003) **A large scale analysis of resistance gene homologues in Arachis.** *Molecular Genetics and Genomics* 270, 34-45.

Radutoiu S, Madsen LH, Madsen EB, Felle HH, Umehara Y, Gronlund M, Sato S, Nakamura Y, Tabata S, Sandal N, and Stougaard J. (2003) **Plant recognition of symbiotic bacteria requires two LysM receptor-like kinases.** *Nature* 425, 585-592.

Madsen EB, Madsen LH, Radutoiu S, Olbryt M, Rakwalska M, Szczyglowski K, Sato S, Kaneko T, Tabata S, Sandal N, and Stougaard J. (2003) **A receptor kinase gene of the LysM type is involved in legume perception of rhizobial signals.** *Nature* 425, 637-640.

Harrison J, de Crescenzo MAP, Sene O, and Hirel B. (2003) **Does lowering glutamine synthetase activity in nodules modify nitrogen metabolism and growth of *Lotus japonicus*?.** *Plant Physiology* 133, 253-262.

VandenBosch KA, and Stacey G. (2003) **Summaries of legume genomics projects from around the globe. Community resources for crops and models.** *Plant Physiology* 131, 840-865.

Perry JA, Wang TL, Welham TJ, Gardner S, Pike JM, Yoshida S, and Parniske M. (2003) **A TILLING reverse genetics tool and a web-accessible collection of mutants of the legume *Lotus japonicus*.** *Plant Physiology* 131, 866-871.

Somers DA, Samac DA, and Olhoft PM. (2003) **Recent advances in legume transformation.** *Plant Physiology* 131, 892-899.

Szczyglowski K, and Amyot L. (2003) **Symbiosis, inventiveness by recruitment?.** *Plant Physiology* 131, 935-940.

Shimada N, Aoki T, Sato S, Nakamura Y, Tabata S, and Ayabe S. (2003) **A cluster of genes encodes the two types of chalcone isomerase involved in the biosynthesis of general flavonoids and legume-specific 5-deoxy(iso)flavonoids in *Lotus japonicus*.** *Plant Physiology* 131, 941-951.

Borisov AY, Madsen LH, Tsyganov VE, Umehara Y, Voroshilova VA, Batagov AO, Sandal N, Mortensen A, Schauser L, Ellis N, Tikhonovich IA, and Stougaard, J. (2003) **The sym35 gene required for root nodule development in pea is an ortholog of nin from *Lotus japonicus*.** *Plant Physiology* 131, 1009-1017.

Tansengco ML, Hayashi M, Kawaguchi M, Imaizumi-Anraku H, and Murooka Y. (2003) **crinkle, a novel symbiotic mutant that affects the infection thread growth and alters the root hair, trichome, and seed development in *Lotus japonicus*.** *Plant Physiology* 131, 1054-1063.

Wienkoop S, and Saalbach G. (2003) **Proteome analysis. Novel proteins identified at the peribacteroid membrane from Lotus japonicus root nodules.** *Plant Physiology* 131, 1080-1090.

Flemetakis E, Dimou M, Cotzur D, Aivalakis G, Efrose RC, Kenoutis C, Udvardi M, and Katinakis P. (2003) **A *Lotus japonicus* beta-type carbonic anhydrase gene expression pattern suggests distinct physiological roles during nodule development.** *Biochimica et Biophysica Acta-Gene Structure and Expression* 1628, 186-194.

Santi C, von Groll U, Ribeiro A, Chiurazzi M, Auguy F, Bogusz D, Franche C, and Pawlowski K. (2003) **Comparison of nodule induction in legume and actinorhizal symbioses, The induction of actinorhizal nodules does not involve ENOD40.** *Molecular Plant-Microbe Interactions* 16, 808-816.

Kato K, Kanahama K, and Kanayama Y. (2003) **Nitrate-independent expression of plant nitrate reductase in *Lotus japonicus* root nodules.** *Plant and Cell Physiology* 44, S39-S39.

Suganuma N, Yamamoto A, Kato T, Okada T, Kawaguchi M, and Kouchi H. (2003) **Gene expression analysis in ineffective nodules of *Lotus japonicus* Fix mutant Ljsym75 with macroarray.** *Plant and Cell Physiology* 44, S79-S79.

Shimada N, Aoki T, Sato S, Kaneko T, Tabata S, and Ayabe S. (2003) **Structures of the genes involved in 5-deoxyisoflavonoid biosynthesis in *Lotus japonicus*.** *Plant and Cell Physiology* 44, S96-S96.

Asamizu E, Nakamura Y, Sato S, and Tabata S. (2003) **Global gene expression analysis of the *Lotus japonicus* nodulation process by SAGE, and generation of novel 3 ' cDNA fragments.** *Plant and Cell Physiology* 44, S131-S131.

Sato S, Nakamura Y, Kaneko T, Asamizu E, Kato T, and Tabata S. (2003) **Progress in *Lotus japonicus* genome sequencing project.** *Plant and Cell Physiology* 44, S132-S132.

Sakai T. (2003) **Comprehensive preparation and utilization of mutants in the model plant of legume, *Lotus japonicus*.** *Plant and Cell Physiology* 44, S148-S148.

Inada S, Takahashi M, Okada K, and Sakai T. (2003) **Molecular genetic analysis of leaf movement induced by blue light in *Lotus japonicus*.** *Plant and Cell Physiology* 44, S195-S195.

Hanyu M, Hattori Y, and Saeki K. (2003) ***Mesorhizobium loti* genes that alter nodulation capacity of *Rhizobium etli* on *Lotus japonicus*.** *Plant and Cell Physiology* 44, S211-S211.

Umehara Y, Chen WL, and Kouchi H. (2003) **The symbiotic mutants of *Lotus japonicus* derived from the regenerated plants.** *Plant and Cell Physiology* 44, S211-S211.

Kumagai H, Shimomura K, Tajima S, and Kouchi H. (2003) **The efficacy of RNAi in *Lotus japonicus* roots and root nodules.** *Plant and Cell Physiology* 44, S211-S211.

Furuya F, Uchiumi T, Suzuki A, and Abe M. (2003) **Expression of leghemoglobin-GFP fusion gene in the hairy root of *Lotus japonicus*.** *Plant and Cell Physiology* 44, S211-S211.

Hayashi M, Tansengco M, Maekawa T, Yano K, Kawaguchi M, and Murooka Y. (2003) **Phenotype analysis of infection thread mutants in *Lotus japonicus*.** *Plant and Cell Physiology* 44, S212-S212.

Kumagai H, and Kouchi H. (2003) **Gene silencing by expression of hairpin RNA in *Lotus japonicus* roots and root nodules.** *Molecular Plant-Microbe Interactions* 16, 663-668.

Asaimizu E, Kato T, Sato S, Nakamura Y, Kaneko T, and Tabata S. (2003) **Structural analysis of a *Lotus japonicus* genome. IV. Sequence features and mapping of seventy-three TAC clones which cover the 7.5 Mb regions of the genome.** *DNA Research* 10, 115-122.

Suganuma N, Nakamura Y, Yamamoto M, Ohta T, Koiwa H, Akao S, and Kawaguchi M. (2003) **The *Lotus japonicus* Sen1 gene controls rhizobial differentiation into nitrogen-fixing bacteroids in nodules.** *Molecular Genetics and Genomics* 269, 312-320.

Kato K, Okamura Y, Kanahama K, and Kanayama Y. (2003) **Nitrate-independent expression of plant nitrate reductase in *Lotus japonicus* root nodules.** *Journal of Experimental Botany* 54, 1685-1690.

Flemetakis E, Dimou M, Cotzur D, Efrose RC, Aivalakis G, Colebatch G, Udvardi M, and Katinakis P. (2003) **A sucrose transporter, LjSUT4, is up-regulated during *Lotus japonicus* nodule development.** *Journal of Experimental Botany* 54, 1789-1791.

Broughto WJ, Hernandez G, Blair M, Beebe S, Gepts P, and Vanderleyden J. (2003) **Beans (Phaseolus spp.) - model food legumes.** *Plant and Soil* 252, 55-128.

Gronlund M, Gustafsen C, Roussis A, Jensen D, Nielsen LP, Marcker KA, and Jensen EO. (2003) **The Lotus japonicus ndx gene family is involved in nodule function and maintenance.** *Plant Molecular Biology* 52, 303-316.

Lombari P, Ercolano E, El Alaoui H, and Chiurazzi M. (2003) **A new transformation-regeneration procedure in the model legume *Lotus japonicus*, root explants as a source of large numbers of cells susceptible to Agrobacterium-mediated transformation.** *Plant Cell Reports* 21, 771-777.

Wretblad S, Bohman S, and Dixelius C. (2003) **Overexpression of a *Brassica nigra* cDNA gives enhanced resistance to Leptosphaeria maculans in *B-napus*.** *Molecular Plant-Microbe Interactions* 16, 477-484.

Akamine S, Nakamori K, Chechetka SA, Banba M, Umehara Y, Kouchi H, Izui K, and Hata S. (2003) **cDNA cloning, mRNA expression, and mutational analysis of the squalene synthase gene of *Lotus japonicus*.** *Biochimica et Biophysica Acta-Gene Structure and Expression* 1626, 97-101.

Kawaguchi M. (2003) **SLEEPLESS, a gene conferring nyctinastic movement in legume.** *Journal of Plant Research* 116, 151-154.

Young ND, Mudge J, and Ellis THN. (2003) **Legume genomes, more than peas in a pod.** *Current Opinion in Plant Biology* 6, 199-204.

Kaneko T, Asamizu E, Kato T, Sato S, Nakamura Y, and Tabata S. (2003) **Structural analysis of a *Lotus japonicus* genome. III. Sequence features and mapping of sixty-two TAC clones which cover the 6.7 Mb regions of the genome.** *DNA Research* 10, 27-33.

Nakagawa T, Izumi T, Banba M, Umehara Y, Kouchi H, Izui K, and Hata S. (2003) **Characterization and expression analysis of genes encoding phosphoenolpyruvate carboxylase and phosphoenolpyruvate carboxylase kinase of *Lotus japonicus*, a model legume.** *Molecular Plant-Microbe Interactions* 16, 281-288.

Harris JM, Wais R, and Long SR. (2003) **Rhizobium-induced calcium spiking in *Lotus japonicus*.** *Molecular Plant-Microbe Interactions* 16, 335-341.

Zhang SL, Sandal N, Polowick PL, Stiller J, Stougaard J, and Fobert PR. (2003) **Proliferating Floral Organs (Pfo), a *Lotus japonicus* gene required for specifying floral meristem determinacy and organ identity, encodes an F-box protein.** *Plant Journal* 33, 607-619.

Suarez R, Marquez J, Shishkova S, and Hernandez G. (2003) **Overexpression of alfalfa cytosolic glutamine synthetase in nodules and flowers of transgenic *Lotus japonicus* plants.** *Physiologia Plantarum* 117, 326-336.

Akashi T, Sawada Y, Shimada N, Sakurai N, Aoki T, and Ayabe S. (2003) **cDNA cloning and biochemical characterization of S-adenosyl-L-methionine, 2,7,4 '-trihydroxyisoflavanone 4'-O-methyltransferase, a critical enzyme of the legume isoflavonoid phytoalexin pathway.** *Plant and Cell Physiology* 44, 103-112.

Iturbe-Ormaetxe I, Haralampidis K, Papadopoulou K, and Osbourn AE. (2003) **Molecular cloning and characterization of triterpene synthases from *Medicago truncatula* and *Lotus japonicus*.** *Plant Molecular Biology* 51, 731-743.

Fei HM, Chaillou S, Hirel B, Mahon JD, and Vessey JK. (2003) **Overexpression of a soybean cytosolic glutamine synthetase gene linked to organ-specific promoters in pea plants grown in different concentrations of nitrate.** *Planta* 216, 467-474.

Imaizumi-Anraku H. (2003) **Legume nodulation and mycorrhiza formation-common genes are involved in both symbiotic interactions.** *Nippon Nogeikagaku Kaishi-Journal of the Japan Society for Bioscience Biotechnology and Agrochemistry* 77, 121-123.

Tawaraya K. (2003) **Signaling in arbuscular mycorrhizal symbiosis.** *Nippon Nogeikagaku Kaishi-Journal of the Japan Society for Bioscience Biotechnology and Agrochemistry* 77, 124-125.

Simon-Rosin U, Wood C, and Udvardi MK. (2003) **Molecular and cellular characterisation of LjAMT2,1, an ammonium transporter from the model legume *Lotus japonicus*.** *Plant Molecular Biology* 51, 99-108.

van Spronsen PC, Tak T, Rood AMM, van Brussel AAN, Kijne JW, and Boot KJM. (2003) **Salicylic acid inhibits indeterminate-type nodulation but not determinate-type nodulation.** *Molecular Plant-Microbe Interactions* 16, 83-91.

Perry AS, Brennan S, Murphy DJ, Kavanagh TA, and Wolfe KH. (2002) **Evolutionary re-organisation of a large operon in adzuki bean chloroplast DNA caused by inverted repeat movement.** *DNA Research* 9, 157-162.

Jiang QY, and Gresshoff PM. (2002) **Shoot control of hypernodulation and aberrant root formation in the har1-1 mutant of *Lotus japonicus*.** *Functional Plant Biology* 29, 1371-1376.

Uchiumi T, Shimoda Y, Tsuruta T, Mukoyoshi Y, Suzuki A, Senoo K, Sato S, Kato T, Tabata S, Higashi S, and Abe M. (2002) **Expression of symbiotic and nonsymbiotic globin genes responding to microsymbionts *on Lotus japonicus*.** *Plant and Cell Physiology* 43, 1351-1358.

Harris J. (2002) **Shedding light on an underground problem.** *Proceeding of the National Academy of Sciences of the United States of America* 99, 14616-14618.

Nishimura R, Ohmori M, Fujita H, and Kawaguchi M. (2002) **A Lotus basic leucine zipper protein with a RING-finger motif negatively regulates the developmental program of nodulation.** *Proceeding of the National Academy of Sciences of the United States of America* 99, 15206-15210.

Aoki T, Kamizawa A, and Ayabe S. (2002) **Efficient Agrobacterium-mediated transformation of *Lotus japonicus* with reliable antibiotic selection.** *Plant Cell Reports* 21, 238-243.

Krusell L, Madsen LH, Sato S, Aubert G, Genua A, Szczyglowski K, Duc G, Kaneko T, Tabata S, de Bruijn F, Pajuelo E, Sandal N, and Stougaard J. (2002) **Shoot control of root development and nodulation is mediated by a receptor-like kinase.** *Nature* 420, 422-426.

Nishimura R, Hayashi M, Wu GJ, Kouchi H, Imaizumi-Anraku H, Murakami Y, Kawasaki S, Akao S, Ohmori M, Nagasawa M, Harada K, and Kawaguchi M. (2002) **HAR1 mediates systemic regulation of symbiotic organ development.** *Nature* 420, 426-429.

Nakamori K, Takabatake R, Umehara Y, Kouchi H, Izui K, and Hata S. (2002) **Cloning, functional expression, and mutational analysis of a cDNA for *Lotus japonicus* mitochondrial phosphate transporter.** *Plant and Cell Physiology* 43, 1250-1253.

Endo M, Hakozaki H, Kokubun T, Masuko H, Takahata Y, Tsuchiya T, Higashitani A, Tabata S, and Watanabe M. (2002) **Generation of 919 expressed sequence tags from immature flower buds and gene expression analysis using expressed sequence tags in the model plant *Lotus japonicus*.** *Genes & Genetic Systems* 77, 277-282.

Pedrosa A, Sandal N, Stougaard J, Schweizer D, and Bachmair A. (2002) **Chromosomal map of the model legume *Lotus japonicus*.** *Genetics* 161, 1661-1672.

Sandal N, Krusell L, Radutoiu S, Olbryt M, Pedrosa A, Stracke S, Sato S, Kato T, Tabata S, Parniske M, Bachmair A, Ketelsen T, and Stougaard J. (2002) **A genetic linkage map of the model legume *Lotus japonicus* and strategies for fast mapping of new loci.** *Genetics* 161, 1673-1683.

Nishimura R, Ohmori M, and Kawaguchi M. (2002) **The novel symbiotic phenotype of enhanced-nodulating mutant of *Lotus japonicus*, astray mutant is an early nodulating mutant with wider nodulation zone.** *Plant and Cell Physiology* 43, 853-859.

Wallace IS, Wills DM, Guenther JF, and Roberts DM. (2002) **Functional selectivity for glycerol of the nodulin 26 subfamily of plant membrane intrinsic proteins.** *FEBS Letters* 523, 109-112.

Orea A, Pajuelo P, Pajuelo E, Quidiello C, Romero JM, and Marquez AJ. (2002) **Isolation of photorespiratory mutants from *Lotus japonicus* deficient in glutamine synthetase.** *Physiologia Plantarum* 115, 352-361.

Stracke S, Kistner C, Yoshida S, Mulder L, Sato S, Kaneko T, Tabata S, Sandal N, Stougaard J, Szczyglowski K, and Parniske M. (2002) **A plant receptor-like kinase required for both bacterial and fungal symbiosis.** *Nature* 417, 959-962.

Nakamura Y, Kaneko T, Asamizu E, Kato T, Sato S, and Tabata S. (2002) **Structural analysis of a *Lotus japonicus* genome. II. Sequence features and mapping of sixty-five TAC clones which cover the 6.5-Mb regions of the genome.** *DNA Research* 9, 63-70.

Podila GK. (2002) **Signaling in mycorrhizal symbioses - elegant mutants lead the way.** *New Phytologist* 154, 541-545.

Novero M, Faccio A, Genre A, Stougaard J, Webb KJ, Mulder L, Parniske M, and Bonfante P. (2002) **Dual requirement of the LjSym4 gene for mycorrhizal development in epidermal and cortical cells of *Lotus japonicus* roots.** *New Phytologist* 154, 741-749.

Colebatch G, Kloska S, Trevaskis B, Freund S, Altmann T, and Udvardi MK. (2002) **Novel aspects of symbiotic nitrogen fixation uncovered by transcript profiling with cDNA arrays.** *Molecular Plant-Microbe Interactions* 15, 411-420.

Pajuelo P, Pajuelo E, Orea A, Romero JM, and Marquez AJ. (2002) **Influence of plant age and growth conditions on nitrate assimilation in roots of *Lotus japonicus* plants.** *Functional Plant Biology* 29, 485-494.

Endo M, Matsubara H, Kokubun T, Masuko H, Takahata Y, Tsuchiya T, Fukuda H, Demura T, and Watanabe M. (2002) **The advantages of cDNA microarray as an effective tool for identification of reproductive organ-specific genes in a model legume, *Lotus japonicus*.** *FEBS Letters* 514, 229-237.

Imaizumi R, Kameya N, Nakamura I, Ayabe S, and Aoki T. (2002) **Construction of activation tag lines of *Lotus japonicus*.** *Plant and Cell Physiology* 43, S73-S73.

Izumi T, Nakagawa T, Banba M, Umehara Y, Kouchi H, Izui K, and Hata S. (2002) **Expression anayses of phosphoenolpyruvate Carboxylases (PEPCs) and a PEPC kinase in *Lotus japonicus*.** *Plant and Cell Physiology* 43, S74-S74.

Shimomura K, Takane K, Nomura M, Kouchi H, and Tajima S. (2002) **Characterization of transgenic *Lotus japonicus* carrying sense and anti-sense uricase CDNA of *Lotus japonicus*.** *Plant and Cell Physiology* 43, S74-S74.

Thu HM, Fuke T, Nomura M, Takegawa K, Asamizu E, Tabata S, and Tajima S. (2002) **Expression analysis of SNARE-like genes in *Lotus japonicus*.** *Plant and Cell Physiology* 43, S74-S74.

Sato S, Nakamura Y, Kaneko T, Kato T, Asamizu E, and Tabata S. (2002) **A large scale genome analysis of *Lotus japonicus* - I. Genome sequencing and mapping.** *Plant and Cell Physiology* 43, S124-S124.

Nakamura Y, Sato S, Kaneko T, Asamizu E, Kato T, and Tabata S. (2002) **A large scale genome analysis of *Lotus japonicus* - II. Annotation pipeline and WWW database.** *Plant and Cell Physiology* 43, S124-S124.

Murakmi Y, Imaizumi-Anraku H, Kawaguchi M, Kawaski S. (2002) **HEGS/AFLP saturation mapping of causal gene of the initial process mutant of nodulation sym70 of *Lotus japonicus*.** *Plant and Cell Physiology* 43, S124-S124.

Asamizu E, Sato S, and Tabata S. (2002) **Gene expression analysis of *Lotus japonicus* nodulation process using the micro-SAGE method.** *Plant and Cell Physiology* 43, S207-S207.

Flemetakis E, Agalou A, Kavroulakis N, Dimou M, Martsikovskaya A, Slater A, Spaink HP, Roussis A, and Katinakis P. (2002) ***Lotus japonicus*** **gene ljsbp is highly conserved among plants and animals and encodes a homologue to the mammalian selenium-binding proteins.** *Molecular Plant-Microbe Interactions* 15, 313-322.

Poulsen C, and Podenphant L. (2002) **Expressed sequence tags from roots and nodule primordia of *Lotus japonicus* infected with *Mesorhizobium loti*.** *Molecular Plant-Microbe Interactions* 15, 376-379.

Genre A, and Bonfante P. (2002) **Epidermal cells of a symbiosis-defective mutant of *Lotus japonicus* show altered cytoskeleton organisation in the presence of a mycorrhizal fungus.** *Protoplasma* 219, 43-50.

Roberts DM, and Tyerman SD. (2002) **Voltage-dependent cation channels permeable to $NH4^{+}$, K^{+}, and Ca^{2+} in the symbiosome membrane of the model legume *Lotus japonicus*.** *Plant Physiology* 128, 370-378.

Kawaguchi M, Imaizumi-Anraku H, Koiwa H, Niwa S, Ikuta A, Syono K, and Akao S. (2002) **Root, root hair, and symbiotic mutants of the model legume *Lotus japonicus*.** *Molecular Plant-Microbe Interactions* 15, 17-26.

Sprent J. (2002) **Knobs, knots and nodules - the renaissance in legume symbiosis research.** *New Phytologist* 153, 2-6.

Colebatch G, Trevaskis B, and Udvardi M. (2002) **Symbiotic nitrogen fixation research in the postgenomics era.** *New Phytologist* 153, 37-42.

Kawaguchi M, Motomura T, Imaizumi-Anraku H, Akao S, and Kawasaki S. (2001) **Providing the basis for genomics in *Lotus japonicus*, the accessions Miyakojima and Gifu are appropriate crossing partners for genetic analyses.** *Molecular Genetics and Genomics* 266, 157-166.

Muller J, Wiemken A, and Boller T. (2001) **Redifferentiation of bacteria isolated from *Lotus japonicus* nroot nodules colonized by Rhizobium sp NGR234.** *Journal of Experimental Botany* 52, 2181-2186.

Orea A, Pajuelo P, Pajuelo E, Marquez AJ, and Romero JM. (2001) **Characterisation and expression studies of a root cDNA encoding for ferredoxin-nitrite reductase from *Lotus japonicus*.** *Physiologia Plantarum* 113, 193-202.

Hayashi M, Miyahara A, Sato S, Kato T, Yoshikawa M, Taketa M, Hayashi M, Pedrosa A, Onda R, Imaizumi-Anraku H, Bachmair A, Sandal N, Stougaard J, Murooka Y, Tabata S, Kawasaki S, Kawaguchi M, and Harada, K. (2001) **Construction of a genetic linkage map of the model legume Lotus japonicus using an intraspecific F-2 population.** *DNA Research* 8, 301-310.

Sato S, Kaneko T, Nakamura Y, Asamizu E, Kato T, and Tabata S. (2001) **Structural analysis of a *Lotus japonicus* genome. I. Sequence features and mapping of fifty-six TAC clones which cover the 5.4 Mb regions of the genome.** *DNA Research* 8, 311-318.

Lohar DP, Schuller K, Buzas DM, Gresshoff PM, and Stiller J. (2001) **Transformation of *Lotus japonicus* using the herbicide resistance bar gene as a selectable marker.** *Journal of Experimental Botany* 52, 1697-1702.

Salvemini F, Marini AM, Riccio A, Patriarca EJ, and Chiurazzi M. (2001) **Functional characterization of an ammonium transporter gene from *Lotus japonicus*.** *Gene* 270, 237-243.

Stougaard J. (2001) **Genetics and genomics of root symbiosis.** *Current Opinion in Plant Biology* 4, 328-335.

Kapranov P, Routt SM, Bankaitis VA, de Bruijn FJ, and Szczyglowski K. (2001) **Nodule-specific regulation of phosphatidylinositol transfer protein expression in *Lotus japonicus*.** *Plant Cell* 13, 1369-1382.

Marsh JF, and Schultze M. (2001) **Analysis of arbuscular mycorrhizas using symbiosis-defective plant mutants.** *New Phytologist* 150, 525-532.

van Spronsen PC, Gronlund M, Bras CP, Spaink HP, and Kijne JW. (2001) **Cell biological changes of outer cortical root cells in early determinate nodulation.** *Molecular Plant-Microbe Interactions* 14, 839-847.

Niwa S, Kawaguchi M, Imaizumi-Anraku H, Chechetka SA, Ishizaka M, Ikuta A, and Kouchi H. (2001) **Responses of a model legume *Lotus japonicus* to lipochitin oligosaccharide nodulation factors purified from *Mesorhizobium loti* JRL501.** *Molecular Plant-Microbe Interactions* 14, 848-856.

Men AE, Meksem K, Kassem MA, Lohar D, Stiller J, Lightfoot D, and Gresshoff PM. (2001) **A bacterial artificial chromosome library of *Lotus japonicus* constructed in an *Agrobacterium tumefaciens*-transformable vector.** *Molecular Plant-Microbe Interactions* 14, 422-425.

Banba M, Siddique ABM, Kouchi H, Izui K, and Hata S. (2001) ***Lotus japonicus* forms early senescent root nodules with *Rhizobium etli*.** *Molecular Plant-Microbe Interactions* 14, 173-180.

Kato T, Kaneko T, Sato S, Nakamura Y, and Tabata S. (2000) **Complete structure of the chloroplast genome of a legume, *Lotus japonicus*.** *DNA Research* 7, 323-330.

Guenther JF, and Roberts DM. (2000) **Functional characterization of two members of the Major Intrinsic Protein family in nodules of the model legume *Lotus japonicus*.** *Molecular Biology of the Cell* 11, 221A-221A.

Roberts DM, and Tyerman SD. (2000) **A voltage-activated cation selective channel permeable to NH_4^+ on the symbiosome membrane of the model legume *Lotus japonicus*.** *Molecular Biology of the Cell* 11, 222A-222A.

Shimada N, Akashi T, Aoki T, and Ayabe S. (2000) **Induction of isoflavonoid pathway in the model legume *Lotus japonicus*, molecular characterization of enzymes involved in phytoalexin biosynthesis.** *Plant Science* 160, 37-47.

Ozawa R, Shimoda T, Kawaguchi M, Arimura G, Horiuchi J, Nishioka T, and Takabayashi J. (2000) ***Lotus japonicus* infested with herbivorous mites emits volatile compounds that attract predatory mites.** *Journal of Plant Research* 113, 427-433.

Ito M, Miyamoto J, Mori Y, Fujimoto S, Uchiumi T, Abe M, Suzuki A, Tabata S, and Fukui K. (2000) **Genome and chromosome dimensions of *Lotus japonicus*.** *Journal of Plant Research* 113, 435-442.

Solaiman MZ, Senoo K, Kawaguchi M, Imaizumi-Anraku H, Akao S, Tanaka A, and Obata H. (2000) **Characterization of mycorrhizas formed by Glomus sp on roots of hypernodulating mutants of *Lotus japonicus*.** *Journal of Plant Research* 113, 443-448.

Kawaguchi M. (2000) **Current development of *Lotus japonicus* research.** Journal of Plant Research 113, 449-449.

Asamizu E, Watanabe M, and Tabata S. (2000) **Large scale structural analysis of cDNAs in the model legume, *Lotus japonicus*.** *Journal of Plant Research* 113, 451-455.

Saeki K, and Kouchi H. (2000) **The lotus symbiont, *Mesorhizobium loti*, Molecular genetic techniques and application.** *Journal of Plant Research* 113, 457-465.

Tajima S, Takane K, Nomura M, and Kouchi H. (2000) **Symbiotic nitrogen fixation at the late stage of nodule formation in *Lotus japonicus* and other legume plants.** *Journal of Plant Research* 113, 467-473.

Hayashi M, Imaizumi-Anraku H, Akao S, and Kawaguchi M. (2000) **Nodule organogenesis in *Lotus japonicus*.** *Journal of Plant Research* 113, 489-495.

Kawasaki S, and Murakami Y. (2000) **Genome analysis of *Lotus japonicus*.** *Journal of Plant Research* 113, 497-506.

Kawaguchi M. (2000) ***Lotus japonicus* 'Miyakojima' MG-20, An early-flowering accession suitable for indoor handling.** *Journal of Plant Research* 113, 507-509.

Imaizumi-Anraku H, Kouchi H, Syono K, Akao S, and Kawaguchi M. (2000) **Analysis of ENOD40 expression in alb1, a symbiotic mutant of *Lotus japonicus* that forms empty nodules with incompletely developed nodule vascular bundles.** *Molecular and General Genetics* 264, 402-410.

Cote F, Roberts KA, and Hahn MG. (2000) **Identification of high-affinity binding sites for the hepta-beta-glucoside elicitor in membranes of the model legumes *Medicago truncatula* and *Lotus japonicus*.** *Planta* 211, 596-605.

Bonfante P, Genre A, Faccio A, Martini I, Schauser L, Stougaard J, Webb J, and Parniske M. (2000) **The *Lotus japonicus* LjSym4 gene is required for the successful symbiotic infection of root epidermal cells.** *Molecular Plant-Microbe Interactions* 13, 1109-1120.

Takane K, Tajima S, and Kouchi H. (2000) **Structural and expression analysis of uricase mRNA from *Lotus japonicus*.** *Molecular Plant-Microbe Interactions* 13, 1156-1160.

Asamizu E, Nakamura Y, Sato S, and Tabata S. (2000) **Generation of 7137 non-redundant expressed sequence tags from a legume, *Lotus japonicus*.** *DNA Research* 7, 127-130.

Endo M, Kokubun T, Takahata Y, Higashitani A, Tabata S, and Watanabe M. (2000) **Analysis of expressed sequence tags of flower buds in *Lotus japonicus*.** *DNA Research* 7, 213-216.

Flemetakis E, Kavroulakis N, Quaedvlieg NEM, Spaink HP, Dimou M, Roussis A, and Katinakis P. (2000) ***Lotus japonicus* contains two distinct ENOD40 genes that are expressed in symbiotic, nonsymbiotic, and embryonic tissues.** *Molecular Plant-Microbe Interactions* 13, 987-994.

Wopereis J, Pajuelo E, Dazzo FB, Jiang QY, Gresshoff PM, de Bruijn FJ, Stougaard J, and Szczyglowski K. (2000) **Short root mutant of *Lotus japonicus* with a dramatically altered symbiotic phenotype.** *Plant Journal* 23, 97-114.

Nukui N, Ezura H, Yuhashi KI, Yasuta T, and Minamisawa K. (2000) **Effects of ethylene precursor and inhibitors for ethylene biosynthesis and perception on nodulation in *Lotus japonicus* and *Macroptilium atropurpureum*.** *Plant and Cell Physiology* 41, 893-897.

Senoo K, Solaiman MZ, Kawaguchi M, Imaizumi-Anraku H, Akao S, Tanaka A, and Obata H. (2000) **Isolation of two different phenotypes of mycorrhizal mutants in the model legume plant *Lotus japonicus* after EMS-treatment.** *Plant and Cell Physiology* 41, 726-732.

Bolte S, Schiene K, and Dietz KJ. (2000) **Characterization of a small GTP-binding protein of the rab 5 family in *Mesembryanthemum crystallinum* with increased level of expression during early salt stress.** *Plant Molecular Biology* 42, 923-936.

Guenther JF, and Roberts DM. (2000) **Water-selective and multifunctional aquaporins from *Lotus japonicus* nodules.** *Planta* 210, 741-748.

Webb KJ, Skot L, Nicholson MN, Jorgensen B, and Mizen S. (2000) ***Mesorhizobium loti*** **increases root-specific expression of a calcium-binding protein homologue identified by promoter tagging in *Lotus japonicus*.** *Molecular Plant-Microbe Interactions* 13, 606-616.

Bras CP, Jorda MA, Wijfjes AHM, Harteveld M, Stuurman N, Thomas-Oates JE, and Spaink HP. (2000) **A *Lotus japonicus* nodulation system based on heterologous expression of the fucosyl transferase NodZ and the acetyl transferase NolL in *Rhizobium leguminosarum*.** *Molecular Plant-Microbe Interactions* 13, 475-479.

Clarkson DT, Carvajal M, Henzler T, Waterhouse RN, Smyth AJ, Cooke DT, and Steudle E. (2000) **Root hydraulic conductance, diurnal aquaporin expression and the effects of nutrient stress.** *Journal of Experimental Botany* 51, 61-70.

Henzler T, Waterhouse RN, Smyth AJ, Carvajal M, Cooke DT, Schaffner AR, Steudle E, and Clarkson DT. (1999) **Diurnal variations in hydraulic conductivity and root pressure can be correlated with the expression of putative aquaporins in the roots of *Lotus japonicus*.** *Planta* 210, 50-60.

Schauser L, Roussis A, Stiller J, and Stougaard J. (1999) **A plant regulator controlling development of symbiotic root nodules.** *Nature* 402, 191-195.

Limami A, Phillipson B, Ameziane R, Pernollet N, Jiang QJ, Poy R, Deleens E, Chaumont-Bonnet M, Gresshof PM, and Hirel B. (1999) **Does root glutamine synthetase control plant biomass production in *Lotus japonicus L.*?** *Planta* 209, 495-502.

Roberts NJ, Brigham J, Wu B, Murphy JB, Volpin H, Phillips DA, and Etzler ME. (1999) **A Nod factor-binding lectin is a member of a distinct class of apyrases that may be unique to the legumes.** *Molecular and General Genetics* 262, 261-267.

Hussain AKMA, Jiang QY, Broughton WJ, and Gresshoff PM. (1999) ***Lotus japonicus*** **nodulates and fixes nitrogen with the broad host range Rhizobium sp NGR234.** *Plant and Cell Physiology* 40, 894-899.

Genschel U, Powell CA, Abell C, and Smith AG. (1999) **The final step of pantothenate biosynthesis in higher plants, cloning and characterization of pantothenate synthetase from *Lotus japonicus* and *Oryza sativum* (rice).** *Biochemical Journal* 341, 669-678.

Stougaard J, Szczyglowski K, de Bruijn FJ, and Parniske M. (1999) **Genetic nomenclature guidelines for the model legume *Lotus japonicus*.** *Trends in Plant Science* 4, 300-301.

Borg S, Podenphant L, Jensen TJ, and Poulsen C. (1999) **Plant cell growth and differentiation may involve GAP regulation of Rac activity.** *FEBS Letters* 453, 341-345.

Jorgensen JE, Gronlund M, Pallisgaard N, Larsen K, Marcker KA, and Jensen EO. (1999) **A new class of plant homeobox genes is expressed in specific regions of determinate symbiotic root nodules.** *Plant Molecular Biology* 40, 65-77.

Martirani L, Stiller J, Mirabella R, Alfano F, Lamberti A, Radutoiu SE, Iaccarino M, Gresshoff PM, and Chiurazzi M. (1999) **T-DNA tagging of nodulation- and root-related genes in *Lotus japonicus*, Expression patterns and potential for promoter trapping and insertional mutagenesis.** *Molecular Plant-Microbe Interactions* 12, 275-284.

Kapranov P, Jensen TJ, Poulsen C, de Bruijn FJ, and Szczyglowski K. (1999) **A protein phosphatase 2C gene, LjNPP2C1, from *Lotus japonicus* induced during root nodule development.** *Proceeding of the National Academy of Sciences of the United States of America* 96, 1738-1743.

Quaedvlieg NEM, Schlaman HRM, Admiraal PC, Wijting SE, Stougaard J, and Spaink HP. (1998) **Fusions between green fluorescent protein and beta-glucuronidase as sensitive and vital bifunctional reporters in plants.** *Plant Molecular Biology* 38, 861-873.

Schauser L, Handberg K, Sandal N, Stiller J, Thykjaer T, Pajuelo E, Nielsen A, and Stougaard J. (1998) **Symbiotic mutants deficient in nodule establishment identified after T-DNA transformation of *Lotus japonicus*.** *Molecular and General Genetics* 259, 414-423.

Wegel E, Schauser L, Sandal N, Stougaard J, and Parniske M. (1998) **Mycorrhiza mutants of *Lotus japonicus* define genetically independent steps during symbiotic infection.** *Molecular Plant-Microbe Interactions* 11, 933-936.

Szczyglowski K, Kapranov P, Hamburger D, and de Bruijn FJ. (1998) **The *Lotus japonicus* LjNOD70 nodulin gene encodes a protein with similarities to transporters.** *Plant Molecular Biology* 37, 651-661.

Quaedvlieg NEM, Schlaman HRM, Admiraal PC, Wijting SE, Stougaard J, and Spaink P. (1998) **Fusions between green fluorescent protein and beta-glucuronidase as sensitive and vital bifunctional reporters in plants.** *Plant Molecular Biology* 37, 715-727.

Szczyglowski K, Shaw RS, Wopereis J, Copeland S, Hamburger D, Kasiborski B, Dazzo FB, and de Bruijn FJ. (1998) **Nodule organogenesis and symbiotic mutants of the model legume *Lotus japonicus*.** *Molecular Plant-Microbe Interactions* 11, 684-697.

Mikami K, Ichimura K, Iuch S, Yamaguchi-Shinozaki K, and Shinozaki K. (1997) **Molecular characterization of a cDNA encoding a novel small GTP-binding protein from *Arabidopsis thaliana*.** *Biochimica et Biophysica Acta-Gene Structure and Expression* 1354, 99-104.

Thykjaer T, Finnemann J, Schauser L, Christense L, Poulsen C, and Stougaard J. (1997) **Gene targeting approaches using positive-negative selection and large flanking regions.** *Plant Molecular Biology* 35, 523-530.

Thykjaer T, Danielsen D, She Q, and Stougaard J. (1997) **Organization and expression of genes in the genomic region surrounding the glutamine synthetase gene Gln1 from *Lotus japonicus*.** *Molecular & General Genetics* 255, 628-636.

Stiller J, Martirani L, Tuppale S, Chian RJ, Chiurazzi M, and Gresshoff PM. (1997) **High frequency transformation and regeneration of transgenic plants in the model legume *Lotus japonicus*.** *Journal of Experimental Botany* 48, 1357-1365.

Szczyglowski K, Hamburger D, Kapranov P, and deBruijn FJ. (1997) **Construction of a *Lotus japonicus* late nodulin expressed sequence tag library and identification of novel nodule-specific genes.** *Plant Physiology* 114, 1335-1346.

Imaizumi-Anraku H, Kawaguchi M, Koiwa H, Akao S, and Syono K. (1997) **Two ineffective-nodulating mutants of *Lotus japonicus* - Different phenotypes caused by the blockage of endocytotic bacterial release and nodule maturation.** *Plant and Cell Physiology* 38, 871-881.

Skot L, Gordon AJ, Timms E, James CL, Webb KJ, and Mizen S. (1997) **Down-regulation of sucrose synthase expression and activity in transgenic hairy roots of *Lotus japonicus*.** *Symbiosis* 22, 241-254.

Skadhauge B, Gruber MY, Thomsen KK, and vonWettstein D. (1997) **Leucocyanidin reductase activity and accumulation of proanthocyanidins in developing legume tissues.** *American Journal of Botany* 84, 494-503.

Kapranov P, deBruijn FJ, and Szczyglowski K. (1997) **Novel, highly expressed late nodulin gene (LjNOD16) from *Lotus japonicus*.** *Plant Physiology* 113, 1081-1090.

Borg S, Brandstrup B, Jensen TJ, and Poulsen C. (1997) **Identification of new protein species among 33 different small GTP-binding proteins encoded by cDNAs from *Lotus japonicus*, and expression of corresponding mRNAs in developing root nodules.** *Plant Journal* 11, 237-250.

Skot L, Minchin FR, Timms E, Fortune MT, Webb KJ, and Gordon AJ. (1996) **Analysis of the two nodulins, sucrose synthase and ENOD2, in transgenic *Lotus* plants.** *Plant and Soil* 186, 99-106.

Jiang QY, and Gresshoff PM. (1997) **Classical and molecular genetics of the model legume *Lotus japonicus*.** *Molecular Plant-Microbe Interactions* 10, 59-68.

Tajima S, Takane K, Ohkawa K, Sugimoto A, and Okazaki K. (1996) **Preparation of a monoclonal antibody against soybean nodule uricase (nod-35), and immunoblot analysis of the expression of nod-35 in tissues of various legumes.** *Plant and Cell Physiology* 37, 1188-1192.

Pillai SV, Funke RP, and Gresshoff PM. (1996) **Yeast and bacterial artificial chromosome (YAC and BAC) clones of the model legume *Lotus japonicus*.** *Symbiosis* 21, 149-164.

Grant WF, and Small E. (1996) **The origin of the Lotus corniculatus (Fabaceae) complex, A synthesis of diverse evidence.** *Canadian Journal of Botany-Revue Canadienne de Botanique* 74, 975-989.

Kawaguchi M, Imaizumi-Anraku H, Fukai S, and Syono K. (1996) **Unusual branching in the seedlings of *Lotus japonicus* - Gibberellins reveal the nitrogen-sensitive cell divisions within the pericycle on roots.** *Plant and Cell Physiology* 37, 461-470.

Waterhouse RN, Smyth AJ, Massonneau A, Prosser IM, and Clarkson DT. (1996) **Molecular cloning and characterisation of asparagine synthetase from *Lotus japonicus*, Dynamics of asparagine synthesis in N-sufficient conditions.** *Plant Molecular Biology* 30, 883-897.

Oger P, Petit A, and Dessaux Y. (1996) **A simple technique for direct transformation and regeneration of the diploid legume species *Lotus japonicus*.** *Plant Science* 116, 159-168.

Stougaard, J and Beuselinck, PR. (1996) **Registration of GIFU B-129-S9 *Lotus japonicus* germplasm.** *Crop Science* 36, 476-476.

Thykjaer T, Stiller J, Handberg K, Jones J, and Stougaard J. (1995) **The maize transposable element Ac is mobile in the legume *Lotus japonicus*.** *Plant Molecular Biology* 27, 981-993.

Schauser L, Christensen L, Borg S, and Poulsen C (1995) **PZF, A cDNA isolated from *Lotus japonicus* and soybean root-nodule libraries, encodes a new plant member of the ring-finger family of zinc-binding proteins.** *Plant Physiology* 107, 1457-1458.

Stougaard J. (1993) **Substrate-dependent negative selection in plants using a bacterial cytosine deaminase gene.** *Plant Journal* 3, 755-761.

Handberg K, and Stougaard J. (1992) ***Lotus japonicus*, an autogamous, diploid legume species for classical and molecular genetics.** *Plant Journal* 2, 487-496.

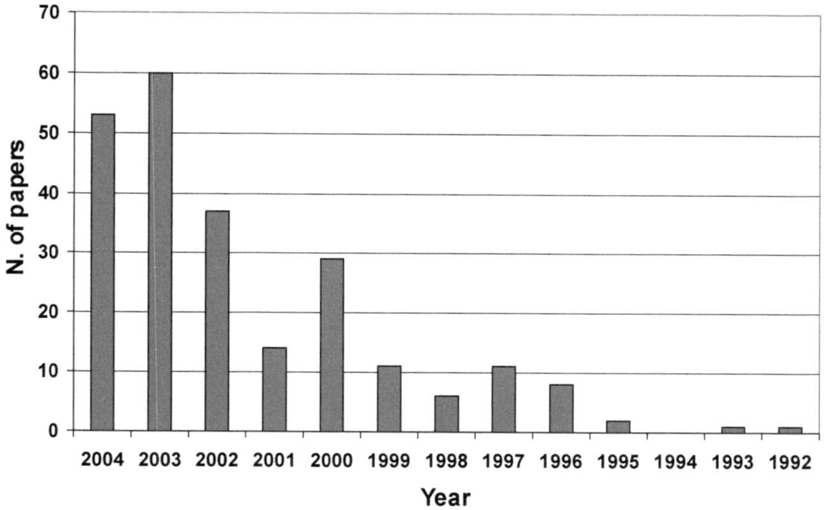

Figure 1. Evolution in the number of Lotus japonicus-*related articles published per year. (Source of information: ISI Web of Knowledge, up to January 2005).*

Index

MIX
Papier aus verantwortungsvollen Quellen
Paper from responsible sources
FSC® C105338

If you have any concerns about our products,
you can contact us on
ProductSafety@springernature.com

In case Publisher is established outside the EU,
the EU authorized representative is:
Springer Nature Customer Service Center GmbH
Europaplatz 3, 69115 Heidelberg, Germany

Printed by Libri Plureos GmbH
in Hamburg, Germany